A Casebook for a First Course in Statistics and Data Analysis

Samprit Chatterjee
Mark S. Handcock
Jeffrey S. Simonoff

New York University

Annotated Edition for Instruction

JOHN WILEY & SONS, INC.
New York • Chichester • Brisbane • Toronto • Singapore

ISBN 0–471–12255–6

Printed in the United States of America

10 9 8 7 6 5 4 3 2

Printed and bound by Malloy Lithographing, Inc.

To all students of statistics, in hope that they will be as intrigued by the subject as the authors have been and continue to be.

——— S.C.

To Martina Morris and David L. Wallace, for imparting a tolerance for ambiguity.

——— M.S.H.

To three great teachers: Sandy Reisman, Judy Tanur and John Hartigan.

——— J.S.S.

Preface

The most effective way to learn statistics is by actively engaging in doing the statistical analysis. This idea drives this casebook. An introductory course in statistics often fails to give the students an idea of the excitement of statistics, and its relevance in the present day world. A considerable amount of material has to be covered, with no complementary time for discussion of real life examples. Students often come away with a blurred impression of formulas, and some words like "mean," "standard deviation," and "regression." The point that statistical analysis is vital to arrive at conclusions in a sensible and rational manner is often neglected. This casebook is an attempt to remedy this deficiency by providing an active resource for classroom use. The book is based on cases that we have developed through almost fifty cumulative years of teaching the introductory statistics course at New York University.

We have attempted in this casebook to present cases representing situations and contexts from a diverse set of fields, where statistical analysis is required to arrive at a meaningful conclusion. Topics covered include eruptions of the "Old Faithful" geyser, the issuance of international adoption visas, the space shuttle Challenger tragedy, patterns in the Dow Jones Industrial Average and Standard and Poor's index, health expenditures of states, random drug and disease testing, baseball free agency, performance of NBA guards, energy consumption of a household, grape yields in a vineyard, and the birth and nursing of a beluga whale calf. All of the datasets are real and complete.

Each case is motivated by a question that needs to be answered, and full background material is presented. The statistical analysis flows naturally from the question. The discussion given in the cases attempts to demonstrate the logic of the analysis and emphasize the interactive and iterative nature of the task. The aim of these cases is to show the reader by example that statistical analysis clarifies and throws light on a complex situation. It enables one to draw useful conclusions. Besides the final conclusion, much is learned about the problem during the analysis. The journey, as well as the arrival, matters.

In addition to investigation of the specific questions raised by a particular case, we hope that the reader also will develop a feel for the kind of approach to data analysis that is likely to be fruitful in general. As statistical software has become generally available, the possibilities of superficial, but inadequate, analysis of data have increased correspondingly. However, if a data analyst is trained to develop a system of general principles in performing a data analysis that are widely applicable, it is much more likely that she will analyze future data sets in a reasonable way. It is our hope that this casebook can be helpful in highlighting the kinds of questions that need to be answered if such a system is being used.

The casebook and your introductory course topics

The material presented in this book can be analyzed using techniques that are almost always taught in an introductory course. The cases should be used concurrently with the statistical methods discussed in a course (or an independent study of introductory statistics). The cases are grouped by broad statistical topics, and are arranged by topics in a sequence that is conventionally followed in a beginning course. The cases can be done in or out of sequence. There is, however, a certain sense of progression in the material. Concepts from data analysis are used in cases dealing with applied probability, and statistical inference. Cases on regression analysis draw heavily on material covered in data analysis and inference (indeed, general issues of data analysis pervade almost all of the cases).

Using the different kinds of cases

The cases presented in the book fall into three categories. Cases in the first category are analyzed completely. These are meant to be models or paradigms for the system of general principles mentioned earlier. We have presented the approach that, to us, appears most simple and direct. There may be other ways of analyzing the same data. We feel confident, however, that the conclusions reached by a valid alternative analysis will be very similar to the ones that we reach. There are many paths to the summit! We would like to hear from any reader who finds a discrepancy between our conclusions and theirs and also welcome suggestions for effective analyses different from the ones that we have presented.

The second category of cases in the book consists of ones that are partially analyzed, in that no analysis is actually performed, but suggestions are made that guide the reader along the path to an effective analysis. This will get the reader gradually involved in direct analysis of the data. In the third category of cases, we present the problem, pose questions, and provide the data. No analysis is suggested; the analyst is on her own. Readers who have worked through the first two categories of cases should find the cases in the third category within their capabilities. The hope of the authors is that, having worked through the cases in the book, the reader will be well prepared to analyze her own data in a real–life setting.

Each case is organized in the same format. Each case describes the background of the data, and poses the questions that are to be answered by statistical analysis. In the cases that are completely analyzed, the analysis is described in detail. The actual sequence of steps followed is noted. The cases are written in an informal style, so that the reader can have the sense of examining the data (using a statistical package) on a computer, with a statistically knowledgeable person looking over their shoulder. Finally, the conclusions of the analysis are summarized in a clear, readable, and non–technical form. Presentation of statistical findings with clarity is essential if they are to be taken seriously. The authors feel that the importance of presenting conclusions of a statistical analysis effectively is not often emphasized, due to the lack of instructive examples accessible to the student. In reality, the inability to communicate effectively often leads to the unfortunate circumstance that little or no attention is paid to the statistical analysis and the findings.

Using the computer

The output, and most plots that appear in the cases (with some editing), were generated using the PC package `STATISTIX 4.0` (a product of Analytical Software). Virtually all of the analyses that we present, however, can be performed using commonly available statistical software. We believe that a statistical package that is not capable of performing the kinds of analyses we describe has serious shortcomings and should be abandoned for a more adequate one.

The casebook probably contains more material than can be covered in a single course. This remark holds particularly concerning the material contained on regression analysis. We have provided the extra material with the hope that it will be used in the teaching of a course in applied regression analysis (the second course). Availability of this case material should considerably ease the learning and understanding of the concepts in regression analysis.

Besides students of an introductory statistics course, this book is likely to prove useful to anyone who is interested in learning how to apply statistical analysis to data encountered in practice. For people who are well versed only in mathematical statistics, the book will provide a useful supplement to theory. Statistics is an applied discipline and justifies itself only in useful application. This casebook is intended to be a pioneer in the line of statistical texts, in that it is designed for the beginner, with this vision and motivation. We would be interested to hear from our readers concerning how well we have succeeded in this enterprise. The authors can be reached electronically on the Internet at the addresses `schatterjee@stern.nyu.edu`, `mhandcock@stern.nyu.edu`, and `jsimonoff@stern.nyu.edu`, respectively.

We are always on the lookout for novel applications and interesting data and welcome any submissions in this area from our readers. To foster the exchange of ideas and material we have provided Internet access to a growing archive of supplementary cases similar in style to those in this book. The archive can be accessed by **gopher** or the World Wide Web (WWW) using a WWW browser (e.g., **mosaic**). We also will maintain a list of useful suggestions sent to us and updated information about the cases in the book. Information on how to access this archive is given in the Appendix.

We would like to thank the many hundreds of introductory statistics students at New York University who have been (unwitting) guinea pigs in the development of these cases. One of the joys of discussing real data with students is that they bring their own backgrounds and experiences to the discussion, with the result that the teacher learns as much as the students do. We would like to thank David Ahlstrom, Orley Ashenfelter, Mark and Barbi Barnhill, Charlie Himmelberg, Jeanne McLaughlin–Russell, Martina Morris, Sundar Polavaram, Tom Pugel and Brooke Squire for providing data sets that are used in the cases. We would also like to thank our colleagues at New York University for teaching from a draft version of this book, and contributing to the final version. Special thanks go to Halina Frydman, Joel Owen and Donald Richter.

SAMPRIT CHATTERJEE
MARK S. HANDCOCK
JEFFREY S. SIMONOFF

Alma Aty, Kazakhstan
Croton–on–Hudson, New York
East Meadow, New York

Using this casebook for an introductory course
Notes for an instructor

Our general pedagogical approach

The cases provided in this book should be used to demonstrate the relevance of the statistical concepts covered in the course to real life applications. To get a detailed idea of the purpose and the philosophy of the casebook, the preface and the introduction to the casebook ("How to use this book") should be consulted. We will not go over that material here. A reading of the preface and the introduction will lead to a more effective utilization of the casebook.

A commonly held view among people who teach data analysis and statistics to students from a specific functional area is that if examples do not come from that area, the students will not be interested, and will not learn what they need to know. So, for example, examples in a statistics course taught in a business school should all relate to aspects of business.

As should be apparent from the table of contents of this book, we strongly disagree with this point of view. Students are not merely future managers, psychologists or biologists; they are citizens of the world. Students bring to class diverse knowledge and interests and the teaching of statistics should take advantage of this. When the context of the case and the students' backgrounds match, the statistical methods will become more memorable from the association. We have found that if a student is interested in the material, they cannot help comprehending the methodology. We were each personally drawn to the field of statistics by (among other things) its wide applicability to virtually all areas of life, and we think that it is important to share this view with our students.

Classroom and assignment suggestions

We have deliberately provided several cases (based on different types of data from different functional areas) for almost all major areas of statistical methodology, so that teachers can choose the case or cases that best reflect the interests of the students and teacher. We do not expect that anyone would, or should, try to cover all of the cases in class (or as homework assignments or group work for the students).

The cases that are fully analyzed (marked by an **F**) should be used concurrently or immediately following the discussion of the statistical concepts in class. For example, "Eruptions of the 'Old Faithful' geyser" will add much to the discussion of statistical summaries in the early part of the course. Similarly the case "Random drug and disease testing" will bring alive the concept of conditional probability. The statistical content of a particular case can be ascertained from the *Topics covered*, and the *Key words* for each case (pointers are also provided in the index). An instructor will almost always find a case fully discussed (an **F** case) for most topics discussed in an introductory course. To provide variety each major topic is discussed in two **F** cases. Your choice of **F** cases can be discussed in class or assigned for reading. We find it useful to do the analysis, "live and on–line" in class. We have provided in the **F** cases all relevant output (both numerical and graphical), so that a full discussion of the case can be carried out without any overhead transparencies or on–line computing. The students must, however, have the book on hand to follow the discussion. The **F** cases are complete in the sense that they can be read independently outside of the class. We find, however, that doing them in class leads to a lively discussion, and generates interest in statistical methods.

The style of the cases is deliberately informal; this book is written for the student rather than the teacher. The cases are written with the objective of simulating a "conversation with the student" so that they can understand the process of data analysis from a relative expert. As such, we have deliberately resisted the urge to post–process the analyses in the light of the final conclusions.

After the **F** cases, the **G** cases can be attempted. They may be assigned as homework and gone over in class. The **G** cases often follow the corresponding **F** cases and should provide an easy transition. It is particularly effective to develop a discussion while going through a **G** case on–line in class. The students are motivated by directing the analysis while the teacher can provide the role of facilitator and sage. We have found that such discussion is a good opportunity to make connections between different concepts. As an alternative, **G** cases are appropriate for on–line use in "smart" classrooms, where the teacher and/or students can individually produce output and plots interactively, allowing the entire class to see how the analysis proceeds.

The **O** cases ask the student to do an independent analysis of an open statistical problem. They can be assigned to be worked individually, or as group projects. Breaking a class into groups of 5 or 6 students, and assigning them an **O** case to work on in a group, can prove to be a useful learning experience. In particular, students with more skill in performing data analyses can act as mentors for less proficient students in their group. The **O** cases do not introduce new technical material and hence do not carry a *Technical terms* section.

We have supplemented each of the **G** and **O** cases in this "Teacher's Edition" with additional guided analysis. This material can be used as a basis for solutions, if the case is given as homework, or as a resource for a recitation session. The additional material is clearly demarcated by gray bars within each case.

Virtually all cases listed under "Statistical Inference" and "Analysis involving Regression" include a large dose of basic data analysis. As such, the categorization of cases in the table of contents under–represents the number of data analysis cases available. Teachers who feel that they would like more data analysis cases can easily use inference or regression cases for that purpose, simply glossing over the technical aspects where appropriate (with the promise of coming back to those details later, perhaps). To help with this process we have provided an expanded categorization of the cases in "How to use this book." The cases that involve material from the more advanced areas are marked with the symbol " $\hookrightarrow$."

A very effective way to ensure that students are exposed to specific kinds of data relevant to a specific functional area, if desired, would be to assign homeworks using data sets drawn from that area. The homework would be similar in style to the **O** cases, but with more direct application to specific needs of the students. The particular data sets would be determined by the teacher's personal interests and the circumstances of the class. We do not wish to impose a restricted number of choices on the teacher at this level for all functional areas, although we do feel that the **F**, **G** and **O** cases given here illustrate the kinds of analyses that are possible.

The casebook and the computer

To get the full benefit of this casebook, access to interactive computing is essential. A personal computer with a reasonable statistical package will enable a student to carry out virtually all of the analyses illustrated in, or required for, the casebook. This is a modest requirement, and is likely to be satisfied by most colleges and universities, including those with a minimum of computing facilities. The **F** cases can be studied and followed even by those with no direct computing facilities, as we have provided all needed computer output, but we do not recommend this. This is a passive learning approach. We would like to encourage a more active "hands on" approach where a student generates the needed computer output. We have found this direct experiential learning to be very useful and motivating for students.

The casebook and your overall course objectives

The use of worked–out, real data examples takes a little getting used to for the teacher who has never done so before. We are not suggesting that the teacher merely read the case aloud to the students. Rather, the context and background of the problem are first described by the teacher, paraphrasing some of the material in the case. Then, this is tied into the statistical material just having been covered in the class, to show how the functional question can be answered using statistical methods. From this point, the specific plots and/or output can be studied and discussed, leading to new questions. Questions and comments from the students might relate to confusion about the form of the computer output, and suggestions of their own about the underlying causes of observed effects. The intention is to develop in the students an understanding of the steps to a good analysis, which can then be applied to any other data situation. Of course, the written version of the case is always available to the student for further individual study.

We have deliberately not included the kinds of cases that are likely to appear in an introductory text. For example, most texts have many examples of the use of the Binomial probability function for the calculation of probabilities, or of the use of normal tables for the calculation of Gaussian–based probabilities. Similarly, most introductory business texts provide examples of the construction of control charts. Such examples are quite useful, of course, but we have not replicated them in this book. Some introductory courses include the analysis of variance (ANOVA). Although we do not construct ANOVA tables, cases that could incorporate an ANOVA perspective are "Eruptions of the 'Old Faithful' geyser," "The performance of stock mutual funds" (one–way) and "Volume and weight from a vineyard harvest" (two–way). Similarly, general issues of quality control (QC) are addressed in "Productivity versus quality in the assembly plant" and "The flight of the space shuttle Challenger."

We also might mention a misconception that could arise from the title of this book. The **F** cases here are not meant to be cases in the sense of the "case method" of instruction pioneered at the Harvard Business School. That is, the intention is not to provide data examples with only a bare skeleton of guidance, which can be used to motivate the introduction of various statistical methods and concepts. Rather, we expect that the methods and concepts have already been discussed, and these cases provide examples of the power of the methods just described.

If the class size is over 100 students, then the casebook should probably be used in recitation sessions. The leader of the recitation can use the casebook as a resource and guide. We have found that this provides structure for each session, which facilitates learning (not to mention reducing the stress on the leader!).

The material in this casebook can be used in many creative ways. We have indicated some ways in which we have successfully used it. We would like to hear about other successful experimentation. We hope that we have provided raw material encouraging the teaching of the introductory statistics course in an innovative and imaginative way. We hope that this casebook will help breathe life into the introductory statistics course, much maligned but absolutely essential in today's complex world.

Contents

Preface *v*

Notes for an instructor *viia*

Introduction

How to use this book *1*

Part 1 **Data Analysis**

Eruptions of the "Old Faithful" geyser *5*

International adoption rates *13*

The performance of stock mutual funds *21*

Predicting the sales and airplay of popular music *23*

Another look at the "Old Faithful" geyser and adoption visas *24*

Productivity versus quality in the assembly plant *25*

Health care spending in the United States *32*

Part 2 **Applied Probability**

The flight of the space shuttle Challenger *33*

Random drug and disease testing *37*

Amniocentesis, blood tests, and Down's syndrome *41*

Perceptions of the New York City subway system: safety and cleanliness *43*

Perceptions of the New York City subway system: other issues *47*

The Central Limit Theorem for census data *48*

The sampling distribution for the median *59*

The sampling distribution for the standard deviation *67*

Racial imbalance in Nassau County public schools *72*

Part 3 **Statistical Inference**

The return on stocks in the Over the Counter market *82*

Volume and weight from a vineyard harvest *86*

The performance of stock mutual funds compared to the market as a whole *99*

The return on stocks in the New York Stock Exchange *100*

Baseball free agency: do teams get what they pay for? *101*

Baseball free agency: another look at player effectiveness *112*

Reporting of sexual partners by men and women *113*
The comparative volatility of stock exchanges *123*
Validity of t–testing for the sexual partners reported by men and women *129*
Subgroups in the "Old Faithful" geyser data *130*
Mortgage rates for different types of mortgages *131*
Condom use and the prevention of AIDS *132*
Air bags and types of automobiles *134*
A further look at the reporting of sexual partners by men and women *137*
Betting on professional football — can you beat the bookies? *142*
Formal investigation of the perceptions of the New York City subway system *144*

Part 4 **Analysis involving Regression**

Emergency calls to the New York Auto Club *145*
Purchasing power parity: is it true? *153*
PCB contamination of U.S. bays and estuaries *164*
Electricity usage, temperature and occupancy *177*
Long and short term performance of stock mutual funds *185*
Estimating rates of return of investments *186*
Predicting international adoption visas from earlier years *189*
The pattern in employment rates over time *190*
Estimating a demand function *191*
The effectiveness of National Basketball Association guards *201*
The possibility of voting fraud in an election *213*
Purchasing power parity and high inflation countries *222*
Prediction of the time interval between "Old Faithful" eruptions *224*
The success of teams in the National Hockey League *227*
The effectiveness of NBA guards: assists per minute *230*
The effectiveness of NBA guards: beyond points and assists *232*
Another look at emergency calls to the New York Auto Club *233*
Subgroups in the electricity consumption data *234*
A closer look at productivity and quality in the assembly plant *235*
Predicting mortgage rates for different types of mortgages *236*
Better prediction of the yield in a small vineyard *237*
Incomes of Long Island communities *247*

Looking more deeply into the average weight per lug from a vineyard	*250*
Predicting incomes of Long Island communities	*252*
Prediction of yields from partial harvesting in a small vineyard	*253*
The birth of a beluga whale	*256*
Estimating a demand function — it's about time	*276*
A comparison of the Dow Jones Industrial Average and the S & P 500 index	*283*
Further investigation of the DJIA and the S & P 500 index	*285*
Appendix: Descriptions of the data files	*286*
Index	*308*

How to use this book

In this casebook you will find examples drawn from many fields, where statistical analysis is needed to answer a particular question. Almost all introductory statistics textbooks concentrate on explaining the methodology, without paying much attention to applications. At the end of such a course a student comes away with a set of tools (techniques) without having a precise idea about how and when they are to be applied. As a consequence the introductory course often appears to be dull, irrelevant, and not worth remembering. This casebook is an attempt to remedy this problem.

The most effective way to use these cases is to study them concurrently with the statistical methodology being learned. To simplify this we have grouped the cases by broad statistical topics. The relevant statistical concepts and techniques pertaining to a particular case are also noted. The cases are arranged in a sequence of topics that we follow at NYU and that is fairly typical of introductory courses.

The cases presented in the book are also classified by the amount of additional analysis needed for the resolution of the case and completion of the study. Some cases are analyzed fully and are meant to be examples or paradigms for effective analysis. These cases are marked below by an **F**. There are other cases where guidance is provided, in that the appropriate analysis is indicated, but not supplied. These cases are marked below by a **G**. The reader is expected to carry out the analysis. There is a third type of case, and these are marked below by an **O**. In these cases, only the data description is provided, along with background information on the context of the problem and nature of the question at issue. The reader is asked to carry out the relevant analysis in an open ended fashion. The data themselves are provided in a corresponding computer file. No analysis is provided. We believe that after having worked through **F** and **G** type cases, a reader will be able to analyze successfully the **O** cases.

The best way to use the **F** cases is to carry out the analysis described in the text interactively on a computer, to verify statements and conclusions made during the analysis. This direct involvement with the analysis will generate the ability and confidence to analyze cases marked **G** and **O**.

The data for each of the cases are contained in the diskette accompanying this book. The files are in text format (ASCII), and can be imported and analyzed using virtually any standard available software. A brief description of each file is given in the appendix "Descriptions of the data files" and should be consulted as a preliminary step in each analysis.

Given below are the cases, grouped by their relevant statistical concepts and classified by the degree of analysis already presented in the case. Cases marked with the symbol " $\hookrightarrow$ " are **F** cases that contain a good deal of material for that general area (e.g., Data Analysis), although they also contain material from more advanced areas (e.g., Statistical Inference). For example, the case "The flight of the space shuttle Challenger" appears in the "Data Analysis" section with the symbol " $\hookrightarrow$ " because it contains a good deal of material appropriate for "Data Analysis," although it is "officially" an **F** case under "Applied Probability."

DATA ANALYSIS

Association between two variables; Comparison of location and variation in subgroups; Data summary; Examination of univariate data; Identifying subgroups; Transformations; Variation over time

F	Eruptions of the "Old Faithful" geyser	*5*
F	International adoption rates	*13*
G	The performance of stock mutual funds	*21*
O	Predicting the sales and airplay of popular music	*23*
O	Another look at the "Old Faithful" geyser and adoption visas	*24*
F	Productivity versus quality in the assembly plant	*25*
O	Health care spending in the United States	*32*
↪	The flight of the space shuttle Challenger	*33*
↪	Volume and weight from a vineyard harvest	*86*
↪	Baseball free agency: do teams get what they pay for?	*101*
↪	Reporting of sexual partners by men and women	*113*
↪	The comparative volatility of stock exchanges	*123*
↪	Emergency calls to the New York Auto Club	*145*
↪	Purchasing power parity: is it true?	*153*
↪	PCB contamination of U.S. bays and estuaries	*164*
↪	Electricity usage, temperature and occupancy	*177*
↪	The effectiveness of National Basketball Association guards	*201*
↪	The possibility of voting fraud in an election	*213*
↪	The birth of a beluga whale	*256*

APPLIED PROBABILITY

Binomial random variable; Central Limit Theorem; Conditional probability; Definitions of probability; Extra–Binomial variation; Gaussian (normal) distribution

F	The flight of the space shuttle Challenger	*33*
F	Random drug and disease testing	*37*
G	Amniocentesis, blood tests, and Down's syndrome	*41*
G	Perceptions of the New York City subway system: safety and cleanliness	*43*
O	Perceptions of the New York City subway system: other issues	*47*
F	The Central Limit Theorem for census data	*48*
F	The sampling distribution for the median	*59*
F	The sampling distribution for the standard deviation	*67*
F	Racial imbalance in Nassau County public schools	*72*
↪	Air bags and types of automobiles	*134*
↪	A further look at the reporting of sexual partners by men and women	*137*

STATISTICAL INFERENCE

Comparison of Binomial proportions; Confidence interval; Contingency tables; Hypothesis testing; Paired sample comparisons; Prediction interval; t–statistic; Tests for difference in location; Tests of independence; Two–sample (unpaired sample) comparisons; z–statistic

F The return on stocks in the Over the Counter market *82*
F Volume and weight from a vineyard harvest *86*
O The performance of stock mutual funds compared to the market as a whole *99*
O The return on stocks in the New York Stock Exchange *100*
F Baseball free agency: do teams get what they pay for? *101*
O Baseball free agency: another look at player effectiveness *112*
F Reporting of sexual partners by men and women *113*
F The comparative volatility of stock exchanges *123*
G Validity of t–testing for the sexual partners reported by men and women *129*
O Subgroups in the "Old Faithful" geyser data *130*
O Mortgage rates for different types of mortgages *131*
F Condom use and the prevention of AIDS *132*
F Air bags and types of automobiles *134*
F A further look at the reporting of sexual partners by men and women *137*
O Betting on professional football — can you beat the bookies? *142*
O Formal investigation of the perceptions of the New York City subway system *144*
↪ Emergency calls to the New York Auto Club *145*
↪ PCB contamination of U.S. bays and estuaries *164*

ANALYSIS INVOLVING REGRESSION

Indicator variables; Lagged variables; Model selection; Multiple regression; Prediction; Regression coefficients; Regression diagnostics; Residuals; Semilog model; Simple regression; Time series data; Transformation; Unusual observations; Weighted regression

F Emergency calls to the New York Auto Club *145*
F Purchasing power parity: is it true? *153*
F PCB contamination of U.S. bays and estuaries *164*
F Electricity usage, temperature and occupancy *177*
G Long and short term performance of stock mutual funds *185*
F Estimating rates of return of investments *186*
O Predicting international adoption visas from earlier years *189*
O The pattern in employment rates over time *190*
F Estimating a demand function *191*
F The effectiveness of National Basketball Association guards *201*
F The possibility of voting fraud in an election *213*
G Purchasing power parity and high inflation countries *222*
G Prediction of the time interval between "Old Faithful" eruptions *224*
G The success of teams in the National Hockey League *227*
G The effectiveness of NBA guards: assists per minute *230*
O The effectiveness of NBA guards: beyond points and assists *232*

O Another look at emergency calls to the New York Auto Club *233*
O Subgroups in the electricity consumption data *234*
O A closer look at productivity and quality in the assembly plant *235*
O Predicting mortgage rates for different types of mortgages *236*
F Better prediction of the yield in a small vineyard *237*
G Incomes of Long Island communities *247*
G Looking more deeply into the average weight per lug from a vineyard *250*
O Predicting incomes of Long Island communities *252*
G Prediction of yields from partial harvesting in a small vineyard *253*
F The birth of a beluga whale *256*
F Estimating a demand function — it's about time *276*
G A comparison of the Dow Jones Industrial Average and the S & P 500 index *283*
O Further investigation of the DJIA and the S & P 500 index *285*

Eruptions of the "Old Faithful" geyser

Topics covered: Bimodal distribution. Data summary. Examination of univariate data. Identifying subgroups. Prediction.

Key words: Boxplot. Histogram. Mean. Median. Mode. Scatter plot. Standard deviation. Side–by–side boxplots. Stem–and–leaf display.

Data Files: `geyser1.dat`, `geyser2.dat`

A geyser is a hot spring that occasionally becomes unstable and erupts hot water and steam into the air. The "Old Faithful" geyser at Yellowstone National Park in Wyoming is probably the most famous geyser in the world. Visitors to the park try to arrive at the geyser site to see it erupt without having to wait too long; the name of this geyser comes from the fact that eruptions follow a relatively stable pattern. The National Park Service erects a sign at the geyser site predicting when the next eruption will occur. Thus, it is of interest to understand and predict the interval time until the next eruption.

The following analysis is based on a sample of 222 intereruption times taken during August 1978 and August 1979 (data source: *Applied Linear Regression, 2nd. ed.*, by S. Weisberg). The first step in any data analysis is simply to look at the data. A **histogram** gives a good deal of information about the distribution of intereruption times, suggesting some interesting structure. Interval times are in the general range of 40 to 100 minutes, but there are apparently two subgroups in the data, centered at roughly 55 minutes and 80 minutes, respectively, with a gap in the middle.

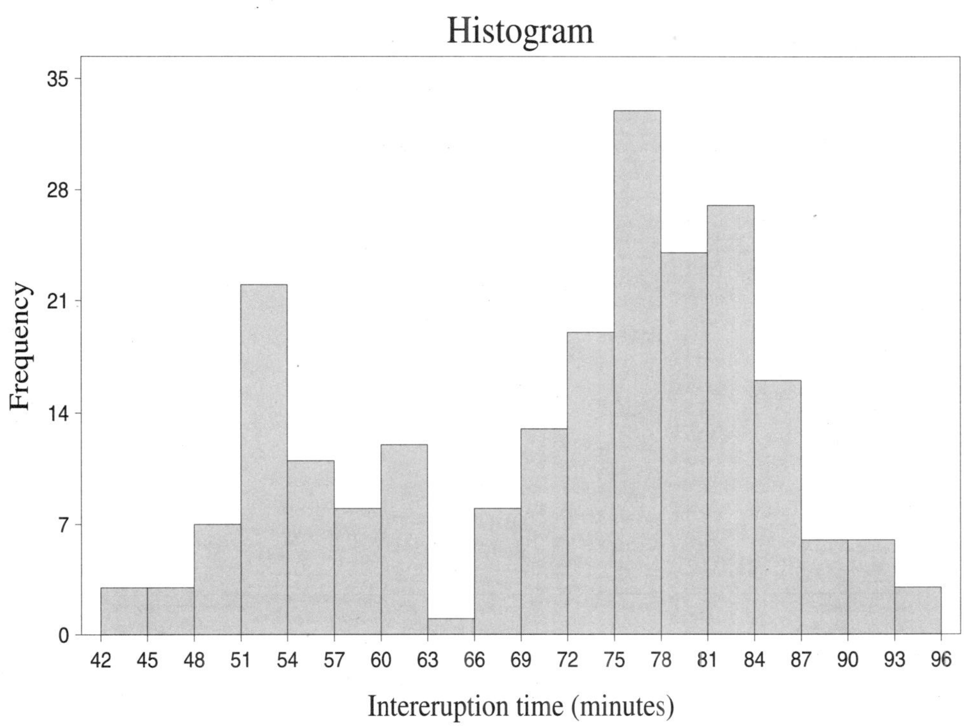

A **stem–and–leaf display** gives a similar impression:

```
STEM AND LEAF PLOT OF INTERVAL Intereruption time (minutes)

   LEAF DIGIT UNIT = 1
   1  2  REPRESENTS 12.

          STEM  LEAVES
     3     4* 234
    11     4. 55788999
    39     5* 0011111111111111222333334444
    54     5. 555566677778889
    67     6* 000011111 2223
    78     6. 66677788999
   107     7* 00000111112222233333333344444
  (44)     7. 55555555555555566666666666677777778888889999
    71     8* 0000000000000111111112222222222233333334444444444
    22     8. 5666666788899
     9     9* 00011134
     1     9. 5

222 CASES INCLUDED     0 MISSING CASES
```

The stem–and–leaf display can give additional information as well. For example, the first column of the plot above is a cumulative frequency column that starts at both ends and meets in the middle. The row that contains the median is marked with parentheses around the count of observations for that row.

Not all exploratory techniques are as effective for this type of data. A **boxplot** (sometimes called a **box–and–whisker** plot) of the interval times shows that interval times are in the general range of 40 to 100 minutes, but the bimodal distribution is hidden by the form of the plot:

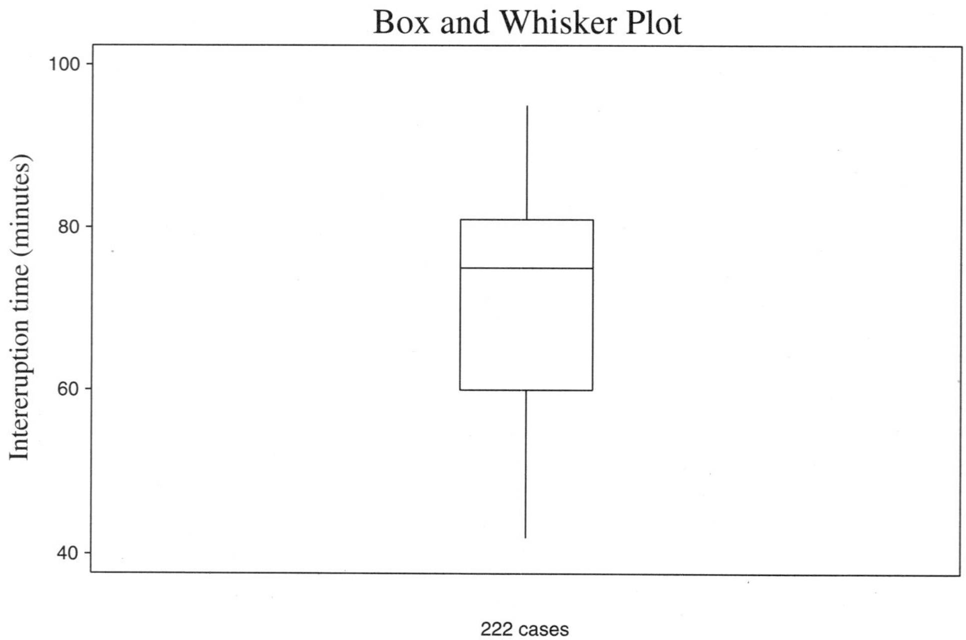

Summary statistics, as given below, look informative, but that is somewhat misleading. For example, the mean intereruption time of about 71 minutes doesn't actually describe a typical result from either subgroup. A useful rule–of–thumb is that roughly 95% of the observations will lie within two standard deviations of the mean, or (here) $71 \pm 25.6 = (45.4, 96.6)$. In fact, all but five of the 222 observations fall in this interval, which is more than we would expect. It is important to remember that summary measures are only trustworthy when data values come from a homogeneous population, which is not the case here.

```
DESCRIPTIVE STATISTICS

                   INTERVAL
N                       222
MEAN                 71.009
SD                   12.799
MINIMUM              42.000
MEDIAN               75.000
MAXIMUM              95.000
```

How, then, can we help the tourists? We need more information. One readily available characteristic of the geyser is the duration of the previous eruption. We could think of our data as pairs of the form (duration of eruption, time until next eruption). Here is a **scatter plot** of those pairs:

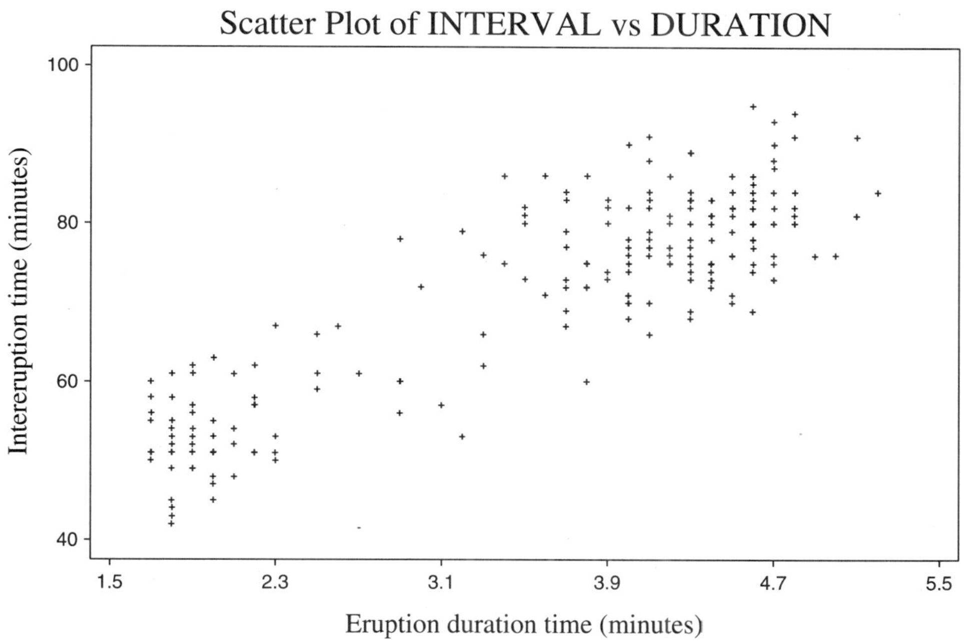

We notice two dominant effects here: there are indeed two distinct subgroups, and a longer eruption tends to be followed by a longer time interval until the next eruption. The existence of two subgroups in this type of data is rare, but not unheard of; J.S. Rinehart, in a 1969 paper in the *Journal of Geophysical Research,* provides a mechanism for this pattern based on the temperature level of the water at the bottom of a geyser tube at the time the water at the top reaches the boiling temperature. That a shorter eruption would be followed by a shorter intereruption time (and a longer eruption would be followed by a longer intereruption time) is also consistent with Rinehart's model, since a short eruption is characterized by having more water at the bottom of the geyser being heated short of boiling temperature, and left in the tube. This water has been heated somewhat, however, so it takes less time for the next eruption to occur.

A long eruption results in the tube being emptied, so the water must be heated from a colder temperature, which takes longer. A. Azzalini and A.W. Bowman provide further discussion of statistical analysis based on this model in a 1990 paper in *Applied Statistics*.

The properties of these two kinds of eruptions can be examined in more detail by separating the observations based on duration of the previous eruption. The following summary statistics are for intereruption times based on separating the eruptions by whether the duration is less than, or greater than, three minutes:

DESCRIPTIVE STATISTICS

FOR ERUPTION DURATION < 3 MINUTES

	INTERVAL
N	67
MEAN	54.463
SD	6.2989
MINIMUM	42.000
MEDIAN	53.000
MAXIMUM	78.000

FOR ERUPTION DURATION > 3 MINUTES

	INTERVAL
N	155
MEAN	78.161
SD	6.8911
MINIMUM	53.000
MEDIAN	78.000
MAXIMUM	95.000

Boxplots also can be stacked side–by–side on the same plot, separated by subgroups, to provide a graphical representation of this pattern:

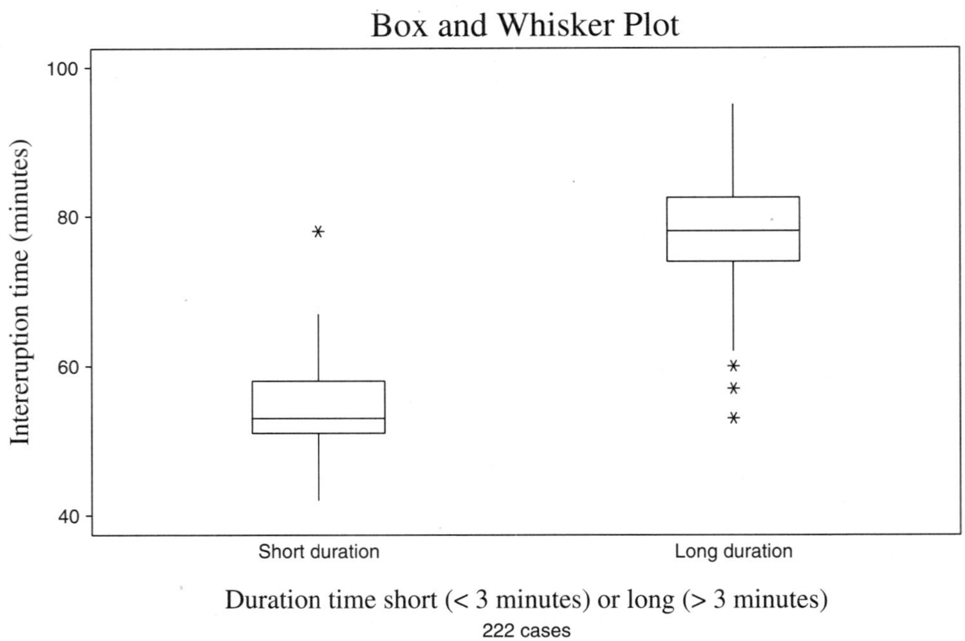

Based on these statistics and plots, a simple prediction rule would be that an eruption of duration less than 3 minutes will be followed by an intereruption interval of about 55 minutes, while an eruption of duration greater than 3 minutes will be followed by an intereruption interval of about 80 minutes. Further, the latter longer intereruption time would be expected to occur about two–thirds of the time.

We can evaluate the effectiveness of this prediction rule by testing it on new data. Azzalini and Bowman, in their 1990 *Applied Statistics* article, gave intereruption and duration times for 296 eruptions in August 1985, which can be used for this purpose (66 of the eruption durations occurred at night, and were recorded as simply "Short" or "Long," which we will treat as less than or greater than 3 minutes, respectively; 2 durations were listed as "Medium," which will be ignored here). Boxplots of the 1985 eruption durations show that apparently the relationship between eruption duration and the previous intereruption time had not changed appreciably in the six intervening years (which is what we would expect), which suggests that the prediction rule will work well:

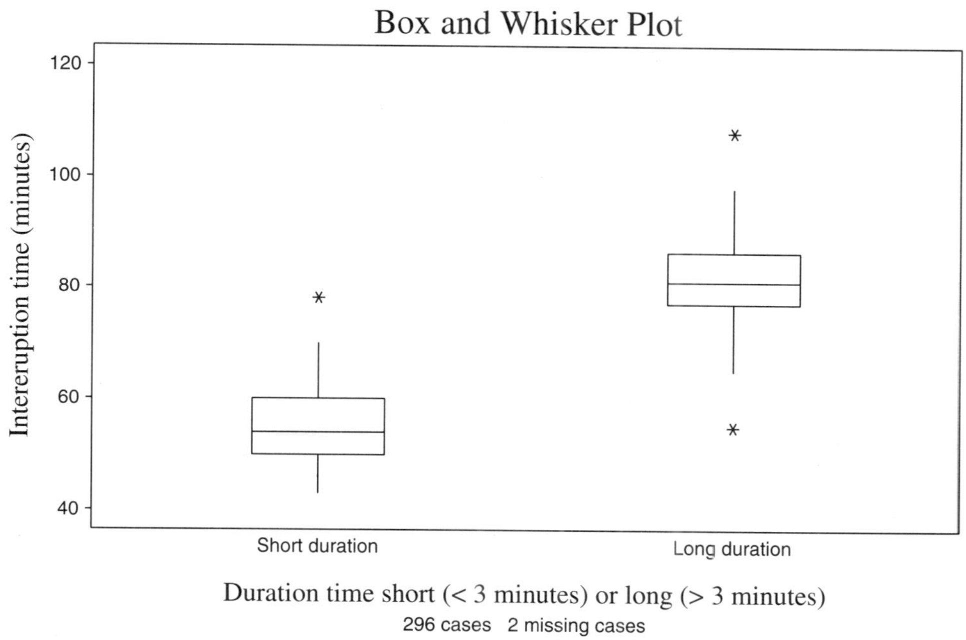

We can examine the error in predicting intereruption time from the simple prediction rule using August 1985 data. For example, the first eruption given lasted 4.0 minutes, and was followed by a 71 minute intereruption time. Since the duration was greater than 3 minutes, we would have predicted the intereruption time to be 80 minutes. Thus, the error made was $71 - 80 = -9$ minutes. This operation is then repeated for all available eruptions. A histogram of the error made in using the prediction rule shows that, except for three unusual eruptions, the rule predicts the intereruption time to within roughly ± 20 minutes (and usually to within ± 10 minutes).

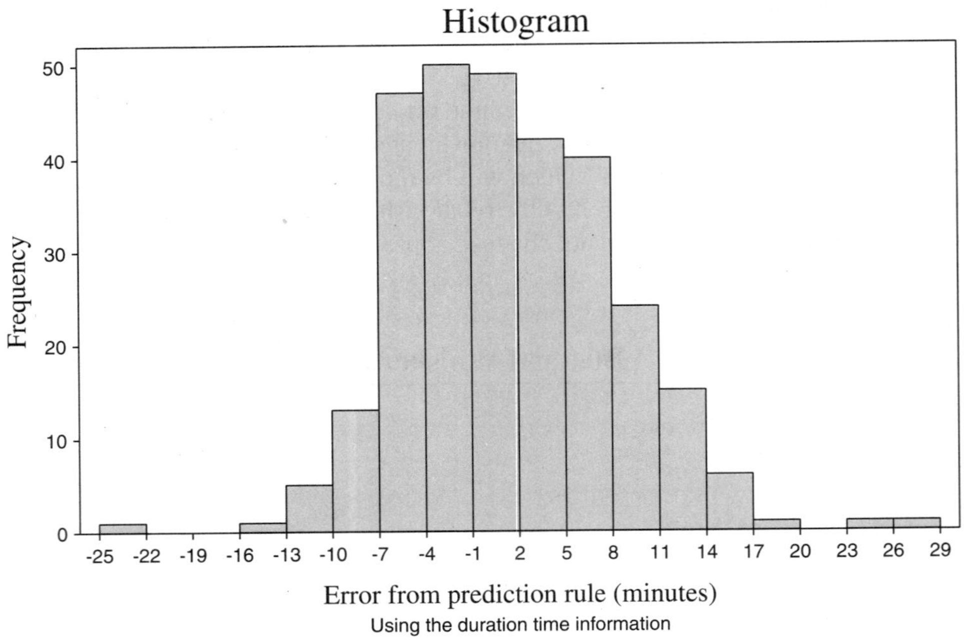

This performance can be compared with making a prediction based just on the distribution of the intereruption times. Recall that the overall median intereruption time for the 1978–1979 data was 75 minutes; a histogram of the errors from using that as a prediction rule for the 1985 data shows that the errors are bimodal, and dramatically larger than if the previous duration is used, even being off by as much as 30 minutes:

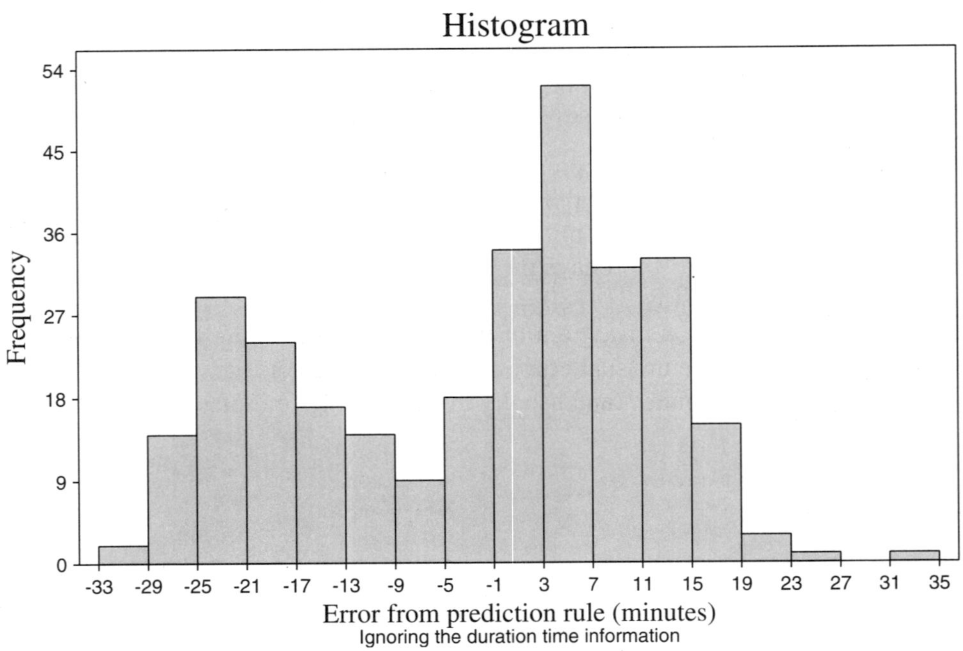

Thus, by using a very simple rule, the National Park Service can try to ensure that visitors to Yellowstone Park will get to see an "Old Faithful" eruption without waiting too long.

Summary

Intereruption times of the "Old Faithful" geyser in Yellowstone National Park are apparently bimodal, with modes centered at around 55 minutes and 75–80 minutes, respectively. These times are directly related to the duration of the previous eruption, with longer eruptions followed by longer intereruption times (and shorter eruptions followed by shorter intereruption times). This pattern is consistent with previously described physical models of geyser eruption. A simple rule that can be used to predict intereruption time is to predict one or the other of the two modal values, based on whether the previous duration was less than, or greater than, three minutes. This rule can be shown to work well on new eruptions of the geyser, supporting its general use.

Technical terms

Boxplot: a graphical device in which a rectangle is used to summarize the distribution of a batch of data. The top and the bottom of the rectangle represent the third and first quartiles, respectively. The line inside the rectangle represents the median. The lines extending from the top and bottom of the rectangle represent either the actual limits of the data, or the limits of the bulk of the data (with unusual observations being represented by individual symbols ["flagged"] if they are further out). The boxplot is particularly useful for comparing the location and variability of several batches of data, as boxes can be plotted side–by–side on one plot.

Histogram: a graphical device to represent the distribution of a batch of data. The data values are usually grouped into mutually exclusive and exhaustive intervals of equal width, and the number of observations in each interval is determined and represented by a vertical bar. In some variations the widths of the intervals are varied, resulting in potentially different appearances in the plot.

Median: an estimate of location determined by ordering the observations in the sample, and then choosing the middle one. If the sample size is even, the median is taken to be the average of the middle two observations. The median has the desirable property of not being greatly influenced by an unusual observation (that is, it is a *robust* estimate).

Mean: an estimate of location determined by adding all of the observations and dividing by the number of observations. It is also known as the average, and is certainly the most commonly used location estimate. An undesirable property is that it can be heavily influenced by even one unusual observation. The usual notation for the sample mean is the use of an overbar, as in $\overline{X}$.

Mode: a value, or range of values, characterized by having a greater likelihood of occurrence than values around it. A distribution might have one mode (being *unimodal*), two modes (*bimodal*), and so on. Another use of the term is as follows: the sample mode is the value that occurs most often in a sample.

Quartile: a value corresponding to a "25% point" in a sample. The first quartile is, roughly speaking, the value such that one–quarter of the observations are less than it, while three–quarters of the observations are greater than it. The third quartile is defined similarly, reversing the quantities "one–quarter" and "three–quarters." The second quartile corresponds to the median. Different statistical packages often use slightly different definitions of the quartiles for a given sample size. A measure of the variability in a set of data that is insensitive to a few unusual values is the difference between the third and first quartiles, called the *inter–quartile range*.

Scatter plot: a method that can be used to study the joint variation of two variables graphically. Each observation is represented by a point on the plot, indexed by the values on the axes. Each axis is used for a different variable. Besides showing how (and whether) two variables are related to each other, scatter plots also can indicate the existence of distinct subgroups in the data.

Standard deviation: an estimate of scale (or variability) determined as the square root of the sum of squared differences between the value of an observation and the sample mean, divided by one less than the sample size.

Stem–and–leaf display: a graphical device to represent the distribution of a batch of data. Very similar to a histogram, it is often accompanied by additional information about the data, such as cumulative frequencies and the position of the median. The plot represents the data values by their numerical values, providing additional information over the histogram, but the grouping intervals are usually chosen based on using round numbers, rather than in an attempt to provide the most effective plot.

International adoption rates

Topics covered: Data summary. Examination of univariate data. Transformations. Variation over time.

Key words: Boxplot. Histogram. Mean. Median. Outliers. Scatter plot. Skewness. Standard deviation. Stem–and–leaf display.

Data File: `adopt.dat`

The issues of adoption in general, and international adoption in particular, have become prominent in recent years. During the years 1991–1993 delegates from 65 countries gathered in The Hague to create a multilateral treaty to facilitate and safeguard intercountry adoptions. Many states and the federal government have instituted legislation recently concerning adoption and the rights of biological parents, adoptive parents, and adopted children. Despite the lack of consistent procedures governing international adoption, American residents have adopted many children from foreign countries, with visas issued for that purpose being consistently between 5000 and 10,000 annually. Clearly, any reasonable attempt to understand the process of international adoption should be based, to some extent, on statistics describing the process. What do the numbers of international adoptions tell us about the process? Is the process stable, or does it vary over time? The data set examined here represents the number of visas issued by the U.S. Immigration and Naturalization Service for the purpose of adoption by U.S. residents (data source: *Ours Magazine*, May/June 1993). The number of visas is categorized by country or region of origin and the fiscal years of 1988, 1991 and 1992. The fiscal year covers October 1 of the previous year through September 30 of the current year. The values for 1991 will be the primary focus here, with the relationships of the 1991 values with those of other years also being examined. Note, by the way, that for most countries, the number of visas issued in any of these years was zero. Analysis will be focused on those countries or regions with nonzero values.

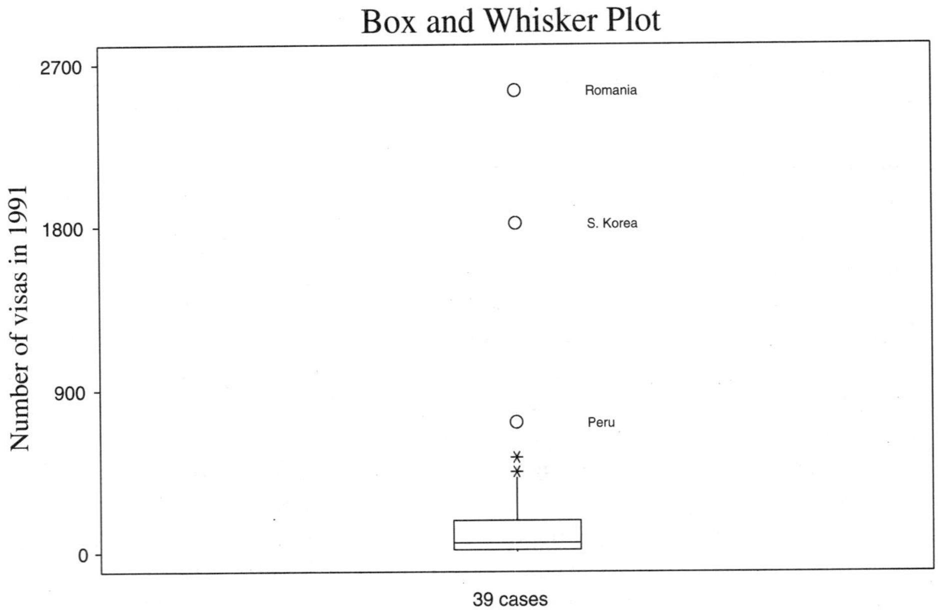

A boxplot, stem–and–leaf display and histogram all show that the distribution of these data has a long right tail (that is, it is not symmetric, but rather positive– or right–skewed). That is, most countries for which any visas were issued had a small number issued, with a few countries having very many more.

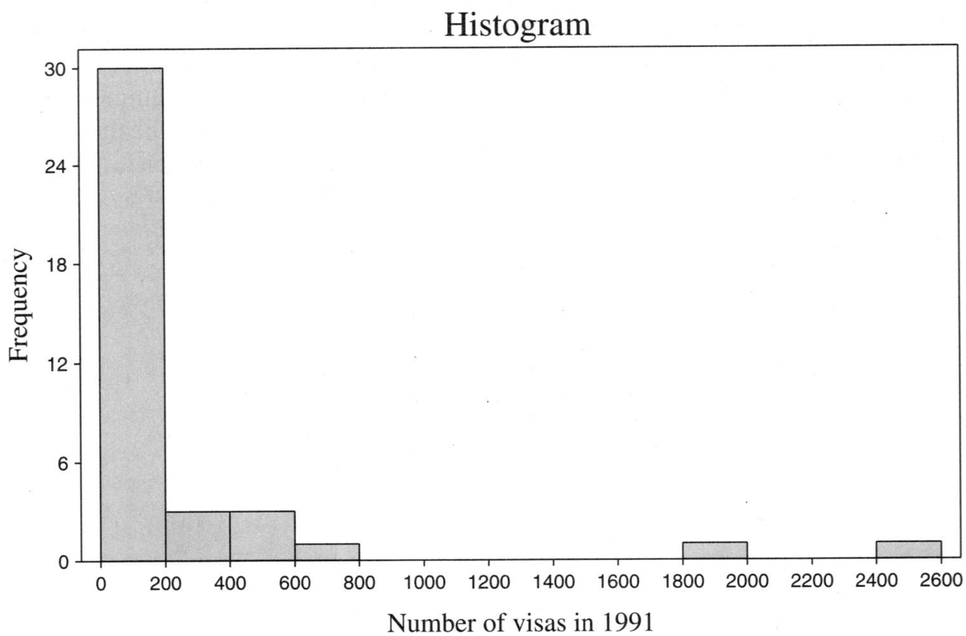

```
STEM AND LEAF PLOT OF VISA91 Number of visas in 1991

   LEAF DIGIT UNIT = 100
   1  2  REPRESENTS 1200.

          STEM  LEAVES
      (30) +0*  000000000000000000000000011111
         9 +0T  223
         6 +0F  445
         3 +0S  7
         2 +0.
         2   1*
         2   1T
         2   1F
         2   1S
         2   1.  8
         1   2*
         1   2T
         1   2F  5

39 CASES INCLUDED     0 MISSING CASES
```

Note, by the way, that the values with leaf value "0" for the stem labeled "0" are not, in fact, exact values of zero, but rather refer to values less than 100.

This pattern to the distribution is not surprising. International adoption is a very political process; governments often decide as a matter of public policy whether they will allow international adoption at all, and at what level. For example, for many years U.S. residents adopted more than 4000 children per year from South Korea. In 1989 the Korean government made the decision to restrict such adoptions in the future, so that the number of visas issued from Korea dropped after that year.

The descriptive statistics are consistent with the long right tail. The mean of 227 is not very meaningful, being inflated by the few unusually large values. Such unusual values are often termed **outliers**, since they lie outside the range of most of the observations. The median of 55 is a much more reasonable value for a "typical" value, since outliers have relatively little effect on it (such statistics are called *robust* statistics).

```
DESCRIPTIVE STATISTICS

                  VISA91
N                     39
MEAN              227.15
SD                496.26
MINIMUM           4.0000
MEDIAN            55.000
MAXIMUM           2552.0
```

These are the kind of data that often benefit from **transformation**; that is, looking at the data through a scale that emphasizes the structure better. One such transformation that is often appropriate for long right–tailed data is the logarithm. Here is a histogram of the logarithm of the data (base 10).

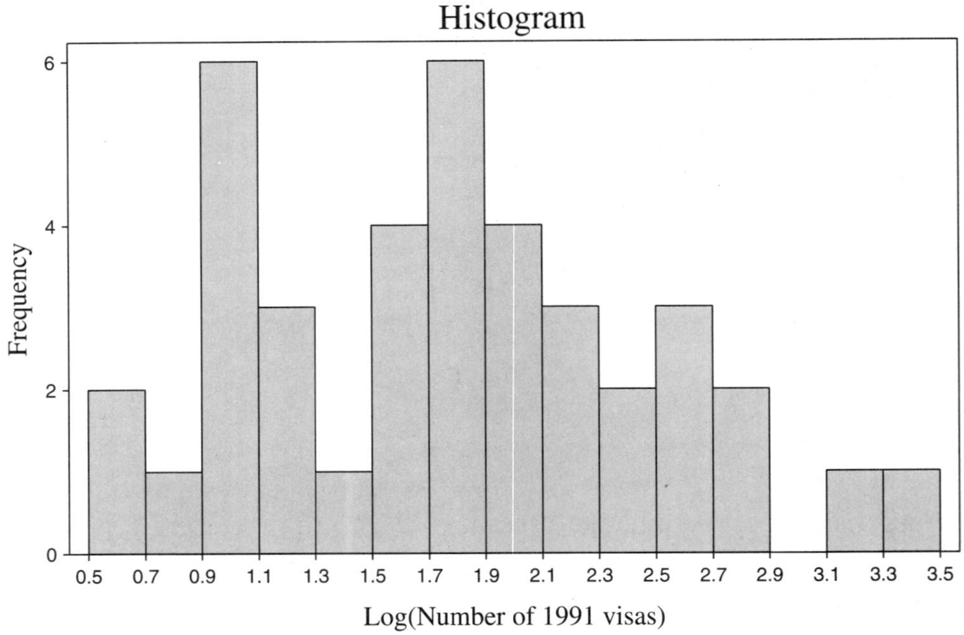

Note that taking logs has brought in the long right tail, and has enabled us to see some interesting structure that was formerly all contained in the tall first bars. There appear to be two groups around 1 (transforming back, that's 10^1, or 10) and a bit less than 2 (10^2, or 100). This might suggest some sort of (perhaps informal, or even unconscious) quota going on at the round numbers of 10 and 100. Another possibility is reporting bias, where numbers are "rounded off" before being reported.

We also should recognize that data of this type can only give us a "snapshot" of the particular moment in time. As was noted earlier, political pressures have a lot to do with the issuance of adoption visas. Indeed, while the U.S. Department of State provides information regarding the status of international adoption in various countries periodically, it also makes the point that these synopses can only "provide a brief snapshot of what is a complex and dynamic process" (see *Ours Magazine*, January/February 1994). Here is a scatter plot of the 1991 visas versus 1988 visas: note that while the pattern was fairly stable overall, we see big shifts in Korea (large 1988, smaller 1991), Romania (small 1988, large 1991) and Peru (small 1988, large 1991), corresponding to shifts in government policy.

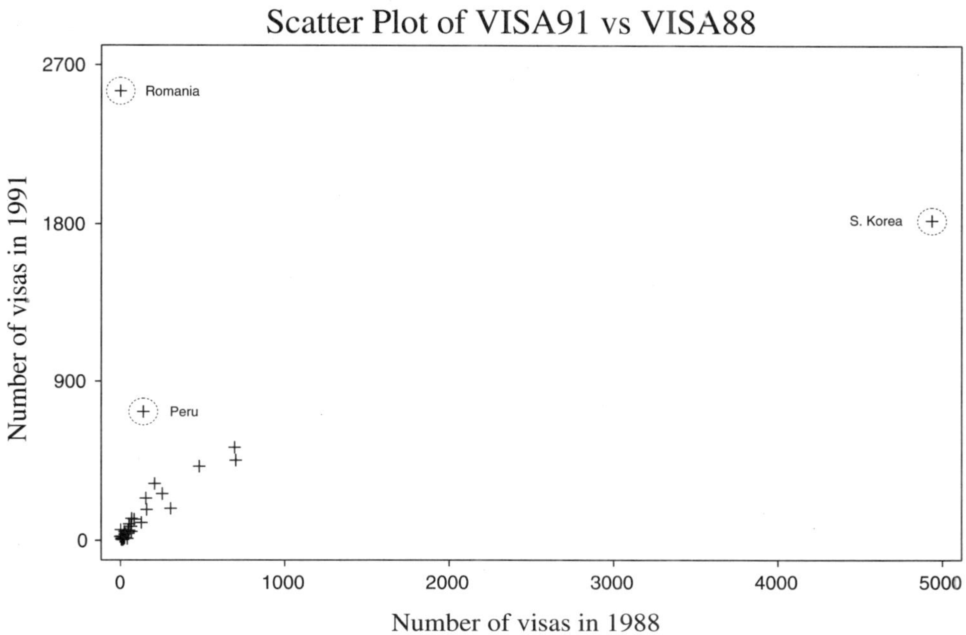

The pattern is even clearer in a plot of the two logged variables (two 1988 visa values are zero, so we set `LOGV88` to zero for them):

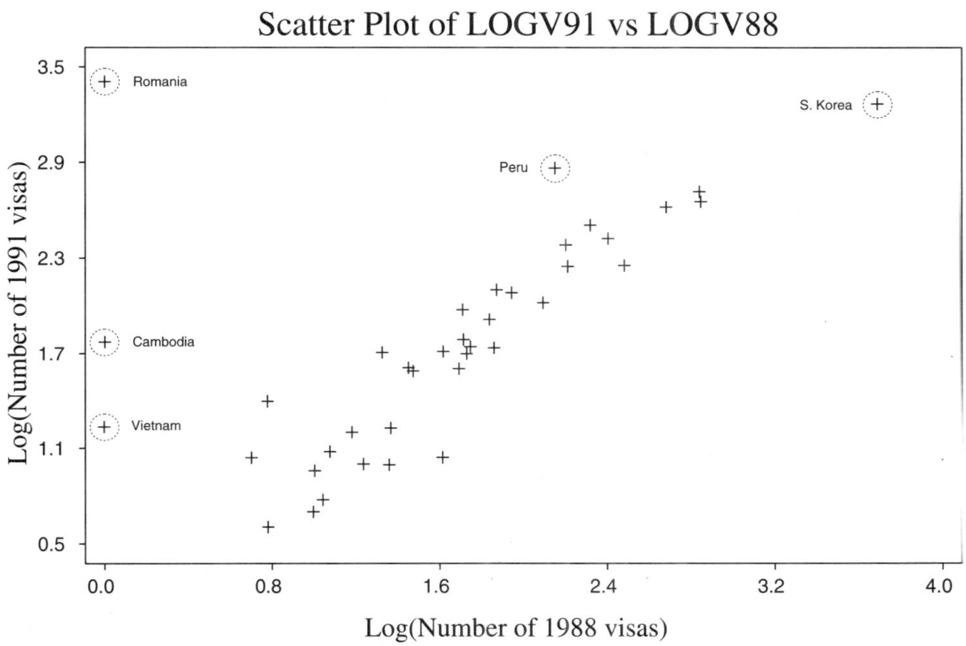

Cambodia also shows up as unusual, going from no visas in 1988 to 59 in 1991. Note also that in this perspective Korea seems more consistent with the general pattern, albeit still an outlier in 1988. Here are corresponding pictures relating 1991 to 1992 visas:

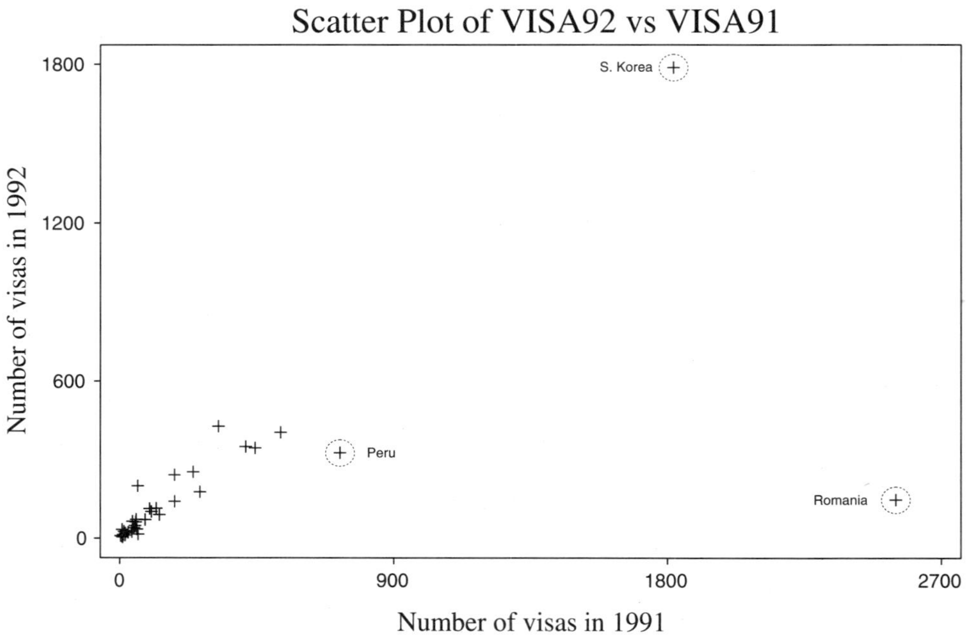

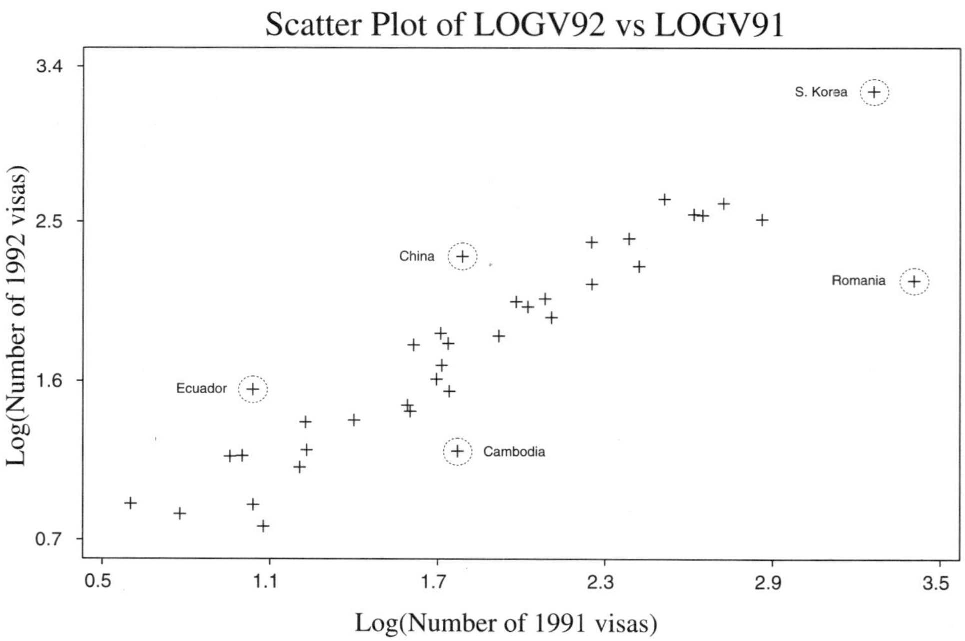

Overall, the pattern in visas issued was fairly stable from 1991 to 1992, but we can see one point in the original scale with very high 1991 visas and moderate 1992 visas. This is Romania, which allowed many international adoptions immediately after the fall of the Communist regime, but then decided to restrict them severely. We also can note that Peru's relatively large number of visas in 1991 was apparently an aberration, since its number dropped by almost 60% in 1992. Thus, while it might be possible to use the number of visas issued in one year to try to predict the number in another year, it must be remembered that such predictions can be very poor if the country being studied has had a shift in its policies towards international adoption.

Finally, we can get an idea of the changes from 1991 to 1992 compared with the changes from 1988 to 1991. We will be focusing on the ratio of the number of visas, a measure of relative change. We also could consider the difference in the number of visas, a measure of absolute change. Here is a scatter plot of the logarithm of the ratio of visas from 1988 to 1991 plotted against the logarithm of the ratio of visas from 1991 to 1992 (one has been added to each case to avoid problems when no visas were issued for a country in a given year).

Scatter Plot of L92ML91 vs L91ML88

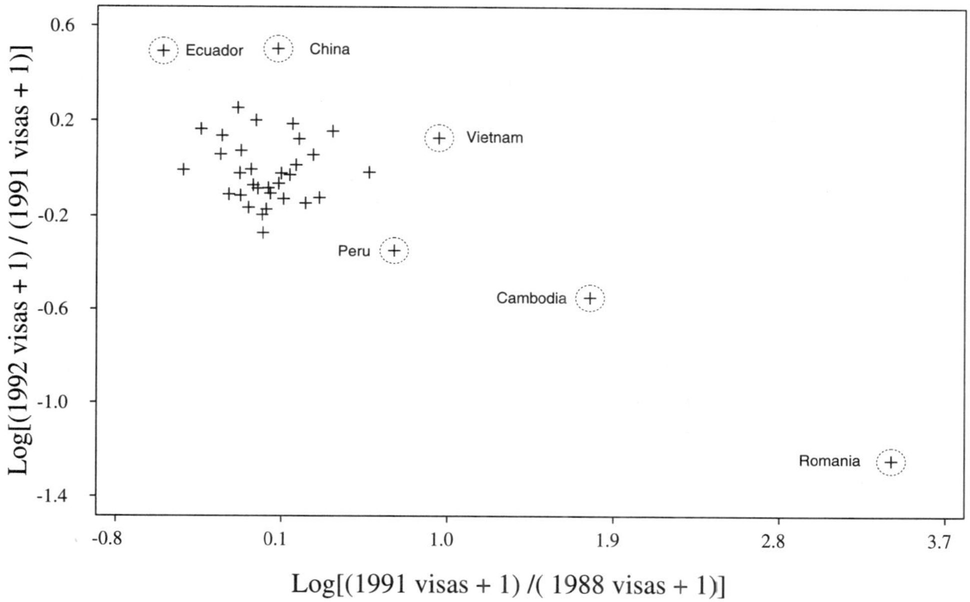

Romania, Cambodia and Peru had large relative increases from 1988 to 1991 and large relative decreases from 1991 to 1992. Vietnam had a big increase from 1988 to 1991 followed by a modest increase from 1991 to 1992. China had a modest increase from 1988 to 1991 followed by a large increase from 1991 to 1992. Ecuador had a big decrease from 1988 to 1991 followed by a big increase from 1991 to 1992. Interestingly, the inverse relationship in the scatter plot suggests that increases tend to be followed by decreases and vice versa.

Summary

While the U.S. Immigration and Naturalization Service issues no visas for the purpose of adoption by U.S. citizens for most countries of origin, certain countries do allow such adoptions. For most such countries, there are few visas issued, although for a few countries the number of visas can be in the thousands for a given year. By transforming the number of visas issued in 1991 using a logarithmic scale, it is possible to see further structure in the data, corresponding to peaks at around 10 and 100 visas. Although the number of visas issued appears to have been fairly stable over the years 1988, 1991 and 1992, there are particular countries that exhibit very large changes from year to year, reinforcing the dynamic nature of the international adoption process. There is also some evidence that increases in visas issued tend to be followed by decreases, and vice versa.

Technical terms

Outlier: an observation whose position is unusual compared with that of the bulk of the data values. Values of this type can have a strong effect on the behavior of different statistics, and are often indicative of complexities in the process being examined that could warrant further study.

Right–skewed data: data where the positions of the values are not symmetrically distributed around the center of the distribution, but rather trail off slowly on the upper side and fall off rapidly on the lower side. The largest few values are called the **right tail** of the distribution. Thus a **long right tail** means that the largest values are spread out. **Positive–skewed** is a synonym, while *left–skewed, left–tailed*, and *short right–tailed* are defined in the natural complementary way.

Transformation: changing the scale of measurement of a variable for the purposes of analysis, rather than using the original scale. Data with a long right–tailed distribution (and a large range of variation) are often analyzed after taking the logarithm (a so–called log transformation), in order to reduce the range of variation, and reveal potential structure in the data values using graphical displays. The logarithm can be taken to any base, such as base 10 (common logarithm) or base e (natural logarithm), although the common logarithm has the advantage of being easily interpreted as the number of digits of the value (log–values between 0 and 1 correspond to single–digit values in the original scale; values between 1 and 2 correspond to double–digit values; and so on). Note that $\ln(x) = \log_{10}(x) \times \ln(10) = 2.3026 \log_{10}(x)$, so that the natural logarithm of a number is proportional to its common logarithm.

The performance of stock mutual funds

Topics covered: Comparison of location of subgroups. Comparison of variation in subgroups.

Key words: Mean. Outliers. Scatter plot. Side–by–side boxplots. Standard deviation.

Data File: `funds.dat`

Mutual funds have become amazingly big business. In 1994, Americans invested \$2 trillion in over 5000 mutual funds. Such funds offer the advantages of providing a diversified investment, but of course at the risk of low (or even negative) returns. The May 1993 issue of *Consumer Reports* provided information about a selection of the largest stock mutual funds available to the general investor, classified by the type of fund (in terms of investment objective):

Type of Fund	Stated Investment Objectives
Balanced	aiming for a "balance" of stocks and bonds, resulting in less risk, but lower returns in a bull market
Equity–Income	investing in dividend–paying stocks, typically those of well–established companies
Growth–and–Income	aiming for potential for growth in share price, as well as regular income from stock dividends
Growth	investing mainly in smaller companies and other stocks with the potential for rapid appreciation
Aggressive–Growth	similar to Growth funds, except that they may use riskier investment strategies
Small–Company	investing in small, young companies with the potential for rapid growth
International–and–Global	International funds invest in stocks outside of the U.S., while Global funds invest in both foreign and U.S. stocks

Is there a difference in performance based on the objective of the fund?

Performance here is measured in two ways. The variable `FIVEYR` represents the five–year performance of the fund (over the years 1988 through 1992) in the following way: the amount an investor would have accumulated by investing \$2000 in the fund at the beginning of each year. Any front–end sales charges and redemption fees have been deducted from the dollar figure. The variable `RETURN92` gives the fund's return (in percent) for 1992, rounded to the nearest integer. Sales charges and redemption fees are not included in the return. Thus, the former variable reflects long–term performance, while the latter variable addresses recent one–year performance.

First, calculate descriptive statistics for the entire sample. What are the mean five–year performance and mean 1992 return? Standard and Poor's 500 Stock Index (commonly called S & P 500) is the widest general–market index of stocks produced by the U.S. credit–rating agency Standard and Poor. The index is constituted by using the prices of securities of 425 U.S. industrial companies and 75 railway and public–utility corporations. Investors regard the S & P 500 as representing the market, the general level of the price of securities in the U.S. The success of an investment strategy is often measured by the degree to which the strategy beats the S & P 500. The reason for this is that by holding a portfolio that represents the S & P 500 (a Market Index Fund), one can do nearly as well as the S & P 500. The S & P 500 is often used as a base level for comparison. The five–year performance and 1992 return for the S & P 500 are $15,440 and 8%, respectively. Taking these as base level performance figures, what does that say about the investment abilities exhibited by the managed mutual funds in the study?

Look at the five–year performance first. Construct side–by–side boxplots, forming groups based on the different types of funds. Note that while the risk of the type of fund tends to increase from left–to–right in the figure, the position of each group on the horizontal axis is somewhat arbitrary. Also, examine descriptive statistics separated by type of mutual fund. Are there differences in five–year performance relating to fund type? Do certain types of funds seem to have been more successful than others over this time period? Are the patterns surprising, given that the market did well during this time period?

Do you note any unusual funds in terms of performance from examination of the boxplots? Note that evaluating any such funds relative to their group, rather than relative to the data set as a whole, is important for these data. The five–year performances of the various fund types are different from each other with respect to both location and scale; that is, both the performance of a typical fund and the variability of that performance from fund to fund changes as the type of fund changes. This is apparent from the different lengths of the boxes in the boxplots. Do you see any relationship between the variability of returns for a given fund type and the performance itself? Given the general investing axiom that higher performance goes with higher risk, what sort of relationship would you expect to exist?

Now shorten the time horizon from five–year to one–year (1992) performance. Repeat the analyses previously done on the five–year return variable. Are there differences related to fund type? Are any observed differences between fund types similar to ones observed for five–year performance? What does that imply about the connection between short– and long–term performance? Another way to investigate this is by constructing a scatter plot of one–year versus five–year performance. Does this plot suggest that funds that do well (or poorly) on one measure necessarily do well (or poorly) on the other?

Comments to the Instructor

The mean five–year performance for the entire sample is $14,940, while the mean 1992 return is 7.35%. Since the corresponding numbers for the S & P 500 index are $15,440 and 8%, respectively, the average performance is below the market average; these numbers are not exactly a ringing endorsement for managed mutual funds.

Let's look at the five–year performance first. Separating by subgroup shows considerable variability between types of fund. In particular, the Aggressive and Small–Company funds have the best performance, the Balanced, Equity–Income, Growth–and–Income and Growth funds are all at the next level of performance, and the International–and–Global funds have relatively poor performance. Apparently over the five year period 1988–1992 an aggressive (risky) strategy was most effective (this is not a big surprise, given that the market did well overall during this time period), and foreign stocks did quite poorly. Note, however, that only the three top types outperformed the S & P. As financial theory would imply, the more aggressive types of funds exhibited more variability than the less aggressive types.

We can also note two apparently unusual funds — one in the Growth–and–Income funds (it is Vista Growth–and–Income, with a five–year performance of $22,331, which is about $4\frac{1}{2}$ standard deviations above the mean for that group), and one in the Growth funds (it's CGM Capital Development, with a five–year performance of $22,934, about $3\frac{1}{2}$ standard deviations above the mean for that group). Each of these funds exhibited particularly strong performance for funds of their type.

The pattern for the 1992 returns is different; all of the fund types seem to be very similar (around the overall median), except that the International–and–Global funds are much worse.

The differences in patterns show that long– and short–term performance can be very different. A scatter plot confirms this. Although there appears to be a small positive slope in the plot, this is actually due completely to the poor performance of the International–and–Global funds for both the long– and short–term. If these funds are omitted from the analysis, there is virtually no relationship between `FIVEYR` and `RETURN92`.

To maintain consistency with the unannotated edition,
the next case begins on the following page.

Predicting the sales and airplay of popular music

Topics covered: Data summary. Examination of univariate data. Prediction.

Key words: Scatter plot.

Data File: `rock.dat`

Each week the music industry magazine *Billboard* publishes lists of the most popular songs from the last week. The songs are ranked in various categories of interest to radio station disk–jockeys, music recording industry executives and, of course, the general public. The "Hot 100 Airplay" list ranks the songs in terms of the amount of airplay that they receive at radio stations over the last week. The data are compiled based on a national sample of 194 stations electronically monitored 24 hours per day. The "Hot 100 Singles Sales" list ranks the songs in terms of the number of sales of singles (records and compact disks) in the last week. The data are compiled from a national sample of point–of–sales equipped retail stores. Both lists provide the ranking of the song on the list from the previous week and the number of weeks that the song has been on the list. The music industry (i.e., recording, radio and television) uses this information to track trends and guide airplay.

Despite the names of these lists, each contains only the highest ranked 75 songs. As we wish to compare the songs in terms of airplay and sales, we will only consider the songs that are on both lists. The data are the 44 songs on both lists for the week ending September 10, 1994.

What is the relationship between the amount of airplay a song receives and the retail sales of that song in a given week? Do some songs appear to be different?

Do songs that receive more airplay in a previous week sell more this week?

One useful measure of the change in sales of a song is the change in ranking of the song on the "Hot 100 Singles Sales" list this week from last week (positive values mean that the song is ranked higher this week than last week). Similarly we can measure the change in airplay of a song by the change in ranking of the song on the "Hot 100 Airplay" list this week from last week.

Is there a relationship between the change in sales of a song and the change in airplay?

How does the change in sales of a song depend on the number of weeks that it has been on the chart? How does the change in airplay of a song depend on the number of weeks that it has been on the chart?

Comments to the Instructor

The plot of sales versus airplay indicates a modest positive relationship, although the variation is quite large. The plots for sales last week versus airplay this week and sales last week versus airplay last week are similar.

The plot of the change in sales versus the change in airplay indicates an extreme outlier with a change in sales of 63 places. This is "At your best (You are love)," by Aaliyah. This was the first week it was on the sales chart and it ranked 13th in sales. The jump in sales is at least this much, as the imputed rank for last week (76) is conservative (that is, it could have been much larger). With this case removed there is still a strong positive association between the change in sales and the change in airplay.

The change in sales decreases steadily as the number of weeks on the sales charts increases. There are a couple of high outliers in the first few weeks on the chart. A similar pattern occurs for the change in airplay as the number of weeks on the airplay charts increases.

To maintain consistency with the unannotated edition,
the next case begins on the following page.

Another look at the "Old Faithful" geyser and adoption visas

Topics covered: Examination of univariate data.

Data Files: `geyser1.dat`, `adopt.dat`

As we saw in the two cases "Eruptions of the 'Old Faithful' geyser" and "International adoption rates," the histogram can be a very useful way to investigate the characteristics of a variable for a set of data. The histogram is (the simplest) example of what is called a **density estimator**. As occurs for all density estimators, the appearance of the histogram is dependent on the degree of smoothing chosen by the data analyst. For the histogram, smoothness is determined by the width of the histogram bins (or, equivalently, the number of bins); large bins imply more smoothness. In addition, the form of the histogram is dependent on the "anchor" of the histogram — that is, the beginning value of the first bin in the histogram.

In this case, the effect of these choices on a histogram estimator is investigated.

For both the "Old Faithful" intereruption times and the (logged) adoption visa counts, construct histogram estimates based on a wide range of bin widths. How stable are the resultant estimates when the bin width is varied? If the appearance of the histogram does not change very much when varying the bin width over a reasonably wide range, then the data analyst can feel confident that any observed patterns are genuine. On the other hand, if the appearance changes in a fundamental way depending on the bin width, any observed patterns when using a particular bin width might just be an accidental result of that choice, and cannot be trusted as much. Do the histograms for these data sets vary enough so as to cast doubt on the impressions discussed in the two earlier cases? Is the appearance of the histogram strongly dependent on the choice of the anchor position? Which choice seems to have a stronger effect on the appearance of the histogram — the bin width or the anchor position?

Comments to the Instructor

Consider first the geyser data. The histograms of both `DURATION` and `INTERVAL` are quite insensitive to the choice of bin width. Bin widths ranging from 0.1 to 1.0 all indicate a bimodal shape for `DURATION`, and bin widths ranging from 1 to 10 all indicate a bimodal shape for `INTERVAL`. The choice of the anchor seems to have virtually no effect at all. Thus, there is every reason to think that the previously noted bimodal shapes of these variables are genuine.

The situation is different for the (logged) adoption visas. Using an anchor of 0.6, the resultant histograms for the logged 1991 visas appear unimodal if bin widths of 0.2 or 0.4 are used, while they appear bimodal if bin widths of 0.3 or 0.5 are used. This casts some doubt on the hypothesis of possible quotas on visas issued. The anchor position has considerably weaker effect on the appearance of the histograms for these data as well. For example, if the bin width is taken to be 0.3, the resultant histograms are quite similar for anchor positions from 0.3 to 0.6.

That the anchor position would have less of an effect on the histogram than the bin width is consistent with statistical theory.

Productivity versus quality in the assembly plant

Topics covered: Association in subgroups. Illusory correlation.

Key words: Boxplot. Regression line. Scatter plot.

Data File: `prdq.dat`

For at least the last fifteen to twenty years, the relative quality of goods manufactured in the United States versus those manufactured in other countries has been a major topic of discussion. One industry where this has been an important issue is the manufacture of automobiles, where the market share of U.S. firms dropped dramatically over this period, in large part as a result of the perception that foreign automobiles (particularly Japanese automobiles) were of higher quality. That this perception was, in fact, a correct one has been supported by many objective criteria, including the ratings of quality produced by various consumer organizations.

A good deal of the success of Japanese manufacturing can be attributed to their willingness to embrace an entirely different way of manufacturing from the classic *mass production* of U.S. and European industry. This approach, which has been termed *lean production*, is based on embracing the notions of statistical quality control, just–in–time manufacturing, and so on.

A common perception in mass production–based industry is that quality and productivity are, in a certain sense, contradictory goals — that is, improved quality can only be achieved at the cost of additional time, effort and cost to the manufacturer, thereby leading to reduced productivity of the workers. If this is true, it casts considerable doubt on the usefulness of lean production, since manufacturers are unlikely to move towards methods to improve quality if they are accompanied by increased costs and reduced productivity. Does there appear to be such a tradeoff for the automobile industry? This would result in an inverse relationship between quality and productivity; is such a relationship actually there?

We examine here data that come from automotive assembly plants in 1989. Values for the number of hours per vehicle spent in production and the number of assembly defects per 100 cars are available for 27 automotive plants. Note that for these measures, high productivity is associated with a low value of hours per vehicle, and high quality is associated with a low value of defects per 100 cars. The source of these data is *The Machine That Changed the World*, by James P. Womack, Daniel T. Jones and Daniel Roos (page 93). On the next page is a scatter plot of quality versus productivity. Superimposed on the plot is a line based on the data that is designed to represent the relationship between the two variables. This line is called a **regression line**.

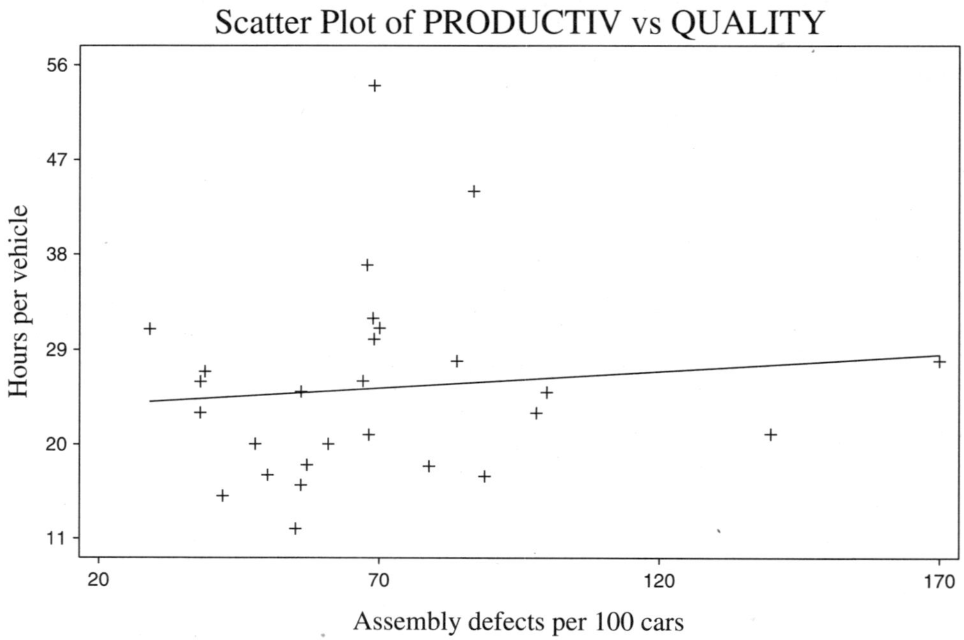

Apparently there is not a tradeoff between quality and productivity, as the observed relationship suggests that productivity increases slightly as the quality increases. This representation is not the most effective one to make, however, since this sample comes from two subpopulations — Japanese (domestic and foreign) facilities and non–Japanese facilities.

It would be more reasonable to look at the quality / productivity relationship for each group separately:

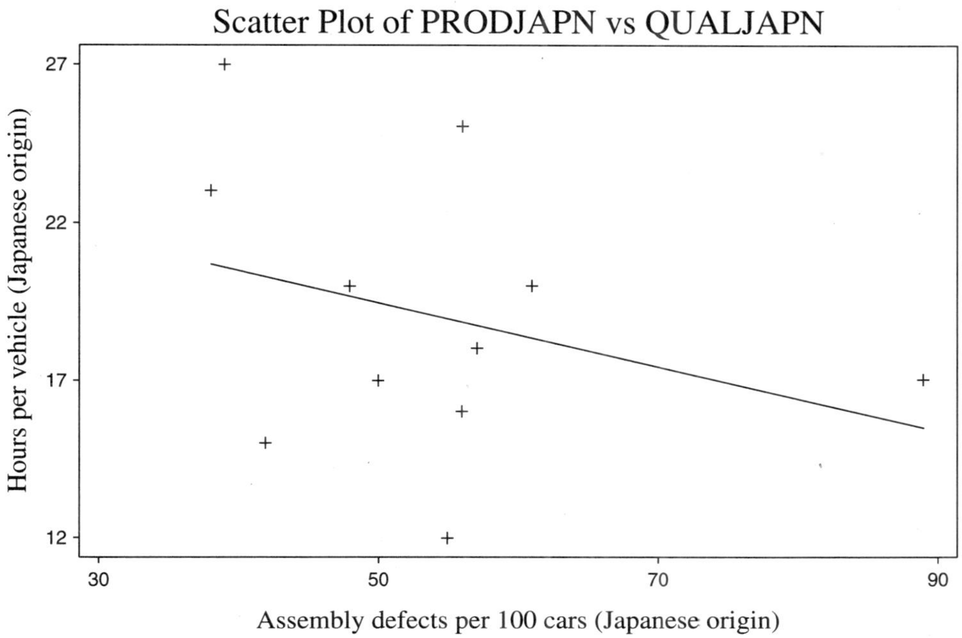

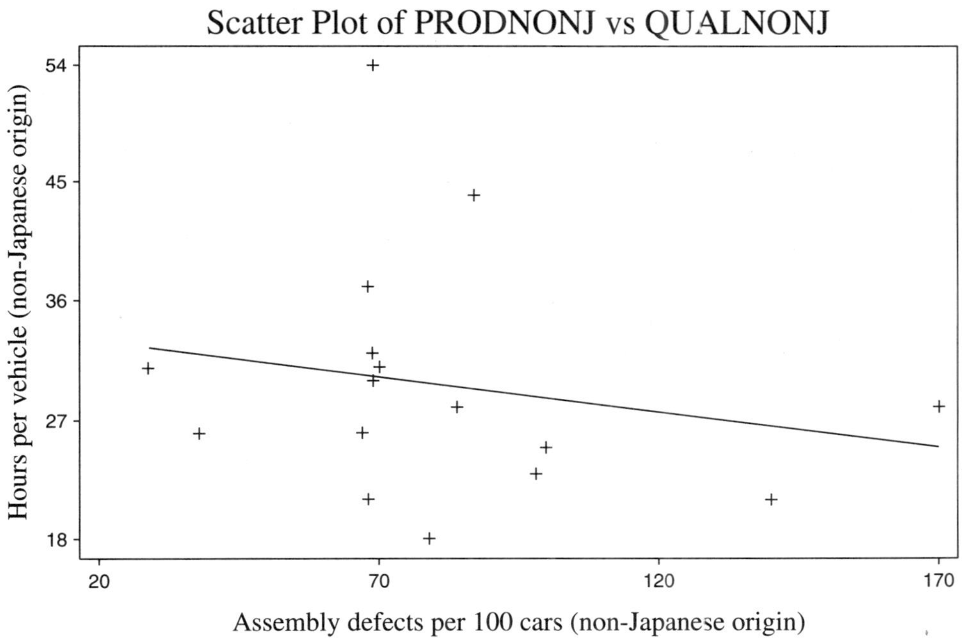

These plots are pretty remarkable. Despite the general *direct* association between productivity and quality, within each group there is an *inverse* association. This can be seen clearly in the following plot:

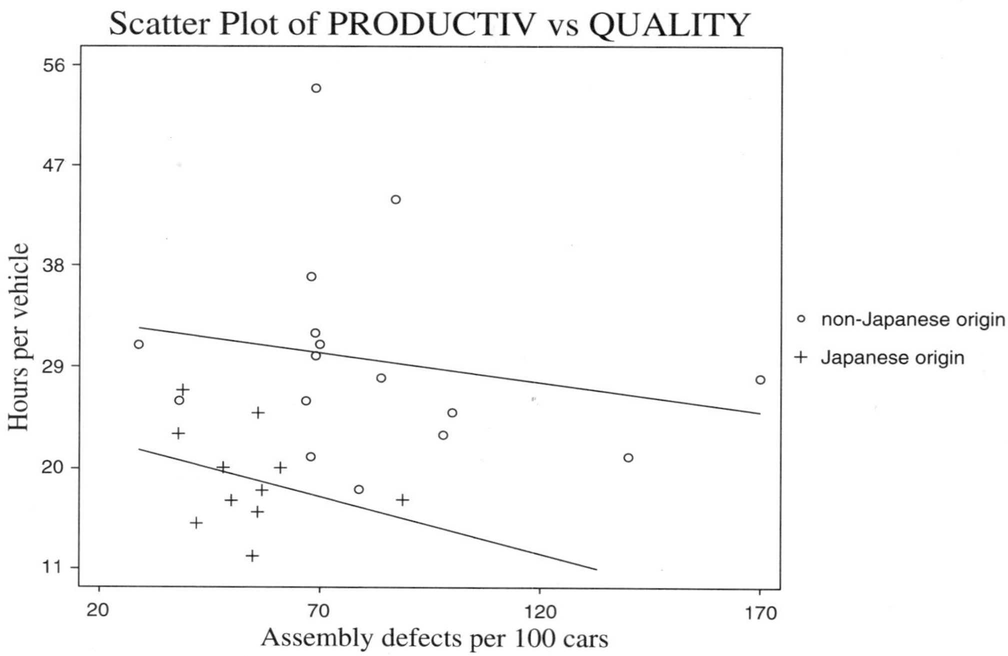

What is going on? This is an example of what is called **illusory correlation**; that is, a situation where the observed association in the entire sample does not reflect the relationship within each subgroup. This is caused (for these data) by the fact that the Japanese plants exhibit both higher quality and higher productivity, while the non–Japanese plants exhibit both lower quality and lower productivity. Side–by–side boxplots illustrate what is happening here:

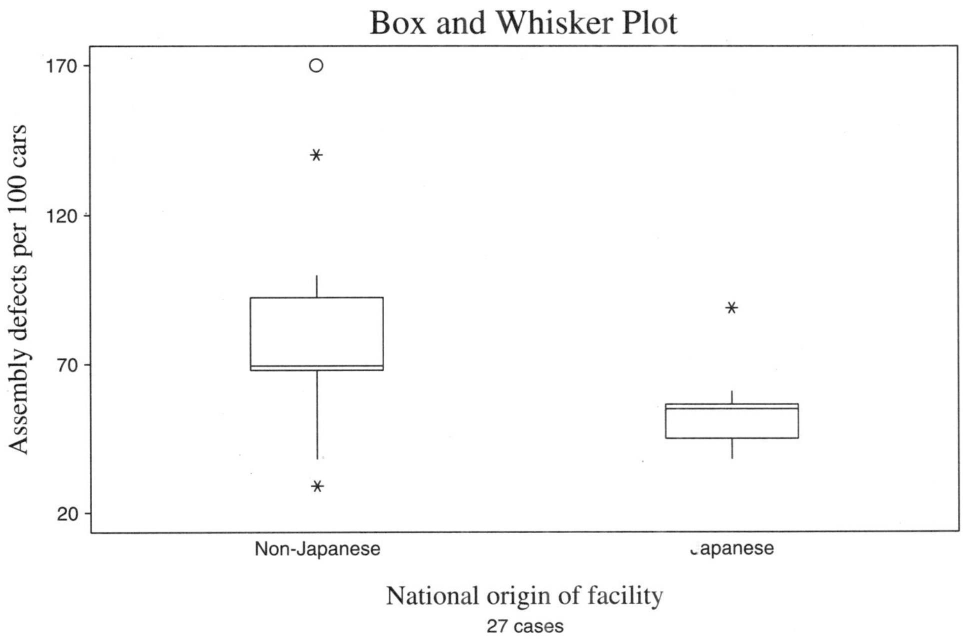

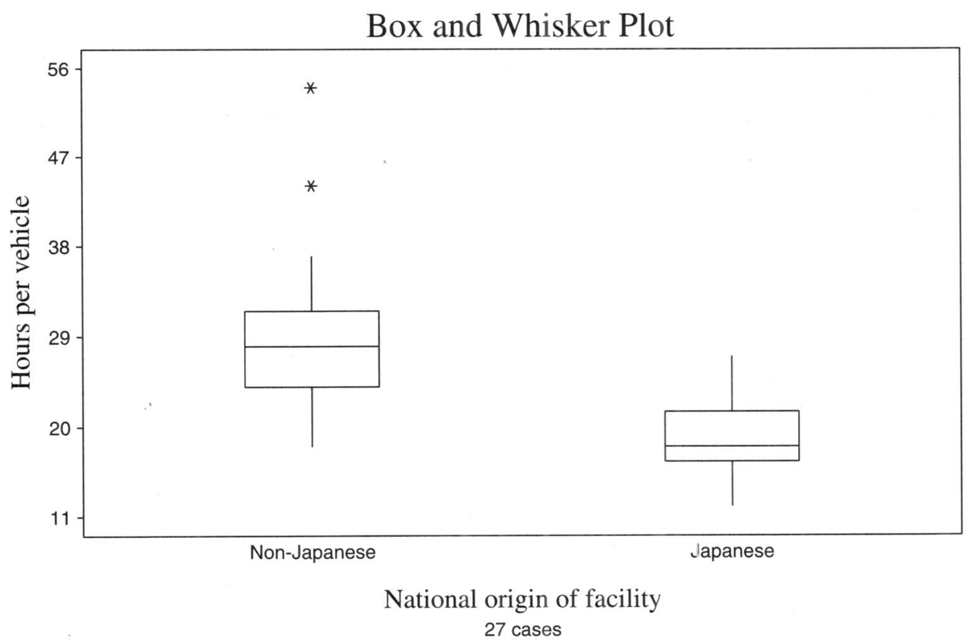

Thus, the relationship between quality and productivity appears to be a direct one, even though, given origin of the plant, the two are inversely related.

Let's go back to those negative correlations we saw before. The one for the non–Japanese firms is the classic quality / productivity tradeoff that occurs in mass production systems, as was discussed earlier. However, the same pattern occurs for the Japanese firms. This directly contradicts the contention that "for lean plants quality is free" (which would imply a positive correlation). The Japanese plants have higher quality and higher productivity, but given those higher general levels, higher quality is, in fact, associated with lower productivity.

If we take a look at the original plots, we might wonder if the somewhat unusual cases in the lower right in each plot have affected the superimposed regression lines. In fact, these points have had little effect:

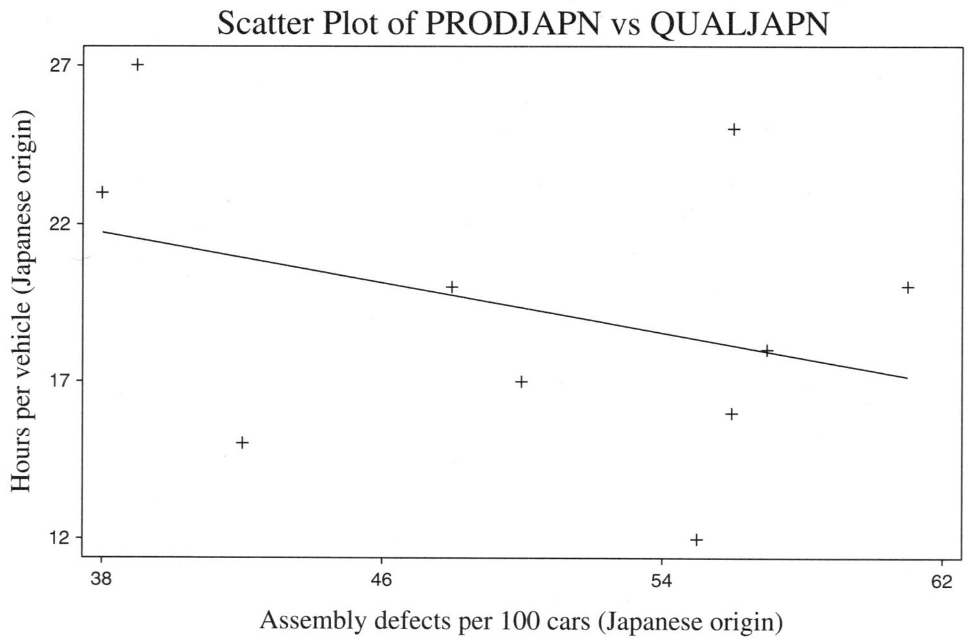

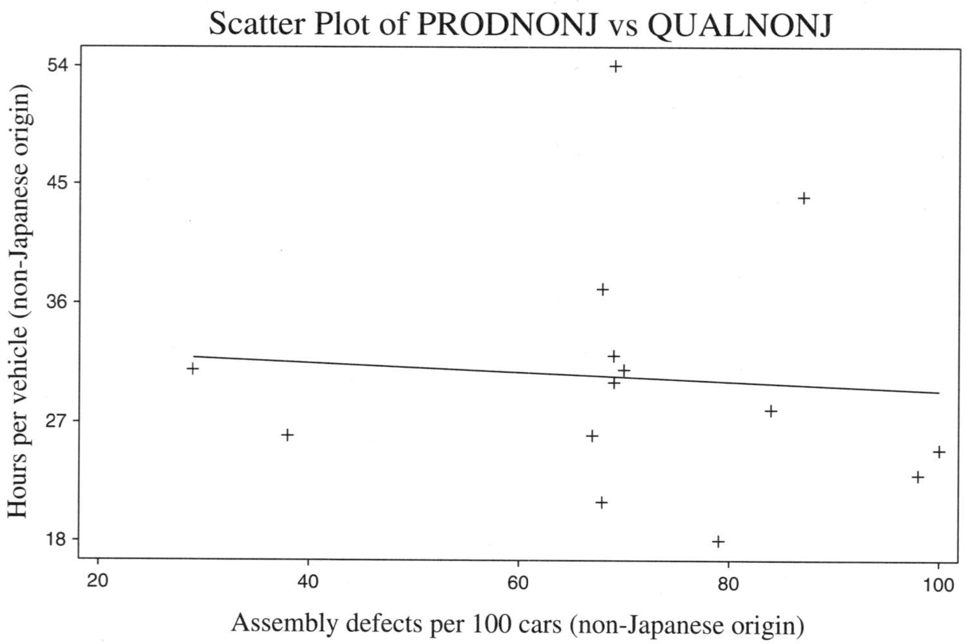

There is a good deal of variability around the regression lines for these data, and we wouldn't want to put too much emphasis on the negative correlations exhibited here.

There is no doubt that the Japanese plants exhibited higher quality and productivity compared with the non–Japanese plants, which could very well be because of "lean production"; however, it would appear that once a plant has achieved those benefits, quality is no longer free.

Summary

The existence of a tradeoff between quality and productivity in manufacturing plants is a common assumption for the mass production system. When examining data about automotive assembly in 1989, this tradeoff does not apparently occur, as quality and productivity are apparently directly related. However, separating the assembly plants by Japanese or non–Japanese origin makes the inverse association characteristic of such a tradeoff apparent. The observed "illusory correlation" occurs because the Japanese plants exhibit both higher quality and higher productivity, while the non–Japanese plants exhibit lower quality and lower productivity.

Technical terms

Correlation: a statistical term referring to the association between two variables. The measure often addresses linear associations in particular; that is, relationships that can be represented on a scatter plot by a straight line. An oft–repeated statement regarding correlation is that "correlation does not imply causation." That is, just because two variables vary together, that does not mean that one causes the other.

Illusory correlation: a situation where the observed association between two variables in the entire sample does not reflect the relationship within individual subgroups. This is often caused by strong association between group membership itself and each of the variables. The anomaly is seen when the correlation is calculated separately for each subgroup.

Regression line: a line determined from data that is designed to represent the relationship between two variables. The values on the line represent the value we would expect for the variable on the vertical axis given the value of the variable on the horizontal axis.

Health care spending in the United States

Topics covered: Comparison of location of subgroups.

Data File: `health.dat`

Health care is currently an important national issue in the United States. Several questions are being discussed and debated. These include the extent of coverage, sources of financing, and modes of delivery. One important question that has been raised concerns the variation of health care spending across the states. Is this variation very large, or could it just be the result of random chance? Is the variation accounted for by specific demographic properties of the states?

Data on per capita health care spending for all 50 states are available for 1991. Per capita spending figures were calculated by dividing the total health spending for a state by its population. The total health spending in a state includes amounts spent in the state by nonresidents, and excludes spending by residents who go to other states for their care. Along with the per capita health spending figures, data were collected on the percentage of per capita income that was spent on health care by each of the states. The corresponding figures for the United States as a whole are $1877 per person and 11.5% of per capita income, respectively. The original data sources were the Department of Health and Human Services and the Census Bureau. The present data come from *The New York Times*, October 15, 1993.

The Census Bureau divides the country into nine regions, with each state falling into one and only one region. These regions are: North East, Middle Atlantic, South Atlantic, East North Central, East South Central, West South Central, West North Central, Mountain and Pacific. Analyze these data for regional variation. Comment on any special features that you note.

Does it appear that health care spending differs across different regions of the country?

It has been suggested that the breakdown of the country into nine subregions is too fine, and that a grouping of six subregions is adequate. Group the states into six meaningful subregions to see if anything important emerges from the altered grouping. Are the conclusions drawn different from those for the data based on nine subgroups? Is there a different number of subgroups that you think might be better than either of these choices for these data?

Comments to the Instructor

The best way to see the regional variation is to construct side–by–side boxplots for per capita and total health spending.

The differences between regions of overall spending on health are most apparent with respect to the Mountain region, where it is relatively low (with median about $1500). The other regions all have median values around $1800. There are, however, considerable differences in the variability within regions, with the East North Central states exhibiting little variability, while the North East and Mountain regions showing considerably more variability.

The region effect is much more pronounced when examining health expenditures as a percentage of per capita income. With respect to this measure, the central regions are considerably higher than either set of coastal regions. This makes sense, since the per capita incomes are lower in the central part of the United States than on either coast. There are also apparent differences in variability between the regions.

Since some of the regions have too few states, a coarser grouping might be useful. One way to reduce the number of regions would be to combine the northern and southern parts of the two Central regions (that is, East North Central with East South Central, and West North Central with West South Central). It is unclear whether this helps in describing these data. It is important to remember, however, that it is only reasonable to combine regions if they are likely to be parts of some relatively homogeneous larger region. Presumably one would also only combine regions that are contiguous geographically.

To maintain consistency with the unannotated edition,
the next case begins on the following page.

The flight of the space shuttle Challenger

Topics covered: Frequency estimates of probabilities. Modeling the probability of failure.

Key words: Sample proportions. Scatter plot.

Data File: `chal.dat`

On January 20, 1986, the space shuttle Challenger took off on the 25 th flight in the National Aeronautics and Space Administration's (NASA) space shuttle program. Less than two minutes into the flight, the spacecraft exploded, killing all on board. A Presidential Commission was appointed to determine the cause of the accident. The Commission was headed by former Secretary of State William Rogers, and is sometimes referred to as the Rogers Commission. The Commission included some distinguished scientists and members of the space exploration community, including astronauts. The late physicist Richard Feynman played a major role in explaining some of the issues. The Commission studied the accident and the events leading up to the fatal launching. They determined the cause of the accident and wrote a two volume *Report of the Presidential Commission on the Space Shuttle Challenger Accident (1986).*

First, a little background information: the space shuttle uses two booster rockets to help lift it into orbit. Each booster rocket consists of several pieces whose joints are sealed with rubber O–rings, which are designed to prevent the release of hot gases produced during combustion. Each booster contains three primary O–rings (for a total of six for the craft). In the 23 previous flights for which there were data (the hardware for one flight was lost at sea), the O–rings were examined for damage.

The forecasted temperature on the launching day of the Challenger was 31° F. The coldest previous launch temperature was 53° F. The sensitivity of the O–rings to temperature was well–known. A warm O–ring will quickly recover its shape after a compression is removed, but a cold one will not. The inability of the O–ring to recover its shape will lead to joints not being sealed and can result in a gas leak. It is the combustion of this leaking gas that resulted in the fiery explosion of the Challenger.

Richard Feynman dramatically demonstrated the weakness of the O–rings at the Rogers Commission hearing. During a lunch break, Feynman went out into Washington D.C. (where the hearings were held) and bought some O–rings. After lunch he took a glass of ice water and dipped the O–rings in it. When he took the O–rings out, it was apparent that there was little resiliency left in them. The culprit was the O–ring and low temperature. Could this have been foreseen?

There was a good deal of discussion among the engineers just before the flight whether the flight should go on as planned or not (an important point is that no statisticians were involved in the discussions). A simplified version of one of the arguments made is as follows. There were seven previous flights where there was damage to at least one O–ring. The table below gives the ambient temperature at launch and the number of primary field joint O–rings damaged during the flight. The entry $\hat{p}$ is the frequency estimate (sample proportion) of the probability of an O–ring being damaged for that flight. That is, it is the proportion of the six O–rings that were damaged for that flight.

Ambient temperature	Number of O–rings damaged	$\hat{p}$
53°	2	.333
57°	1	.167
58°	1	.167
63°	1	.167
70°	1	.167
70°	1	.167
75°	2	.333

If we look at the table, there is no apparent relationship between temperature and the probability of damage; higher damage occurred at both lower and higher temperatures. Thus, that it was cold on the day of the flight doesn't imply that the flight should have been scrubbed.

Unfortunately, this analysis is completely inappropriate. It completely ignores the 16 flights where there was no O–ring damage, acting as if there is no information in those flights. This is obviously not reasonable! If flights with high temperatures **never** had O–ring damage, for example, that would clearly tell us a lot about the relationship between temperature and O–ring damage! In fact, here is a scatter plot of the O–ring damage versus temperature for **all** 23 of the flights for which information is available:

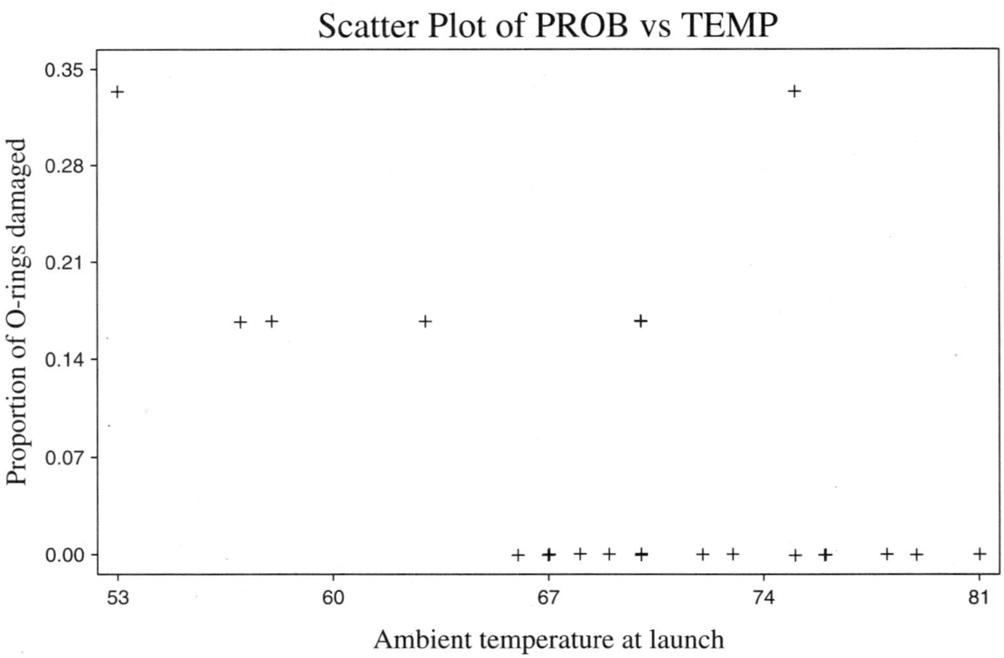

The picture is very different now. To show the dramatic difference we give the two scatter plots adjacent to each other:

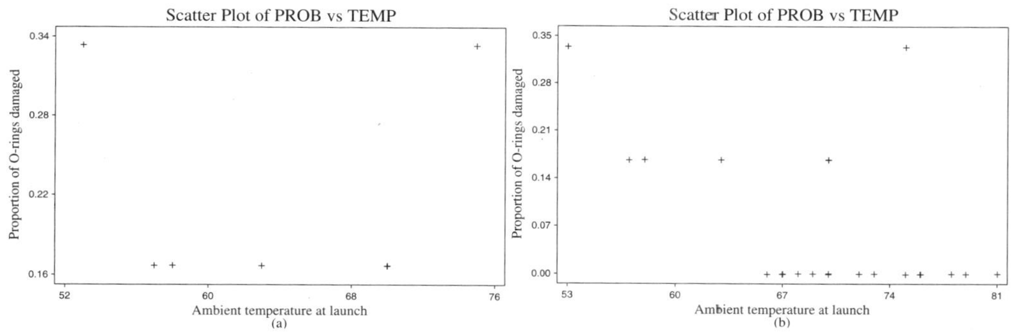

The scatter plot (a) gives the frequency estimate of the probability of O–ring damage versus temperature for launches in which there was O–ring damage, while (b) gives the same for all launches. Except for the one observation in the upper right of the plot, there is a clear inverse relationship between the probability of O–ring damage and the ambient temperature — lower temperature is associated with higher probability of damage (the unusual observation is the flight of the Challenger from October 30 through November 6, 1985). A plot of this kind would certainly have raised some alarms about the advisability of launching the shuttle. Unfortunately, such a plot was never constructed. The basic flaw in the analysis of the thermal distress carried out before the launching was the failure to include flights in which there was no O–ring damage. It is the conclusion to be drawn, as Sherlock Holmes said, from the "dog that did not bark." The Rogers Commission concluded that "a careful analysis of the flight history of O–ring performance would have revealed the correlation of O–ring damage in [sic] low temperature."

The analysis given here is crude, and can be made more precise. It is possible to build a model predicting the probability of O–ring damage as a function of temperature from these data using the method of *logistic regression*. If that is done here, the estimated probability of O–ring damage given an ambient temperature of 31° is **.96**! Indeed, with the benefit of hindsight, it can be seen that the Challenger disaster was not at all surprising, **given data that were available at the time of the flight**. As a result of its investigations, one of the recommendations of the Rogers Commission was that a statistician be part of the ground control team for all flights. A complete (and more exhaustive) discussion of this material can be found in the 1989 paper "Risk Analysis of the Space Shuttle: Pre–Challenger Prediction of Failure," by S.R. Dalal, E.B. Fowlkes and B.A. Hoadley, which appeared in *Journal of the American Statistical Association*.

Summary

Examination of data available at the time of the flight of the space shuttle Challenger on January 20, 1986 indicates that the subsequent explosion of the spacecraft was not unexpected. A scatter plot of the observed proportions of failed O–rings versus ambient temperature at the time of the flight shows a clear inverse relationship between the two variables, indicating that a flight at the low temperature on January 20 was highly inadvisable.

Technical terms

Frequency estimate: a method of estimating a probability, based on calculating the observed sample proportion of occurrences consistent with the event in question from a set of long–run random occurrences.

Probability: a mathematical representation of the likelihood of a particular outcome (or set of outcomes) occurring. Probabilities must satisfy certain basic properties, such as being between 0 and 1 (inclusive). There are several different ways of defining probabilities, including methods based on long–run proportions and personal (subjective) choices.

Random drug and disease testing

Topics covered: Conditional probability.

Key words: False negative. False positive. Probability. Sensitivity. Specificity.

A controversial political issue in recent years has been the possible implementation of random drug and/or disease testing (e.g., drug testing of transportation workers, or testing medical workers for HIV [human immunodeficiency virus, which causes AIDS]). Such testing inevitably involves the tradeoff of the benefit to society of detecting drug abuse or disease versus the cost to the individual of potential invasion of privacy.

Consider, as an example, HIV testing. The standard test is the Wellcome Elisa test. For any diagnostic test, the two key attributes are summarized by **conditional probabilities**:

(1) the *sensitivity* of the test:

$$\text{sensitivity} = P(\text{Positive test result} \quad | \quad \text{Person is actually HIV} - \text{positive})$$

(2) the *specificity* of the test:

$$\text{specificity} = P(\text{Negative test result} \quad | \quad \text{Person is actually not HIV} - \text{positive})$$

According to the Food and Drug Administration, the sensitivity of the Elisa test is approximately .993 (so only .7% of the people who are truly HIV–positive would have a negative test result), while the specificity is approximately .9999 (so only .01% of the people who are truly HIV–negative would have a positive test result).

That sounds pretty good. However, these are not the only numbers to consider when evaluating the appropriateness of random testing. A person who tests positive is interested in a different conditional probability: P (person actually is HIV–positive | a positive test result). That is, what proportion of people who test positive (which would result in great mental distress, possible employment ramifications, etc.) actually are HIV–positive? If the incidence of the disease is low, most positive results could be *false* positives.

W.O. Johnson and J.L. Gastwirth, in a 1991 paper in *Journal of the Royal Statistical Society, Series B*, estimate the incidence of HIV–positive in the general population of people without known risk factors to be .000025. Consider a group of 10,000,000 people without known risk factors. We would expect that about 250 of these people are actually HIV–positive. Based on the sensitivity figure of .993, $(250)(.993) \approx 248$ of these people would have positive Elisa tests. Now think about the 9,999,750 true HIV–negative people. Based on the specificity figure of .9999, $(9,999,750)(.0001) \approx 1000$ false positive results would come from this group. That is, **80% of the positive test results would actually come from people who were not HIV–positive**!

Now, before we decide that diagnostic tests must be a waste of time, we should note that the test *has* dramatically increased the chance of detection; while only .0025% of all people with no known risk factors are HIV–positive, almost 20% of those people with a positive Elisa test are HIV–positive (an increase by a factor of almost 8000!). Still, there are many false positives. This is the problem with random testing for the presence of a rare characteristic. Public policy must consider the potential damage to people who are falsely labeled as having a condition with negative public perception, such as drug abuse, alcohol abuse, or infectious diseases. This question will only become more important in the future; according to a survey of 630 major companies undertaken by the American Management Association in 1992, the proportion of companies conducting drug testing in the workplace rose from 22% in 1987 to 75% in 1991 to 85% in 1992. The survey also found that the proportion of people testing positive dropped during that time. This suggests that fewer drug users are in the workplace or applying for jobs, but it also suggests that therefore a larger proportion of the positive test results are actually false positives. Firms must be careful about how they apply such tests, however; in July 1993 a woman fired from her job because of a (subsequently determined to be false) positive drug test sued to get her job back, claiming that her firing was illegal under the Americans with Disabilities Act. For this reason, most private firms follow Department of Health and Human Services guidelines for testing federal employees, which require rigorous test standards with backup medical corroboration for someone who tests positive.

A study in the United Kingdom in the late 1980's confirms the numbers given here. Out of 3,122,556 blood samples taken from people without known risk factors for HIV, 373 tested positive for HIV based on the Elisa test. These samples were then retested using the much more specific (and expensive) Western Blot test, and 64 cases were confirmed. That is, 83% of the positive test results were false positives.

It should be reemphasized, by the way, that the HIV–positive incidence rate given above is for people without known risk factors (roughly speaking, people who are permitted to volunteer to give blood at most blood centers). The HIV–positive incidence rate is much higher in the general population; for example, the World Health Organization estimated it (in 1991) to be .00143 for women and .0133 for men, respectively, in the general population in North America; .002 for women and .008 for men, respectively, in Latin America; .0007 for women and .005 for men, respectively, in Western Europe; and .025 for both men and women in sub–Saharan Africa. Thus, there would be a smaller proportion of false positives in random testing from the general population than from testing only those people with no known risk factors. Still, false positives could still be a problem: in North America, in general random testing of 10,000,000 people, there would be about 73,650 true HIV–positive people, with approximately 73,134 being correctly identified as HIV–positive; there would be 9,926,350 true HIV–negative people, with approximately 993 false positives from this group. Thus, over 1% of the positives would still be false positives. Of course, it also would be very expensive to administer tests to the entire population, but that's not a directly probabilistic issue.

We also must recognize that the numbers presented here are for the United States and United Kingdom, two countries with relatively sophisticated medical establishments. The British medical journal *The Lancet* reported in June 1992 that the mass screening for HIV that is a major component of AIDS control strategy in Russia has led to gigantic false positive rates: approximately 99.4% in 1990, and 99.8% in 1991. Multiple administration of the less specific screening tests only worsens the problem (since a positive result in *any* of the tests is considered an overall positive result): in pregnant women, for whom two HIV tests during pregnancy are mandatory, the false positive rate exceeded 99.9%.

As medical technology advances, these kinds of issues will become more important. For example, the February 4, 1993, issue of *The New England Journal of Medicine* included an article about a simple new test for HIV infection that is appropriate for newborn babies (the usual tests, based on looking for antibodies to the virus, are not accurate for infants; only one–third of infants born with maternal antibodies are actually infected, and it takes more than a year for the antibody tests to be accurate). The new test (called an *immune–complex–dissociated HIV p24 antigen assay*), is inexpensive (about $80), and various states are considering whether to use it in routine testing of pregnant women and/or infants. Note, however, that unless the specificity of the test is **extraordinarily** high (or the test would only be applied to children of mothers who have been confirmed as being HIV–positive), there are likely to be far more false positives than true positives in this kind of testing. Considering that it might be impossible (or very expensive) to separate out accurately the true positives from the false positives (at least until the baby is a year old), routine application of the test could lead to great emotional distress for parents, and possibly incorrect (or dangerous) treatment of children. The article did refer to one study of the test conducted at UCLA, where 100% of babies in the study that were born of infected women were correctly classified as to whether or not they were infected. Unfortunately, the test only had 29 babies in it, so it cannot be considered to have resolved this question completely.

The issues here are not limited to HIV testing. For example, a study done at a Lyme disease clinic in Boston in 1992 found that of 800 patients who had been referred to the clinic because of positive results on blood tests, 45% were false positives — people who did not, in fact, have Lyme disease. Considering the virtual impossibility of preventing Lyme disease, short of eradicating it in the ticks that are its host, this translates into many patients receiving inappropriate medical care all over the country. Another example of this type of situation is the prostate specific antigen (PSA) blood test for prostate cancer in men. Autopsy studies suggest that about half of all men over 50 have cancerous cells in their prostate, but only about 2.4% die of prostate cancer. The PSA test has high false positive and false negative rates (e.g., about 50% of the men with high PSA readings do not have prostate cancer); when a person tests positive, there are other tests to do, which are often inconclusive. Given that even if the patient has cancerous prostate cells, it might never pose a health threat, it is a very difficult decision to make whether PSA tests should be given routinely or not. The current recommendation by the American Cancer Society is that men over the age of 50 should have an annual PSA test, along with a digital rectal exam, although a statistical analysis published in the September 14, 1994 issue of the *Journal of the American Medical Association* suggests that the benefits of such screening are marginal at best.

By the way, the same sort of confusion of conditional probabilities that occurs when mistaking the sensitivity of a test for the proportion of positive test results that correspond to people who actually have the disease (true positives) occurs in other contexts as well. Consider, for example, the use of DNA "fingerprinting" as evidence in a criminal trial. Laboratory (forensic) evidence addresses the question "What is the probability that the defendant's DNA profile matches that of the crime sample, given that the defendant is not guilty?". This is not, however, the question of interest to the jury, which is "What is the probability that the defendant is not guilty, given that the DNA profiles of the defendant and the crime sample match?". Confusion of these two probabilities is commonly called "the prosecutor's fallacy." The statistician Peter Donnelly made the argument that the two probabilities can sometimes be quite different in a successful appeal of a rape conviction in the United Kingdom in 1993.

Summary

Application of random drug and disease testing has been suggested in recent years as a way to address the presence of drug abuse and infectious disease in the population. Unfortunately, if the attribute being tested for is rare in the population, even very accurate tests are likely to produce a large proportion of false positive results. For example, it can be estimated that in random testing for the HIV virus in people with no known risk factors, over 80% of the positive test results are, in fact, false positives. This reinforces the notion that random testing for a rare characteristic can have important public policy implications, in terms of invasion of privacy, and needs to be carefully considered before implementation.

Technical terms

Conditional probability: a probability value that involves restricting attention to the subset of possibilities defined to be consistent with the occurrence of a specific condition. The notation $P(A \mid B)$, read "the probability of A given B," is defined to represent the probability of the event A occurring, given that the event B has occurred (or will occur).

Amniocentesis, blood tests, and Down's syndrome

Topics covered: Conditional probability.

Key words: False negative. False positive. Probability. Sensitivity. Specificity.

Down's syndrome is a genetic disorder, caused by the inheritance of an extra copy of chromosome 21, which leads to retarded mental and physical development. It is well established that the risk of having a Down's syndrome baby rises with the mother's age. While the overall risk is 1 in 700, it rises to 1 in 270 for mothers at age 35 (and is even more probable for older mothers).

Because of this, amniocentesis, a procedure used to withdraw amniotic fluid containing fetal cells from the mother's uterus, is routinely recommended for pregnant women over age 35. This procedure is virtually 100% sensitive in its ability to detect Down's syndrome, but it carries a small risk of causing miscarriage. It is also not inexpensive, with an average cost of about $1000 in the United States. An alternative procedure, chorionic villus sampling, also carries the risk of miscarriage.

An article in the April 21, 1994, issue of *The New England Journal of Medicine* described the results of a study of the effectiveness of blood tests to detect Down's syndrome. These blood tests are not believed to create any risk of miscarriage. According to the article, a study in which 5385 pregnant women over age 35 were administered both the blood tests (which cost about $75) and amniocentesis indicated that the sensitivity of the blood tests was .89, while the specificity was .75 (see the case "Random drug and disease testing" for the definitions of these terms). A "positive" result for the blood tests corresponds to an estimated probability of Down's syndrome exceeding 1 in 200 from the blood tests, which would then be followed by an amniocentesis.

We will use the definition of conditional probability to learn more about these blood tests. First, recall the definition of conditional probability:

$$P(A \mid B) = \frac{P(A \text{ and } B)}{P(B)},$$

for events A and B. From the definitions of sensitivity and specificity, we therefore know that

$$\begin{aligned} &P(\text{Positive test result} \mid \text{Presence of Down's syndrome}) \\ &\qquad = \frac{P(\text{Positive test result and Presence of Down's syndrome})}{P(\text{Presence of Down's syndrome})} \\ &\qquad = .89 \end{aligned}$$

and

$$\begin{aligned} &P(\text{Negative test result} \mid \text{Absence of Down's syndrome}) \\ &\qquad = \frac{P(\text{Negative test result and Absence of Down's syndrome})}{P(\text{Absence of Down's syndrome})} \\ &\qquad = .75 \end{aligned}$$

An important question is as follows: what proportion of women over age 35 would test positive on the blood tests? Note that such women fall into two mutually exclusive groups: those who test positive and for whom Down's syndrome is present, and those who test positive and for whom Down's syndrome is absent. The first group comprises 0.33% of the population of women over age 35, while the second group comprises 24.91% of the population. (Why? Here's a hint: you'll need to use the overall incidence of Down's syndrome in the population, and the sensitivity [for the first group] and specificity [for the second group] of the blood tests.) Pooling these values, the probability of a positive test result is .2524.

What proportion of these positive results are *false* positives? That is, what is

$$P(\text{Absence of Down's syndrome} \mid \text{Positive test result})?$$

This can be calculated easily (based on what we've already calculated) as .98694. That is, almost 98.7% of the positive test results are false positives. Since any positive test would be followed by an amniocentesis, this is not necessarily a problem, as long as the mother knew that even after a positive blood test the risk of a Down's syndrome baby is still less than 1 in 75.

Another point of concern is the rate of false negative results. What proportion of women over age 35 who test negative on the blood tests (and might therefore not undergo an amniocentesis) would in fact give birth to a Down's syndrome baby? We know that 74.76% of all blood tests would be negative. Using the same kind of calculations that lead to the false positive rate, we can determine the false negative rate to be 0.0546%, or less than 1 in 1800. Thus the use of the blood test to "screen" women ensures that amniocentesis is only applied to higher risk women (a 1 in 75 chance of Down's syndrome) rather than the general population of women over age 35 (a 1 in 270 chance of Down's syndrome). The blood test reduces the health costs and risk of miscarriage for 3 out of 4 women, while leaving a 1 in 1800 chance for the birth of a Down's syndrome baby.

Comments to the Instructor

The proportion of women who test positive where Down's syndrome is present can be calculated as follows:

$$\begin{aligned} P(&\text{Positive test result and presence of Down's syndome}) \\ &= P(\text{Positive test result} \mid \text{Presence of Down's syndrome}) \\ &\quad \times P(\text{Presence of Down's syndrome}) \\ &= (.89)(.003704) \\ &= .003296. \end{aligned}$$

Similarly,

$$\begin{aligned} P(&\text{Positive test result and absence of Down's syndome}) \\ &= P(\text{Positive test result} \mid \text{Absence of Down's syndrome}) \\ &\quad \times P(\text{Absence of Down's syndrome}) \\ &= (1-.75)(.996296) \\ &= .24907. \end{aligned}$$

Thus, the probability of a positive test result is the sum of .00396 and .24907, or .25237.

The proportion of positive test results that are false positives is determined as in the "Random drug and disease testing" case. Since

$$\begin{aligned} &P(\text{Presence of Down's syndrome} \mid \text{Positive test result}) \\ &\qquad = \frac{P(\text{Presence of Down's syndome and positive test result})}{P(\text{Positive test result})} \\ &\qquad = \frac{.003296}{.25237} \\ &\qquad = .01306, \end{aligned}$$

the required probability is $1 - .01306$, or .98694.

The false negative rate is calculated the same way:

$$\begin{aligned} &P(\text{Presence of Down's syndrome} \mid \text{Negative test result}) \\ &\qquad = \frac{P(\text{Presence of Down's syndome and negative test result})}{P(\text{Negative test result})} \\ &\qquad = \frac{.003704 - .003296}{.74763} \\ &\qquad = .000545. \end{aligned}$$

To maintain consistency with the unannotated edition,
the next case begins on the following page.

Perceptions of the New York City subway system: safety and cleanliness

Topics covered: Independence. Joint distribution. Marginal distribution.

Key words: Cross–classified data. Random variable. Sample proportions.

Data File: `subway.dat`

The New York City subway system is one of the largest and most complex rapid transit systems in the world. It provides service to 469 stations, 24 hours a day, seven days a week, using 714 miles of track. In recent years, the Metropolitan Transit Authority (MTA), the quasi–governmental organization that operates the subways (through its subsidiary agency the New York City Transit Authority) has spent millions of dollars to try to improve customer service and the perceptions of the subway system by New Yorkers. Considering that the operating deficit of the MTA was projected to be $5 billion in 1994, the financial viability of the MTA is obviously predicated on its ability to attract and retain new and current riders. In order to do this, the agency needs to know what the perceived strengths and weaknesses of the subways are, so that it can develop appropriate strategies to address these perceptions.

A survey of public perception of the New York City rapid transit system engaged by the MTA is the basis of the data set analyzed here. The sample consists of 62 graduate students at the Leonard N. Stern School of Business at New York University, who were sampled in the Spring of 1994. The data were provided by Sundar Polavaram and L. Brooke Squire.

Attention will be focused here on four questions relating to the respondent's satisfaction level of the cleanliness of the stations (`CLNSTAT`), the cleanliness of the trains (`CLNTRAIN`), the safety in the stations (`SAFSTAT`) and the safety on the trains (`SAFTRAIN`). The responses are coded into a five–point scale, ranging from "Very unsatisfactory" (1) to "Very satisfactory" (5). Each of these variables is a **random variable**, in that it represents a rule that assigns a number to each outcome of a random process.

It might be expected that an individual's responses related to cleanliness would be similar to each other, as would be their responses related to safety. Is that the case? One way to answer this question is to construct the **marginal distribution** of each of the variables; that is, a table that reports the sample proportions that fall into each category of the variable.

Construct marginal distributions for the two cleanliness–related variables `CLNSTAT` and `CLNTRAIN`. Do the distributions look similar? Actually, there are noticeable differences, with many more people expressing dissatisfaction with the cleanliness of the stations than of the trains.

Now construct marginal distributions for the two safety–related variables `SAFSTAT` and `SAFTRAIN`. Do these distributions look similar? In fact, they do look similar to each other, and to that of `CLNTRAIN`.

Do these results imply that the safety–related variables are more associated with each other than are the cleanliness–related variables? Or that the safety–related variables are associated with `CLNTRAIN`? Not really. The reason is that the association between different variables cannot be seen from marginal distributions, which only examine the variables one at a time. Rather, it is necessary to examine the **joint distribution** of the two variables, which summarizes their joint relationship (this is analogous to the difference in information that can be gleaned from two histograms versus a scatter plot).

Are the perceptions of cleanliness and safety related? Construct the joint distribution of the variables CLNSTAT and SAFSTAT. What do you notice about the pattern of sample proportions in the table? The probabilities are higher for pairs of values that are close to each other; that is, when the respondent rated cleanliness low (for example), he or she also tended to rate safety low. Respondent perceptions about cleanliness and safety in stations appear to be directly related to each other.

This pattern is suggestive, but it could be that it is actually not very surprising, even if there was no relationship between the variables. When two random variables are completely unrelated (that is, knowledge of the value of one gives no information about the value of the other), they are said to be **independent**; what would the joint distribution of CLNSTAT and SAFSTAT look like if they were independent?

Mathematical theory can help us to answer this question. Another way of saying that two events A and B are independent is that the conditional probability of A occurring given B occurs is the same as the unconditional probability that A occurs (without knowing anything about whether B occurs); that is, $P(A \mid B) = P(A)$. But, by the definition of conditional probabilities, if A and B are independent, then

$$\begin{aligned} P(A \mid B) \equiv \frac{P(A \text{ and } B)}{P(B)} &= P(A) \\ \Rightarrow P(A \text{ and } B) &= P(A) \times P(B). \end{aligned}$$

This mathematical identity can be used to compare the observed joint distribution of CLNSTAT and SAFSTAT to the distribution that would be expected under independence of the two random variables. For example, under the assumption of independence,

$$P(\texttt{CLNSTAT} = 1 \text{ and } \texttt{SAFSTAT} = 1) = P(\texttt{CLNSTAT} = 1) \times P(\texttt{SAFSTAT} = 1);$$

the two marginal probabilities on the right side of the equation can then be estimated from the appropriate sample proportions from the marginal distributions to yield an estimated joint probability under independence.

Construct the estimated joint distribution of CLNSTAT and SAFSTAT under independence. Does it look similar to the observed joint distribution of the two variables you constructed earlier? The answer is no. The observed probabilities are too large for close values of the two variables, and too small for values that are more different. The following two plots illustrate the difference in the two distributions. Note that under independence each row or column follows a similar pattern of increase or decrease, while the observed joint distribution shows the "piling up" of probabilities for similar values of CLNSTAT and SAFSTAT.

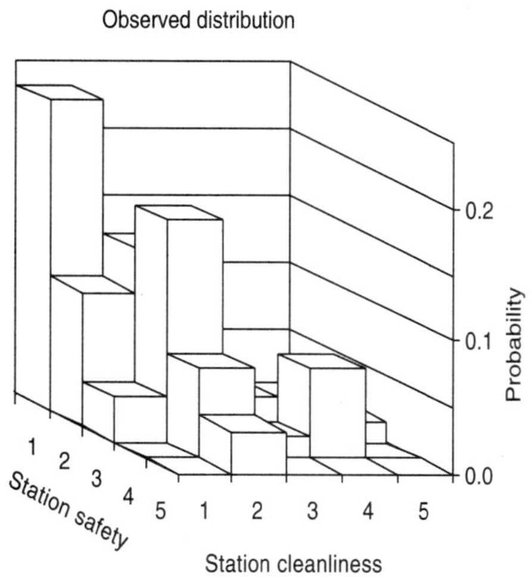

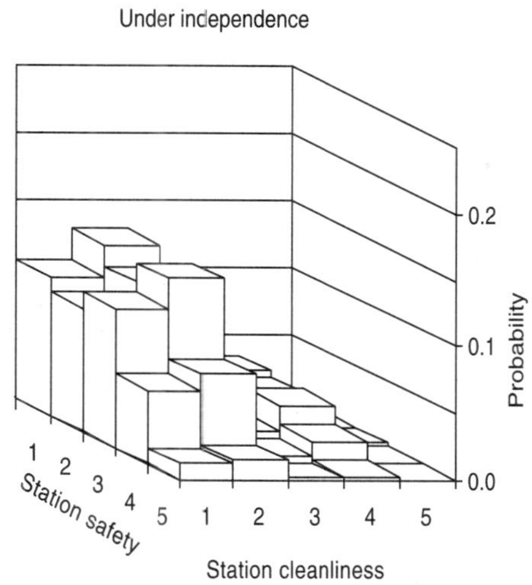

Thus, it appears that there is a real association between the perceptions of cleanliness and safety in subway stations for this sample. Are cleaner stations truly safer, or do riders just *feel* safer in cleaner stations? We cannot answer that question from these data, but subway crime statistics might shed some light on this question.

Now construct the observed joint distribution of CLNTRAIN and SAFTRAIN, the variables referring to trains (rather than stations). Compare this to the estimated joint distribution under independence of the two variables. Do the variables appear to be independent?

We might expect that the two cleanliness–related variables would be strongly related, as would the two safety–related variables. Is that the case here?

Technical terms

Independence: when knowledge of the occurrence of one event gives no information about the occurrence of another event. Two random variables are **independent** if knowledge of the observed value of one random variable gives no information about what the value of the other random variable is (or will be). The existence of independence of two events A and B can be defined as $P(A \mid B) = P(A)$, or, equivalently, $P(A \text{ and } B) = P(A) \times P(B)$. For contingency tables (tables of counts), independence is consistent with the probability in a cell equaling the product of the marginal probabilities of falling in that row and column; that is, $P(\text{Value falls in row } i \text{ and column } j) = P(\text{Value falls in row } i) \times P(\text{Value falls in column } j)$.

Joint distribution: a set of numbers that represents the probabilities of the possible outcomes for two random variables taken as a pair. That is, the joint distribution assigns probabilities to events of the form $\{$ Variable 1 equals x and Variable 2 equals $y\}$.

Marginal distribution: a set of numbers that represents the probabilities of each possible outcome occurring for one random variable.

Random variable: a rule that assigns a number to each possible outcome of some random process. The numbers can have physical meaning (e.g., the number of people arriving at a location in a given time period), or can be assigned arbitrarily (e.g., 1 if the person arriving is female, 0 if the person is male).

Comments to the Instructor

Here are the marginal distributions for the four variables:

Response	CLNSTAT	CLNTRAIN	SAFSTAT	SAFTRAIN
1	.377	.258	.279	.311
2	.443	.290	.246	.197
3	.049	.226	.279	.361
4	.115	.177	.164	.082
5	.016	.048	.033	.049

The joint distribution of the variables CLNSTAT and SAFSTAT is as follows:

	CLNSTAT				
SAFSTAT	1	2	3	4	5
1	.233	.050	.000	.000	.000
2	.100	.133	.000	.017	.000
3	.033	.167	.033	.033	.017
4	.000	.067	.017	.067	.000
5	.000	.033	.000	.000	.000

In contrast, the joint distribution under independence is as follows:

	CLNSTAT				
SAFSTAT	1	2	3	4	5
1	.104	.127	.014	.033	.005
2	.092	.113	.013	.029	.004
3	.104	.127	.014	.033	.005
4	.055	.068	.008	.018	.003
5	.012	.015	.002	.004	.001

Similar constructions can be made for CLNTRAIN and SAFTRAIN, as well as by pairing the two cleanliness variables and the two safety variables. In all cases, there is a clear association between the pairs; that is, they are not independent.

Perceptions of the New York City subway system: other issues

Topics covered: Independence. Joint distribution. Marginal distribution.

Data File: `subway.dat`

In the case "Perceptions of the New York City subway system: safety and cleanliness," it was shown that the perceptions of the riders in the sample regarding the cleanliness of the subways and the safety of the subways were apparently related to each other.

In this case, you will examine other perceptions of the riders, and how they relate to each other. It is natural to expect that variables that are paired (as cleanliness in the stations and cleanliness on the trains are) would be related to each other. Explore if this is the case by comparing the observed joint distribution to the distribution that would be expected under independence. What about other "natural" pairs, such as police presence and safety?

Other pairs are easier to hypothesize as being independent. For example, it is hard to see why ease of token purchase would be related to safety on trains. Explore the observed joint distribution to see if this is the case.

Examine the possible associations of any variables that seem interesting, comparing the observed joint distribution to the distribution that would be expected under independence. Are the results as you would expect? Are any surprising?

Comments to the Instructor

Most "natural" pairs of variables appear to be related; for example, police presence on trains and police presence in stations, crowding on trains and crowding in stations, convenience of train stops and convenience of schedule, and ease of token purchase and ease of token collection all appear to be directly related to each other.

Other apparent relationships that are interesting are the (not surprising) association between perceptions on safety and on police presence; the association between perceptions of the crowdedness of stations and cleanliness of trains and stations (presumably reflecting a tendency for people to associate crowds with clutter); and convenience of schedules and frequency of trains.

Most of the other associations are apparently weak or not present at all. For example, perception of in–station announcements is apparently unrelated to all other variables, except perception of on–train announcements.

The Central Limit Theorem for census data

Topics covered: Central Limit Theorem. Computer experiment. Distributional shape. Random samples. Sampling distributions.

Key words: Histogram. Long–tailed data. Mean. Median. Standard deviation. Standard error.

Data Files: `census1.dat`, `census2.dat`, `census3.dat`

Every 10 years, beginning in 1790, the United States Census Bureau has undertaken a decennial census, for the purpose of describing the people and economy of the United States. This census, which is required by the U.S. Constitution, yields data that are used for many purposes, including congressional apportionment and the allocation of federal funds.

The data set that is the subject of this case is a subpopulation of 5000 workers chosen from the U.S. population. They are full–time year–round white female workers between 16 and 65 years of age. They were chosen as part of a larger study of the structure of the U.S. workforce. Here we will regard the 5000 women as the population of interest, rather than as a sample from the U.S. population as a whole.

The characteristic under study for these women is their income from wages (measured in thousands of dollars). First let's look at some descriptive statistics:

```
DESCRIPTIVE STATISTICS

                    INCOME
N                     5000
MEAN                19.897
SD                  12.573
MINIMUM             0.2920
1ST QUARTILE        12.123
MEDIAN              16.990
3RD QUARTILE        24.419
MAXIMUM             220.84
```

We see that the mean income was a little under \$20,000, with a standard deviation of $\sigma = \$12,573$. This standard deviation is large, relative to the mean. The median is a little under \$17,000 and is about \$3,000 less than the mean, suggesting that the distribution is right–skewed. The range of wage incomes is quite large; from a minimum of \$292 to a maximum of \$220,840. There is obviously a great deal of diversity in income in this population.

To get a better understanding of the distribution of the incomes for these 5000 people, let's look at a histogram:

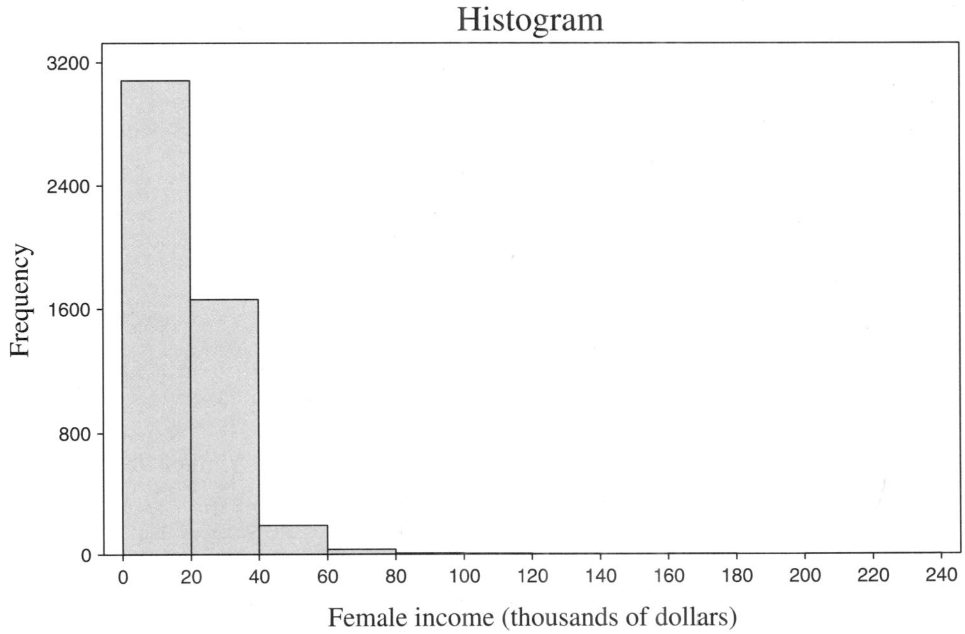

The skewness in the incomes is very apparent, as is the extreme right tail.

Suppose that these income data were not available, and we needed an estimate of the mean income of this population of women. We can see from the above summary that the population mean happens to be $\mu = \$19,897$. How would we estimate it? The usual thing to do is to survey the women and ask them their incomes. If we had the time to survey all 5000 women we would have reconstructed the data set and would know the mean exactly. But this could take a long time and be quite costly. It would be nice if we could think of a way that would enable us to only re–survey a few of the members of the group. To obtain a representative sample, the natural thing to do is to choose randomly the women to be re–surveyed from the original 5000. Based on this sample we could estimate the population mean income by the sample mean income. We know that the sample mean will not be exactly equal to the population mean, but we think that it will be close.

The purpose of this case is to explore how close the estimates based on re–surveys are to the true population value. We also want to examine the distribution of the estimates. We will conduct an experiment on a computer to generate many more samples than would be possible in practice.

Each value of the variable `SIZE10` is the sample mean from a sample of size 10 from the original population. The variable has 400 cases, each corresponding to a different sample. For example, the first value of `SIZE10` is 19.5413. This means that the first sample of 10 randomly chosen women had a mean income of $19,541.3. The second sample of 10 randomly chosen women had a mean of $12,507.5, and so on. Each of the sample means is different, and not necessarily equal to the population mean ($19,897). Some are close, as for the first sample, and some are far away, as for the second sample. The closeness of the estimate is literally the luck of the draw!

However, there is a pattern, based on the distribution of the sample means. How are these sample means distributed about the population mean? We can use the same data analytic tools to analyze the results of our computer experiment as we used to analyze the original population.

First, let's look at some descriptive statistics for `SIZE10`:

```
DESCRIPTIVE STATISTICS

                  SIZE10
N                    400
MEAN              19.887
SD                 4.320
MINIMUM           11.877
1ST QUARTILE      17.192
MEDIAN            19.210
3RD QUARTILE      21.881
MAXIMUM           47.149
```

We see that the overall variability of the sample means is much less than the population values. The standard deviation is 4.320, compared with 12.573. The average of the averages is quite close to the population average, so the sample mean does not seem to be systematically too high or too low. This is the property of **unbiasedness** that is often talked about.

Next, let's look at a histogram of `SIZE10`. This represents the distribution of the estimate based on the sample (called the **sampling distribution of the estimate**).

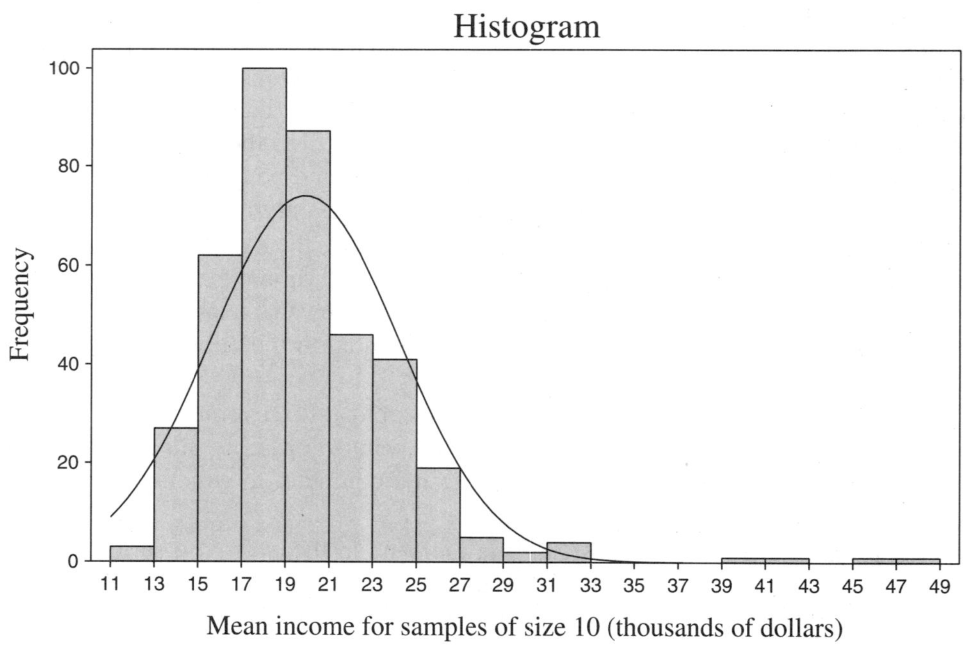

We see that the values are still quite right–skewed, although not as right–skewed as the population values are. The distribution of the normal (Gaussian) random variable with the same mean and variance as that of `SIZE10` is superimposed on the histogram. We can see that the sampling distribution of the sample mean for samples of size 10 is not close to normal, but is closer to a normal distribution than the distribution of the population values is.

We can think of the values of SIZE10 as a random sample from the sampling distribution of the sample mean. We are using the sample (i.e., statistics) to get a handle on the form of the underlying distribution. In that sense we are using statistics to understand probability.

What do we know about the sampling distribution of $\overline{X}$, the sample mean?

(1) The probability distribution of $\overline{X}$ has expected value $\mu_{\overline{X}} = \mu = \$19,897$. It also has standard deviation

$$\sigma_{\overline{X}} = \frac{\sigma}{\sqrt{10}} = \frac{\$12,573}{\sqrt{10}} = \$3,976.$$

This standard deviation, being the standard deviation of a statistic (rather than of a random variable from a population) has a special name; it is called the **standard error** of the mean. The above sample mean (\$19,887) and sample standard deviation (\$4,320) both seem close to these figures. Note that they are not the same because they are also sample statistics!

(2) According to the **Central Limit Theorem,** the probability distribution of $\overline{X}$ will appear to be more and more normal in shape as the sample size increases. We can see this in the above two histograms. There is still a way to go, but the distribution of the sample mean is much closer in shape to the normal than that of the population.

(3) The Central Limit Theorem applies to a very wide range of populations. The values that the population takes can even be discrete (e.g., number of jobs). All that is required is that the standard deviation of the population be nonzero and finite. The theorem also makes it clear that if the distribution of the population is itself normal, then the distribution of the sample mean will be normal for *any* sample size.

The reason that the Central Limit Theorem is so important from a practical point of view is that it leads to greatly simplified probability calculations. If the distribution of the sample mean is known, we can easily use a table or a computer program to calculate probabilities of interesting events. In addition, if the distribution of the sample mean is known to be very close to a normal distribution, we can use the standard normal tables. These are usually readily available and are easy to use.

How large does the sample size need to be to use the normal approximation?

Let's take a look at other sample sizes. The variables SIZE1, SIZE2, SIZE4, SIZE25, SIZE50, SIZE100, and SIZE625 are the sample means from samples of size 1, 2, 4, 25, 50, 100 and 625, respectively. As was the case for SIZE10, for each variable we have taken 400 random samples.

First, here are some descriptive statistics for each variable:

```
DESCRIPTIVE STATISTICS

                 SIZE1        SIZE2        SIZE4        SIZE10       SIZE25
N                  400          400          400           400          400
MEAN            20.731       20.349       19.392        19.887       19.840
SD              12.922       8.8827       5.3992        4.3196       2.2264
MINIMUM         0.2920       6.8405       7.9120        11.877       14.267
1ST QUARTILE    12.813       14.315       15.384        17.192       18.252
MEDIAN          17.680       18.498       18.699        19.210       19.635
3RD QUARTILE    25.773       24.646       22.625        21.881       21.053
MAXIMUM         114.72       78.455       41.394        47.149       27.390

                SIZE50      SIZE100      SIZE625
N                  400          400          400
MEAN            19.754       19.813       19.921
SD              1.6359       1.1979       0.5035
MINIMUM         15.565       16.816       18.440
1ST QUARTILE    18.599       18.996       19.573
MEDIAN          19.616       19.700       19.930
3RD QUARTILE    20.755       20.474       20.259
MAXIMUM         26.329       23.864       21.653
```

The samples of size 1 (i.e., SIZE1) are just a random sample from the population. We note that the mean and standard deviation for SIZE1 are close to those of the population.

Based on the Central Limit Theorem, we expect that as the sample size increases, the sample means will become closer to the population mean. We can see this effect in the means of these variables. Similarly, as the sample size increases, the sample standard deviation decreases, as we would expect.

Here are the theoretical standard deviations for each sampling distribution compared to the values we have observed. Note that these theoretical standard deviations are the standard errors of the sample mean, while the observed values are estimates of these standard errors:

Sample size n	True standard deviation of the sampling distribution $\sigma_{\overline{X}} = \frac{\sigma}{\sqrt{n}}$	Estimated standard deviation of the sampling distribution
1	12.573	12.922
2	8.8905	8.8827
4	6.2865	5.3992
10	3.9759	4.3196
25	2.5146	2.2264
50	1.7781	1.6359
100	1.2573	1.1979
625	0.5029	0.5035

In each case the observed value is close to the expected value from the theory. There is some variation about the expected values, as we expect because of random fluctuation (from the act of drawing a random sample). The point is that the sampling distributions have **exactly** the standard deviations (standard errors) given in the second column of the table. The sample means are, in fact, samples from distributions with those standard deviations.

How does the distributional shape change as the sample size increases? Let's look at histograms of the sample means for each sample size:

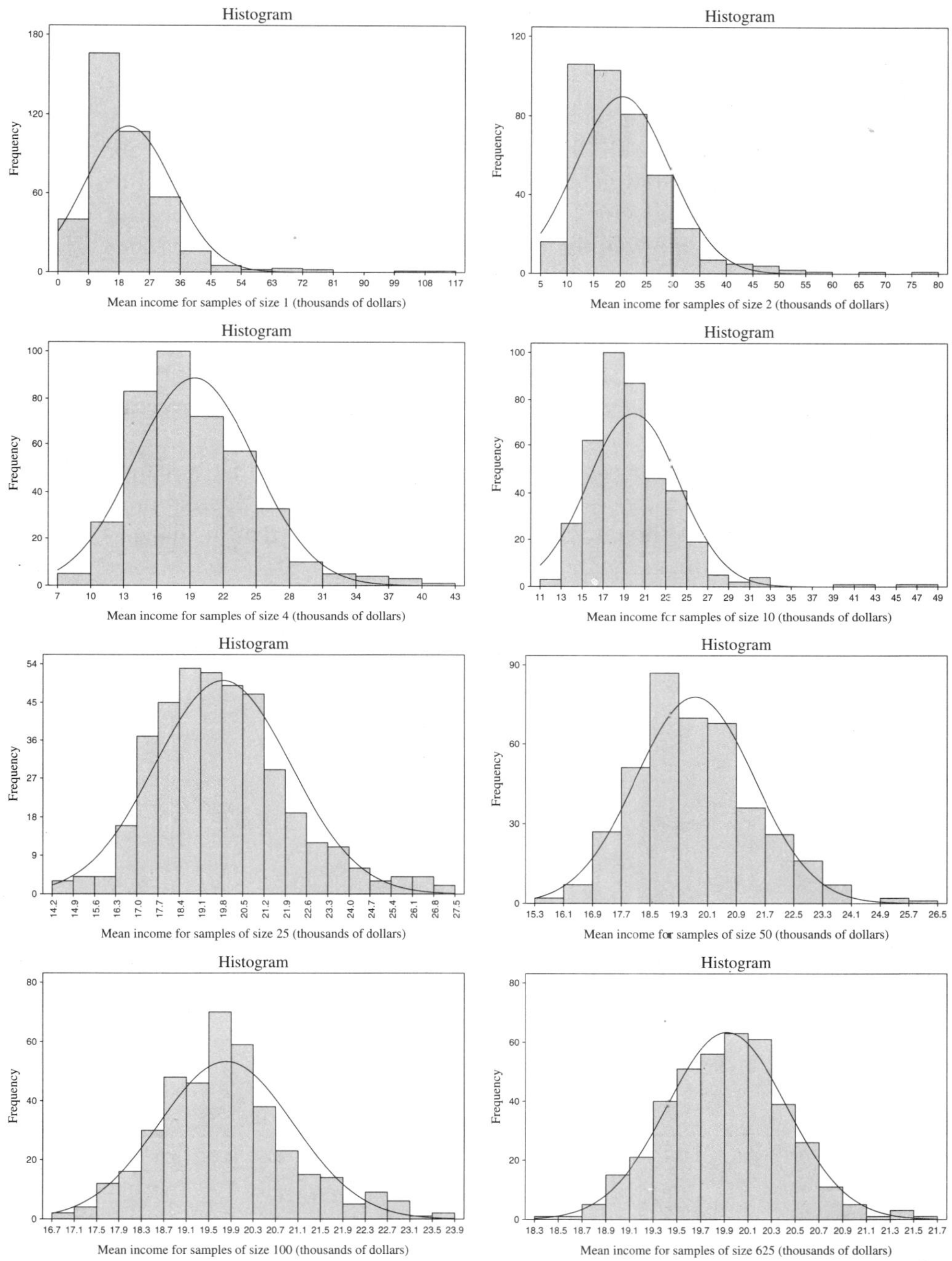

The first histogram, which is just a sample from the population, can be used as a baseline for comparison. After looking at the histograms for a while, we can make some observations:

(1) As the sample size increases, the distributions become closer and closer to the corresponding normal distribution in shape. In doing so, the skewness apparent in the population becomes progressively less noticeable.

(2) The variation of the sample means progressively decreases as the size of the sample increases. This can be seen from the range of values on the horizontal axis.

(3) At what point is the distribution sufficiently normal, so that the normal approximation is adequate? Well, this depends on what we need the probabilities for. Usually we want to know the range of values that contains some specified proportion (say, 95%) of the values. The distribution of the sample mean for samples of size 10 still has a few outlying values in the \$40,000–\$50,000 range. Hence, approximate probabilities based on the normal approximation will be poor. The approximation for a sample size of 25 looks better, and the approximations for sample sizes of 50 and above look good.

How much of a difference does the shape of the distribution of the population make on how large the sample size needs to be for the normal approximation to be adequate?

For these very right–skewed wage incomes it appears that a sample size of 25–50 is necessary. Let's try a transformation to reduce the population skewness. The sampling distribution should then be much closer to normal in shape. Here is a histogram of the logarithm of the income:

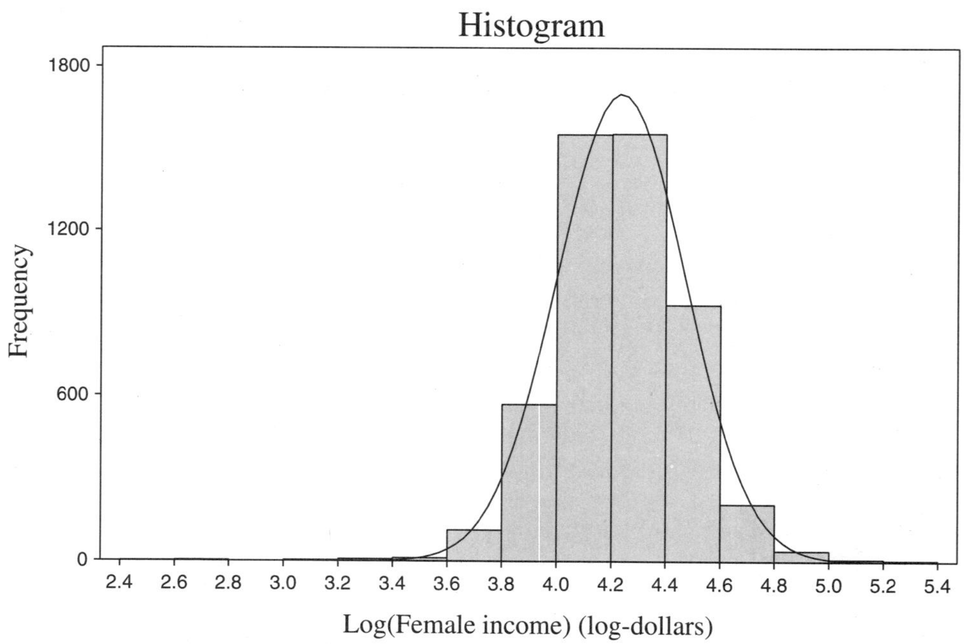

Except for a few lower outliers, the distribution of the population now looks quite normal. The minimum income of \$292 is very low for a full–time full–year worker, which is emphasized in the log scale.

Here are some descriptive statistics:

```
DESCRIPTIVE STATISTICS

                   LINCOME
N                     5000
MEAN                4.2351
SD                  0.2347
MINIMUM             2.4654
1ST QUARTILE        4.0836
MEDIAN              4.2302
3RD QUARTILE        4.3877
MAXIMUM             5.3441
```

The median and mean are very close, reflecting the dramatically reduced skewness. The reason that the logarithm–transformed income appears to be so normal in shape is probably because of the multiplicative nature of incomes. People typically receive wage increases determined as a percentage of their current wage, rather than fixed dollar amounts. Hence, on the log–scale these incremental changes appear as *additive* changes. Indeed, we can then view a person's wage as the *sum* of these changes. In this sense a person's wage is subject to the Central Limit Theorem, just as sample averages (and sums) are. Thus, there appears to be a real–life Central Limit Theorem applying to the individual wages. This is the reason that salaries are often modeled as following a lognormal distribution; this distribution gets its name from the fact that the logarithm of such a random variable is normally distributed.

How does the shape of the sampling distribution of the sample mean for the transformed incomes change as the sample size increases?

As for the untransformed income, we can randomly choose samples of size 1, 2, 4, 25, 50, 100 and 625. The sample means for each one of these samples are then saved in `SIZE1L`, `SIZE2L`, `SIZE4L`, `SIZE25L`, `SIZE50L`, `SIZE100L`, and `SIZE625L`, respectively. As before, each variable represents 400 random samples.

Let's look at histograms of the sample means for each sample size:

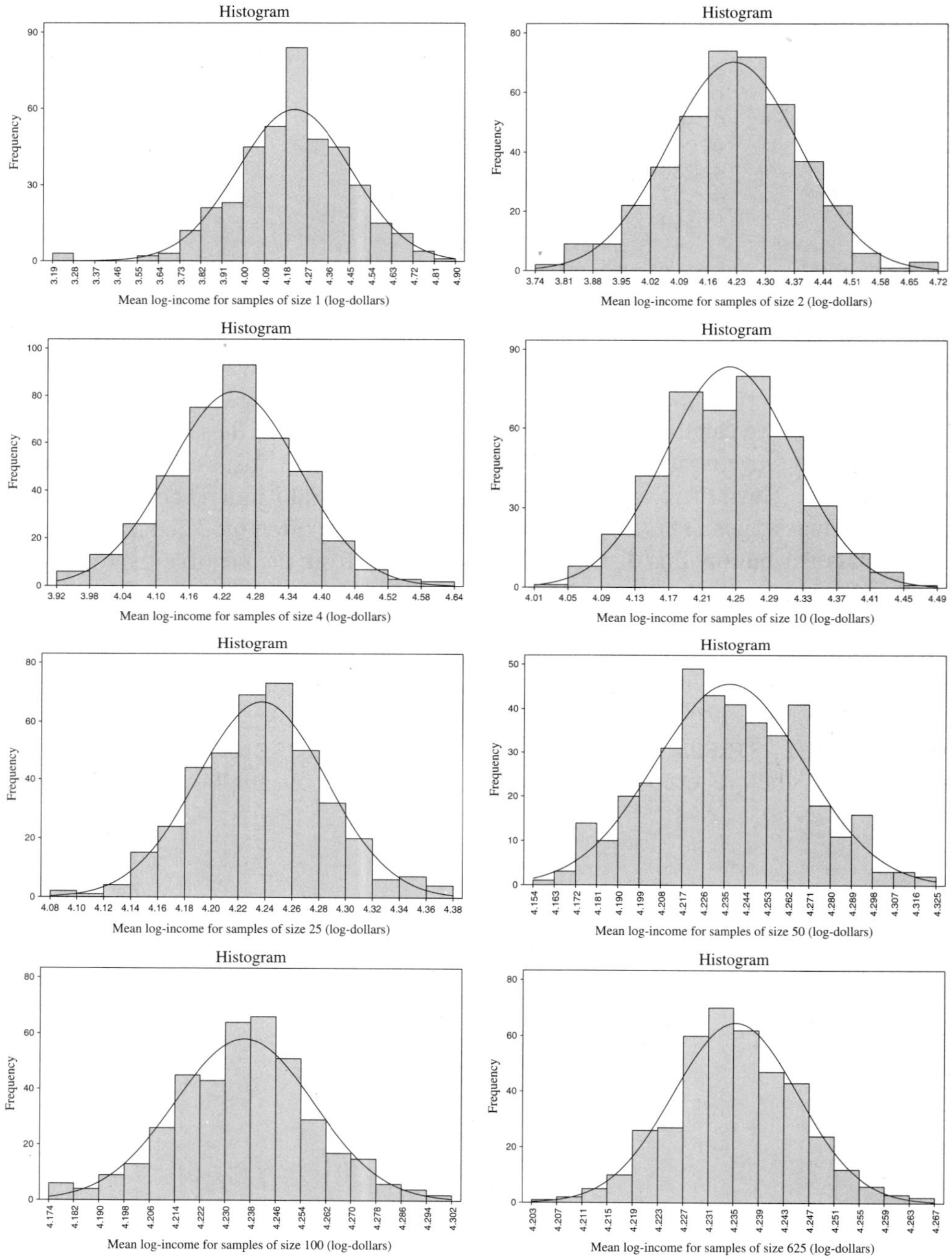

The first histogram, which is just a sample from the population, can be used as a baseline for comparison. After looking at the histograms for a while, we can make some observations. In this case, the distribution of the sample mean for samples of size 2 looks reasonably normal. The approximation for sample sizes of 4 and above looks very good. Remember that these are sample values; some variation is expected, and we do not expect the distribution to match the normal curve exactly.

As before, the variation of the sample means progressively decreases as the size of the sample increases. The distributions are increasingly concentrated about the population mean of 4.2351 log–dollars.

From the descriptive statistics given earlier, we see that the population standard deviation is $\sigma = 0.2347$ log–dollars. Here are the theoretical standard deviations (standard errors) for each sampling distribution compared to the values we obtain from the descriptive statistics:

Sample size n	True standard deviation of the sampling distribution $\sigma_{\overline{X}} = \frac{\sigma}{\sqrt{n}}$	Estimated standard deviation of the sampling distribution
1	0.2347	0.2408
2	0.1660	0.1589
4	0.1174	0.1172
10	0.0742	0.0764
25	0.0469	0.0479
50	0.0332	0.0315
100	0.0235	0.0220
625	0.0094	0.0099

As was the case for the untransformed values, in each case the observed value is close to the expected value from theory. They do vary, though, as we also expect from sample values. So, this computer experiment has verified the theoretical results.

Summary

The Central Limit Theorem is used as justification for the approximation of the distribution of the sample mean by the normal distribution. In this case, we conduct an experiment on the computer to explore how applicable the Central Limit Theorem is for income data from the 1990 census.

The distribution of the income values is extremely right–skewed and non–normal. Based on the experiment, we find that sample sizes of 25–50 are necessary for the Central Limit Theorem approximation to be accurate.

The log–transformed income values appear to be nearly normal in shape. For these data the experimental results suggest that sample sizes of 2–4 are sufficient for the approximation to be accurate.

The experimental results are also consistent with the formula for the standard deviation (standard error) of $\overline{X}$, $\sigma_{\overline{X}} = \frac{\sigma}{\sqrt{n}}$, which is usually derived from theoretical considerations.

Technical terms

Central Limit Theorem: a theoretical result that states that as the sample size increases, the sampling distribution of virtually all statistics can be progressively better approximated by the normal distribution. The result is true under very weak conditions on the shape of the underlying population, as long as the data constitute a random sample from the population.

Gaussian random variable: a continuous random variable, often called a "normal" random variable. It is completely characterized by its mean and standard deviation. Its distribution is symmetric around its mean, and follows a "bell shape." Many observed processes produce normally distributed random variables. The normal random variable with mean 0 and standard deviation 1 is called the *standard* normal random variable.

Population: the complete set of objects that is the subject of study. Often it is too time–consuming or expensive to try to examine the entire population of interest. In this case a sample is often used as a surrogate.

Sample: a subset of the population, which is then the subject of investigation. In order for patterns observed in the sample to be representative of patterns in the population, the sample must be constructed so as to be "typical" of the population. In addition, most inferential statistics assume that the sample is drawn randomly from the population.

Standard error: the standard deviation of a statistic (estimate). It is based on the sampling distribution of the statistic, which is itself dependent on the underlying population distribution of the data and the sample size, as well as the form of the statistic itself.

The sampling distribution for the median

Topics covered: Central Limit Theorem. Computer experiment. Distributional shape. Independent samples. Sampling distributions.

Key words: Histogram. Long–tailed data. Median. Standard deviation. Standard error.

Data Files: `census4.dat`, `census5.dat`

In the case "The Central Limit Theorem for census data," we explored the applicability of the Central Limit Theorem for income data from the 1990 census. In that case we conducted an experiment on the computer to study the sampling distribution of the sample mean. In this case we will undertake a similar experiment based on the sample median.

From the descriptive statistics of the population we know that the median income is \$16,990. The mean income is \$19,897. These values, of course, represent different characteristics of the population. If we rank the women from lowest to highest wage income, then the median income is the income of the person in the middle. As such, the median captures the characteristics of the middle person. The mean is the total income received divided by the number of people. Thus, it summarizes the overall income level of the population. Because the median counts *people,* rather than *dollars,* it usually better represents of the level of income the typical person has. This is particularly true here, where there are some people with very large incomes.

The purpose of this case is to explore how close the sample medians based on re–surveys are to the true population value. We also want to examine the distribution of the estimates.

As in the case "The Central Limit Theorem for census data," let's first generate samples of size 10. Each value of the variable `SIZE10M` is the sample median from a sample of size 10 from the population. The variable has 400 cases, each corresponding to a different sample. For example, the first value of `SIZE10M` is 14.712. This means that the first sample of 10 randomly chosen women had a median income of \$14,712. The second sample of 10 randomly chosen women had a median of \$15,071, and so on. Each of the sample medians is different, and not necessarily equal to the population median (\$16,990). Both of these values are less than the population median. This raises the question: how are these sample medians distributed about the population median?

First, let's look at some descriptive statistics for `SIZE10M`:

```
DESCRIPTIVE STATISTICS

                 SIZE10M
N                    400
MEAN              17.672
SD                3.5018
MINIMUM           9.4165
1ST QUARTILE      15.163
MEDIAN            17.034
3RD QUARTILE      19.671
MAXIMUM           28.545
```

We see that the variability of the sample medians is generally much less than for the population values. The standard deviation is 3.5018, compared with 12.573. The average of the sample medians is almost $800 above the population median, although the median of the sample medians is close to the population median.

Next, let's look at a histogram of SIZE10M. This represents the sampling distribution of the sample median:

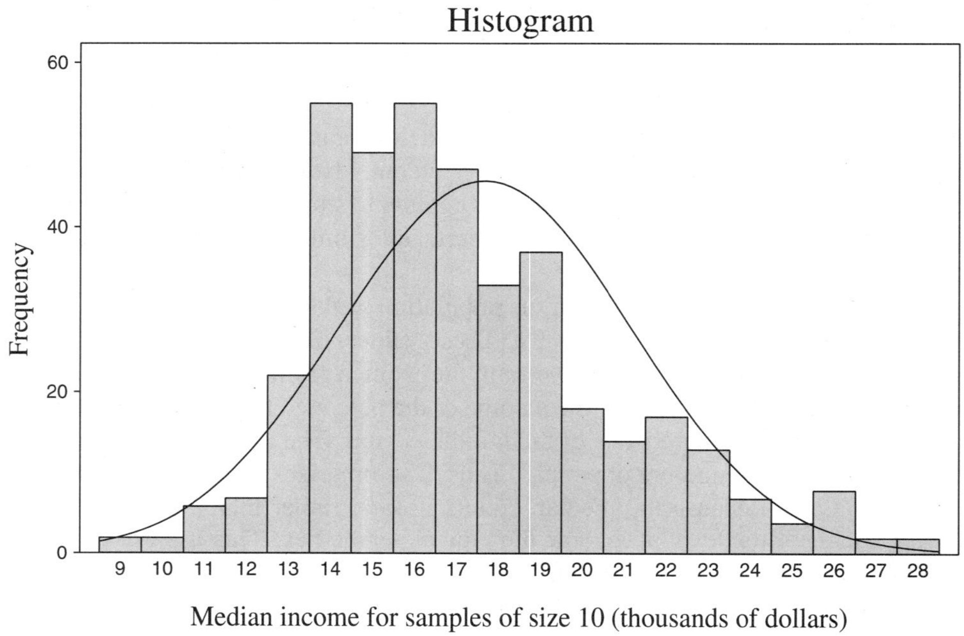

We see that the values are still right–skewed, although not as right–skewed as the population values are. The distribution of the normal (Gaussian) variable with the same mean and variance as that of SIZE10M is superimposed on the histogram. We can see that the sampling distribution of the sample median for samples of size 10 is not close to normal, but is closer to a normal distribution than the distribution of the population values is.

Let's take a look at other sample sizes. The variables SIZE1M, SIZE2M, SIZE4M, SIZE25M, SIZE50M, SIZE100M, and SIZE625M are the sample medians from samples of size 1, 2, 4, 25, 50, 100 and 625, respectively. As was the case for SIZE10M, for each variable we have taken 400 random samples.

First, here are some descriptive statistics for each variable:

DESCRIPTIVE STATISTICS

	SIZE1M	SIZE2M	SIZE4M	SIZE10M	SIZE25M
N	400	400	400	400	400
MEAN	19.426	20.347	18.206	17.672	17.081
SD	10.558	8.9035	5.0559	3.5018	2.1841
MINIMUM	0.2920	6.1070	7.9500	9.4165	11.651
1ST QUARTILE	12.203	14.524	14.711	15.163	15.569
MEDIAN	17.255	18.695	17.790	17.034	16.890
3RD QUARTILE	24.128	24.806	20.683	19.671	18.293
MAXIMUM	89.585	98.545	34.880	28.545	23.805

```
                  SIZE50M    SIZE100M     SIZE625M
N                     400         400          400
MEAN               17.174      16.998       16.982
SD                 1.6519      1.0690       0.4483
MINIMUM            13.611      14.203       15.663
1ST QUARTILE       15.990      16.171       16.681
MEDIAN             17.049      17.014       16.960
3RD QUARTILE       18.233      17.780       17.324
MAXIMUM            23.569      20.328       18.328
```

The samples of size 1 (i.e., SIZE1M) are just a random sample from the population. We note that the mean and standard deviation for SIZE1M are close to those of the population.

Based on the variant of the Central Limit Theorem that applies to the sample median, we expect that as the sample size increases, the sample medians will become closer to the population median. We can see this effect in the means of these variables; as the sample size increases the average sample median approaches the population median. The median of the sample medians also approaches the population median. Similarly, as the sample size increases, the sample standard deviation decreases, as we would expect. We also can note that, compared with the corresponding values given for the sample mean in the case "The Central Limit Theorem for census data," the sample median is, on average, farther from the population median than the sample mean is from the population mean (for a given sample size). Further, the sample median is systematically above the population median (on average), indicating that the estimate is positively biased.

How does the variation of the sample median compare to the variation of the sample mean for the same sample size? Here are the theoretical standard deviations for the sampling distribution of the sample mean compared to the sample standard deviations of the sample medians we have observed. Note that these theoretical standard deviations are the standard errors of the sample mean, while the observed values are estimates of the standard errors of the sample medians:

Sample size n	True standard deviation of the sampling distribution of the sample mean $\sigma_{\overline{X}} = \frac{\sigma}{\sqrt{n}}$	Estimated standard deviation of the sampling distribution of the sample median s_{median}	Ratio $\frac{s_{median}}{\sigma_{\overline{X}}}$
1	12.573	10.558	0.840
2	8.8905	8.9035	1.001
4	6.2865	5.0559	0.804
10	3.9759	3.5018	0.880
25	2.5146	2.1841	0.869
50	1.7781	1.6519	0.929
100	1.2573	1.0690	0.850
625	0.5029	0.4483	0.891

After the first few values, the ratios appear to stabilize at a value of about 0.89. This suggests that the standard error of the median appears to be inversely proportional to $\sqrt{n}$ for large values of n (since we know that the standard error of the mean is). There is some variation in the ratios, as we expect because of random fluctuation (from the act of drawing a random sample). The data suggest that the standard deviation of the sampling distribution of the sample median is

$$\sigma_{median} \approx \frac{0.89\sigma}{\sqrt{n}} \quad \text{for large } n.$$

Thus for these long–tailed data, the sample median is less variable than the sample mean, although it is biased.

How does the distributional shape of the sampling distribution change as the sample size increases? Let's look at histograms of the sample medians for each sample size (they are on the next page).

The first histogram, which is just a sample from the population, can be used as a baseline for comparison. After looking at them for a while, we can make some observations:

(1) As the sample size increases, the distributions become closer and closer to the corresponding normal distribution in shape. In doing so, the skewness apparent in the population becomes progressively less noticeable.

(2) The variability of the sample medians progressively decreases as the size of the sample increases. This can be seen from the range of values on the horizontal axis.

(3) At what point is the distribution sufficiently normal, so that the normal approximation is adequate? The distribution of the sample median for samples of size 50 still exhibits some skewness. Hence, approximate probabilities based on the normal approximation will be poor. The approximations for sample sizes of 100 and above look better.

In the case "The Central Limit Theorem for census data," we saw that the shape of the distribution of the population made a difference on how large the sample size needs to be for the normal approximation to be adequate for the sample mean. The median is supposedly less sensitive to outlying values. Is the sampling distribution of the sample median sensitive to the shape of the population distribution?

Let's consider again the logarithm of the income. We can consider the sample medians of the logarithm of income for each sample size. Each value of the variable `SIZE10ML` is the sample median for a sample of size 10 from the population of log–incomes (i.e., from `LINCOME`). The variables `SIZE1ML`, `SIZE2ML`, `SIZE4ML`, `SIZE25ML`, `SIZE50ML`, `SIZE100ML`, and `SIZE625ML` are the sample medians from samples of size 1, 2, 4, 25, 50, 100 and 625, respectively. For each variable we have taken 400 random samples.

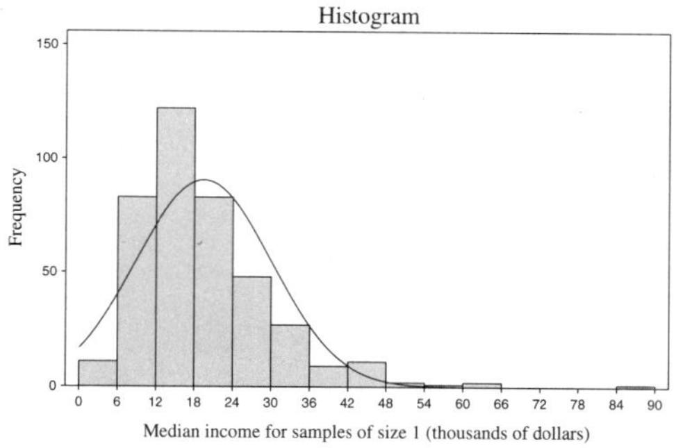
Histogram
Frequency
0
50
100
150
0 6 12 18 24 30 36 42 48 54 60 66 72 78 84 90
Median income for samples of size 1 (thousands of dollars)

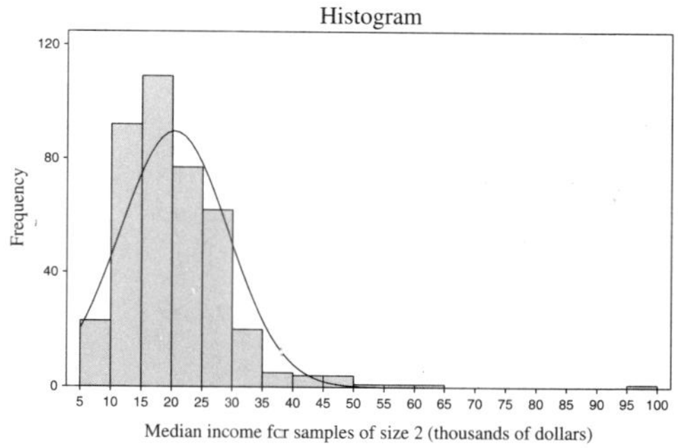
Histogram
Frequency
0
40
80
120
5 10 15 20 25 30 35 40 45 50 55 60 65 70 75 80 85 90 95 100
Median income for samples of size 2 (thousands of dollars)

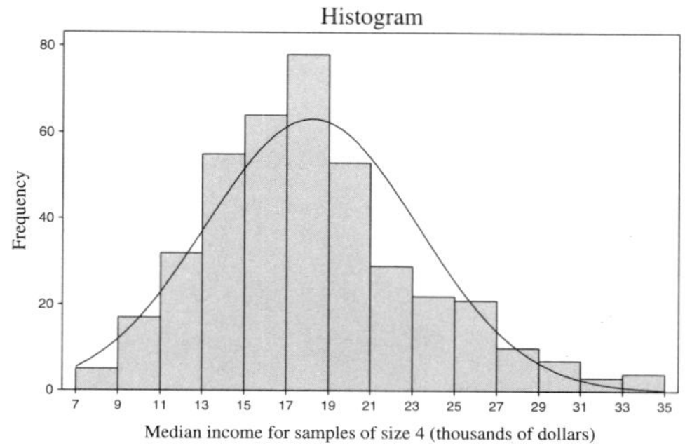
Histogram
Frequency
0
20
40
60
80
7 9 11 13 15 17 19 21 23 25 27 29 31 33 35
Median income for samples of size 4 (thousands of dollars)

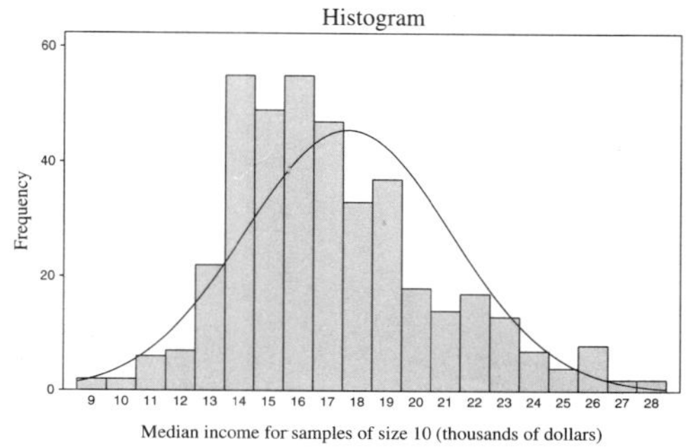
Histogram
Frequency
0
20
40
60
9 10 11 12 13 14 15 16 17 18 19 20 21 22 23 24 25 26 27 28
Median income for samples of size 10 (thousands of dollars)

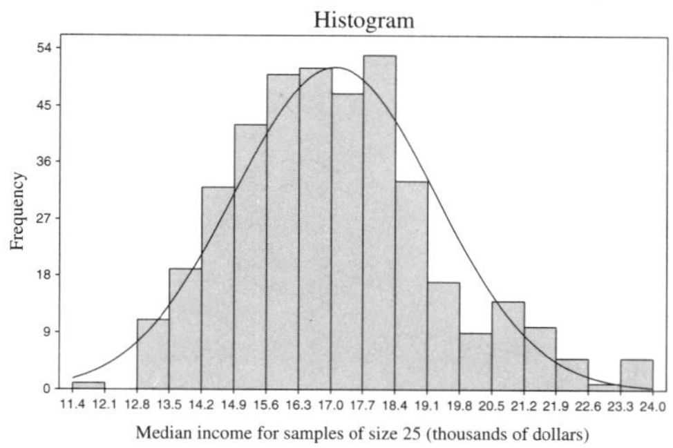
Histogram
Frequency
0
9
18
27
36
45
54
11.4 12.1 12.8 13.5 14.2 14.9 15.6 16.3 17.0 17.7 18.4 19.1 19.8 20.5 21.2 21.9 22.6 23.3 24.0
Median income for samples of size 25 (thousands of dollars)

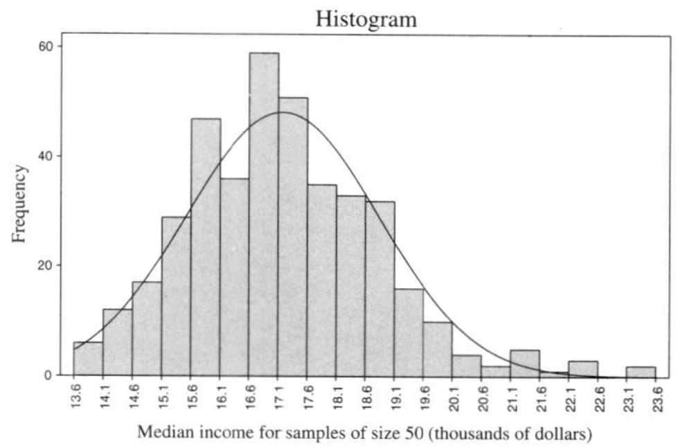
Histogram
Frequency
0
20
40
60
13.6 14.1 14.6 15.1 15.6 16.1 16.6 17.1 17.6 18.1 18.6 19.1 19.6 20.1 20.6 21.1 21.6 22.1 22.6 23.1 23.6
Median income for samples of size 50 (thousands of dollars)

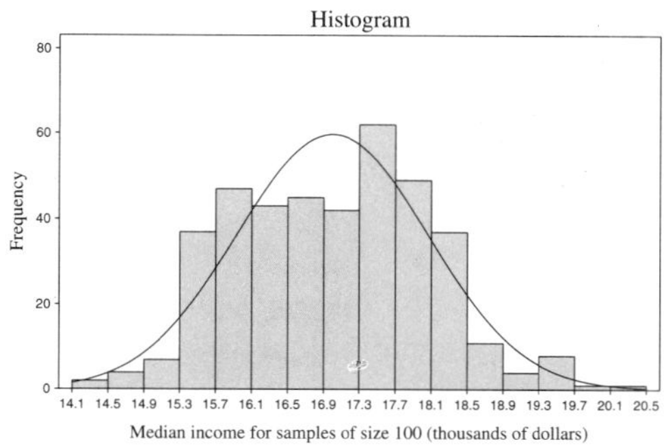
Histogram
Frequency
0
20
40
60
80
14.1 14.5 14.9 15.3 15.7 16.1 16.5 16.9 17.3 17.7 18.1 18.5 18.9 19.3 19.7 20.1 20.5
Median income for samples of size 100 (thousands of dollars)

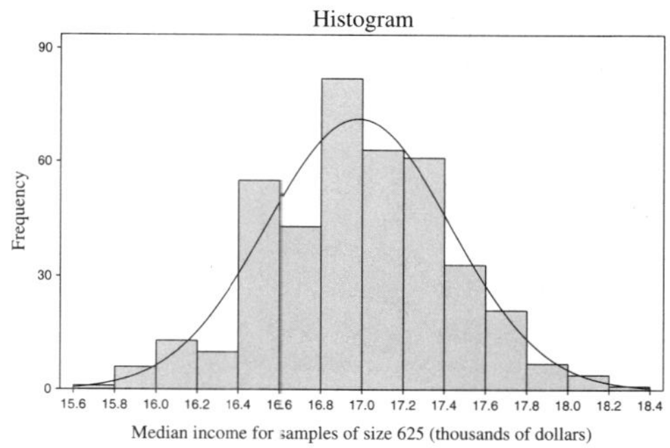
Histogram
Frequency
0
30
60
90
15.6 15.8 16.0 16.2 16.4 16.6 16.8 17.0 17.2 17.4 17.6 17.8 18.0 18.2 18.4
Median income for samples of size 625 (thousands of dollars)

Let's look at histograms of the sample medians of the log–income for each sample size:

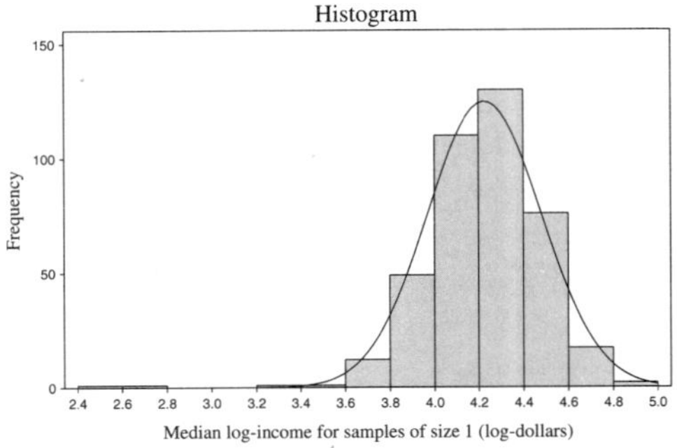

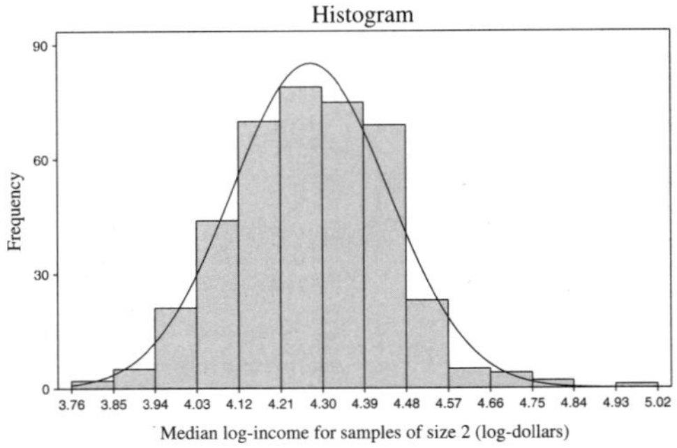

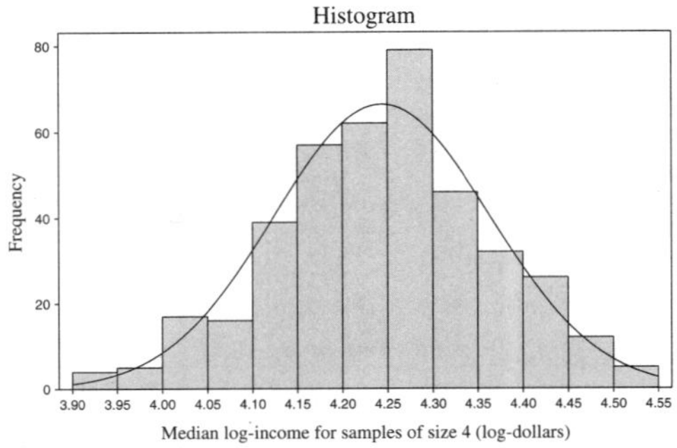

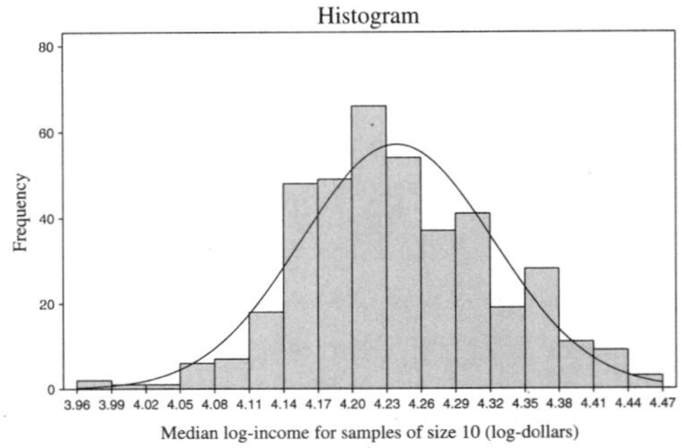

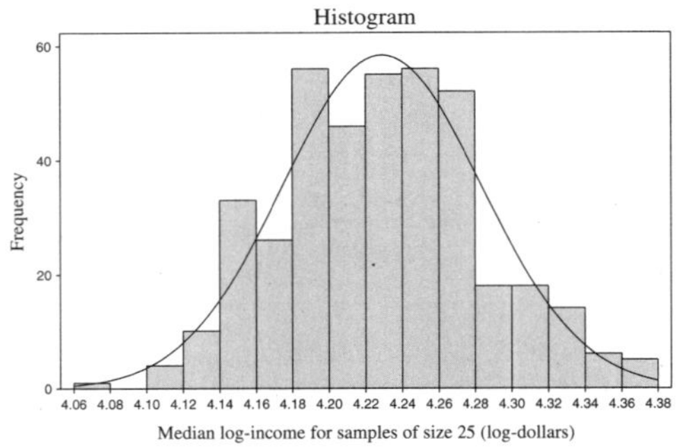

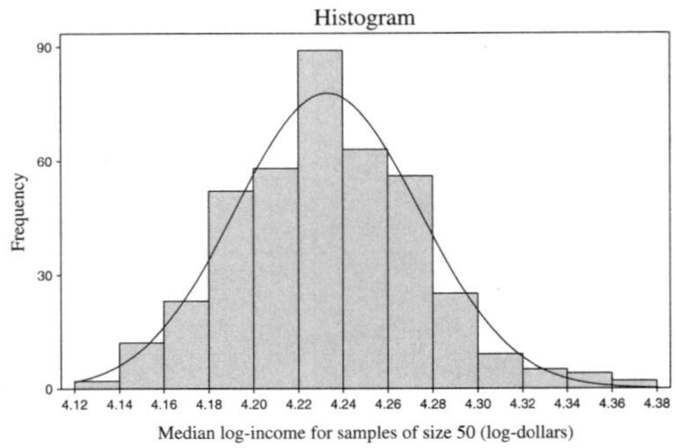

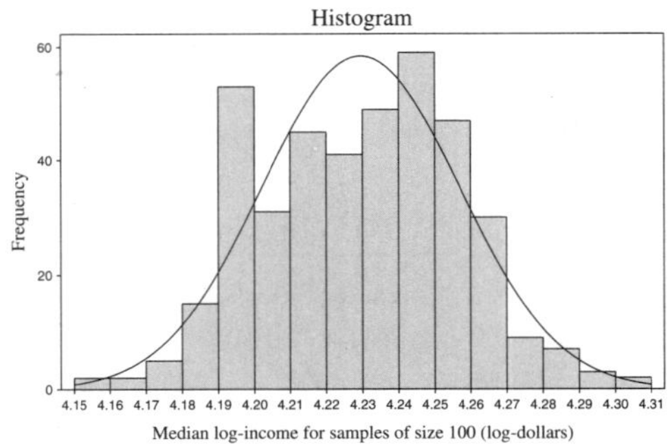

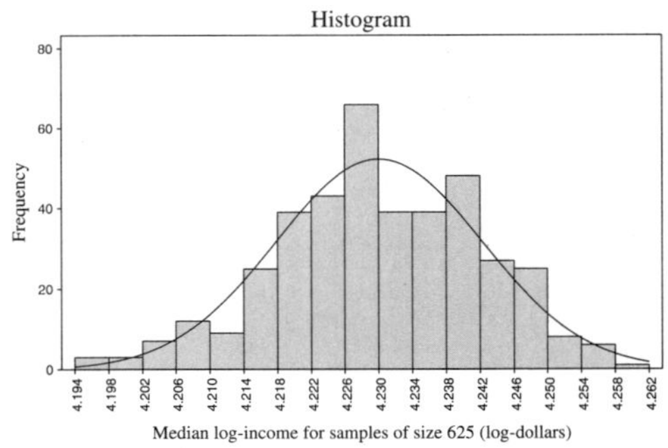

The first histogram, which is just a sample from the population, can be used as a baseline for comparison. After looking at the histograms for a while, we can make some observations. Here, the distribution of the sample median for samples of size 2 looks reasonably normal. The approximation for sample sizes of 4 and above looks very good. Remember that these are sample values; some variation is expected, and we do not expect the distribution to match the normal curve exactly.

As before, the variation of the sample medians progressively decreases as the size of the sample increases. The distributions are increasingly concentrated about the population median of 4.2302 log–dollars. Further, for this nearly symmetric population, the sample medians do not consistently overestimate or underestimate the population value. That is, the estimate appears to be unbiased.

From the descriptive statistics given in the earlier case, we see that the population standard deviation is $\sigma = 0.2347$ log–dollars. Here are the theoretical standard deviations (standard errors) for each sampling distribution for the sample mean compared to the values we obtain from the descriptive statistics for the sample median:

Sample size n	True standard deviation of the sampling distribution of the sample mean $\sigma_{\overline{X}} = \frac{\sigma}{\sqrt{n}}$	Estimated standard deviation of the sampling distribution of the sample median s_{median}	Ratio $\frac{s_{median}}{\sigma_{\overline{X}}}$
1	0.2347	0.2556	1.089
2	0.1660	0.1687	1.016
4	0.1174	0.1202	1.024
10	0.0742	0.0838	1.129
25	0.0469	0.0547	1.166
50	0.0332	0.0411	1.238
100	0.0235	0.0273	1.162
625	0.0094	0.0122	1.298

As happened for the untransformed values, after the first few values the ratios appear to stabilize, this time at a value of about 1.25. This suggests that the standard error of the median is inversely proportional to $\sqrt{n}$ for large values of n. In fact, it suggests that the standard deviation of the sampling distribution of the sample median is

$$\sigma_{median} \approx \frac{1.25\sigma}{\sqrt{n}} \quad \text{for large } n.$$

Note that for this symmetric (nearly normal) population, the sample median is apparently more variable than the sample mean.

Summary

In this case, we conduct an experiment on the computer to explore how applicable the Central Limit Theorem is for the sample median from income data from the 1990 census.

The sample median is subject to a variation of the standard Central Limit Theorem; the sampling distribution of the sample median will appear to be more and more normal in shape as the sample size increases. The mean of the sampling distribution will approach the population median and the standard deviation of the sampling distribution will be roughly inversely proportional to $\sqrt{n}$ for large sample sizes.

The results of the experiment verify these results.

The income values are very right–skewed and non–normally distributed. Based on the experiment, we find that sample sizes of 50 or more are necessary for the Central Limit Theorem approximation to be accurate. We also find that the variation of the sample median is less than the variation in the sample mean of the same sample size. However, on average, the sample median seems to be farther from the population median than the sample mean is away from the population mean, and biased upwards.

The log–transformed income values appear to be nearly normal in shape. For these data the experimental results suggest that sample sizes of 2–4 are sufficient for the approximation to be accurate. For these nearly normal data the variation in the sample median is consistently greater than the variation in the sample mean for the same sample size.

The sampling distribution for the standard deviation

Topics covered: Central Limit Theorem. Computer experiment. Distributional shape. Independent samples. Sampling distributions.

Key words: Histogram. Long–tailed data. Standard deviation. Standard error.

Data Files: `census6.dat`

In the cases "The Central Limit Theorem for census data" and "The sampling distribution for the median" we explored the applicability of the Central Limit Theorem to the sample mean and the sample median for income data from the 1990 census. We found that in both cases the sampling distributions approached a normal distribution as the sample size increased. As the distribution of the income population was very non–normal, the sample size needed to be at least 30 before the sampling distribution was approximately normal. In both cases the standard deviation of the estimator appeared to be inversely proportional to the square root of the sample size.

In this case we will undertake a similar computer experiment on the sample standard deviation.

From the descriptive statistics of the population we know that the standard deviation of the income is $12,573.

The purpose of this case is to explore how close the sample standard deviations based on re–surveys are to the true population value. We also want to examine the distribution of the estimates.

As in the original cases, let's first generate samples of size 10. Each value of the variable `SIZE10S` is the sample standard deviation from a sample of size 10 from the population. The variable has 400 cases, each corresponding to a different sample. For example, the first value of `SIZE10S` is 10.023. This means that the first sample of 10 randomly chosen women had a sample standard deviation of $10,023. The second sample of 10 randomly chosen women had a standard deviation of $13,363, and so on. Each of the sample standard deviations is different, and not necessarily equal to the population standard deviation. Some are close, as for the second sample, and some are far away. However, there is a pattern, based on the distribution of the sample standard deviations. How are these sample standard deviations distributed about the population standard deviation?

First, let's look at some descriptive statistics for `SIZE10S`:

```
DESCRIPTIVE STATISTICS

                 SIZE10S
N                    400
MEAN              11.278
SD                7.9036
MINIMUM           4.1918
1ST QUARTILE      7.6971
MEDIAN            9.6901
3RD QUARTILE      12.200
MAXIMUM           83.763
```

The average of the sample standard deviations is more than $1000 below the population standard deviation. The median value is even lower. Thus, the sample standard deviation is apparently negatively biased as an estimator of the population standard deviation for these data.

Next, let's look at a histogram of SIZE10S. This represents the sampling distribution of the sample standard deviation.

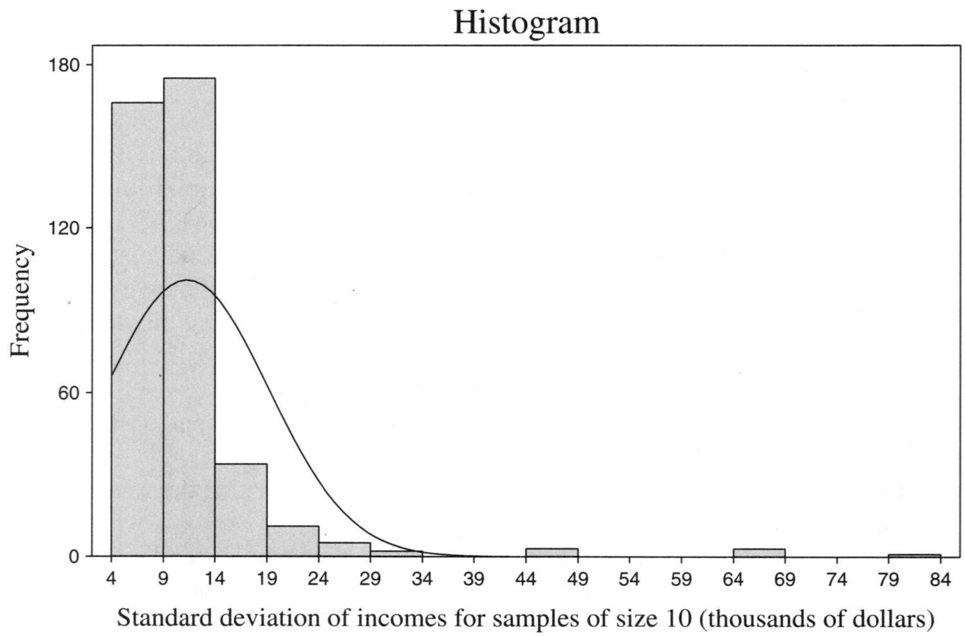

We see that the values are very right–skewed. The distribution of the normal (Gaussian) variable with the same mean and variance as that of SIZE10S is superimposed on the histogram. We can see that the sampling distribution of the sample standard deviation for samples of size 10 is not at all close to normal.

Let's take a look at other sample sizes. The variables SIZE2S, SIZE4S, SIZE25S, SIZE50S, SIZE100S, and SIZE625S are the sample standard deviations from samples of size 2, 4, 25, 50, 100 and 625, respectively. As happened for SIZE10S, for each variable we have taken 400 random samples.

First, here are some descriptive statistics for each variable:

```
DESCRIPTIVE STATISTICS

                   SIZE2S      SIZE4S      SIZE10S     SIZE25S     SIZE50S
N                     400         400          400         400         400
MEAN               8.5726      10.135       11.278      11.284      11.878
SD                 9.5806      8.8313       7.9036      4.9315      4.3466
MINIMUM         7.071E-04      1.1483       4.1918      5.8642      5.5331
1ST QUARTILE       2.4895      5.3738       7.6971      8.7583      9.3826
MEDIAN             6.0210      8.3951       9.6901      10.052      10.536
3RD QUARTILE       11.949      12.371       12.200      12.049      12.695
MAXIMUM            99.908      101.92       83.763      41.316      31.295

                 SIZE100S    SIZE625S
N                     400         400
MEAN               12.104      12.464
SD                 3.2475      1.5527
MINIMUM            7.6178      9.5074
1ST QUARTILE       10.067      11.300
MEDIAN             11.238      12.156
3RD QUARTILE       12.996      13.406
MAXIMUM            24.555      19.820
```

How does the variation of the sample standard deviation compare with the variation of the sample mean for the same sample size? Here are the theoretical standard deviations for the sampling distribution of the sample mean compared with the sample standard deviations of the sample standard deviations we have observed. Note that these theoretical standard deviations are the standard errors of the sample mean, while the observed values are estimates of the standard errors of the sample standard deviations:

Sample size n	True standard deviation of the sampling distribution of the sample mean $\sigma_{\overline{X}} = \frac{\sigma}{\sqrt{n}}$	Estimated standard deviation of the sampling distribution of the sample standard deviation s_{sd}	Ratio $\frac{s_{sd}}{\sigma_{\overline{X}}}$
2	8.8905	9.5806	1.078
4	6.2865	8.8313	1.405
10	3.9759	7.9036	1.988
25	2.5146	4.9315	1.961
50	1.7781	4.3466	2.445
100	1.2573	3.2475	2.583
625	0.5029	1.5527	3.087

The ratios appear to be increasing steadily. This shows that although the standard error of the standard deviation appears to be steadily decreasing, it decreases much more slowly than the standard error for the sample mean.

How does the distributional shape of the sampling distribution change as the sample size increases? Let's look at histograms of the sample standard deviations for each sample size:

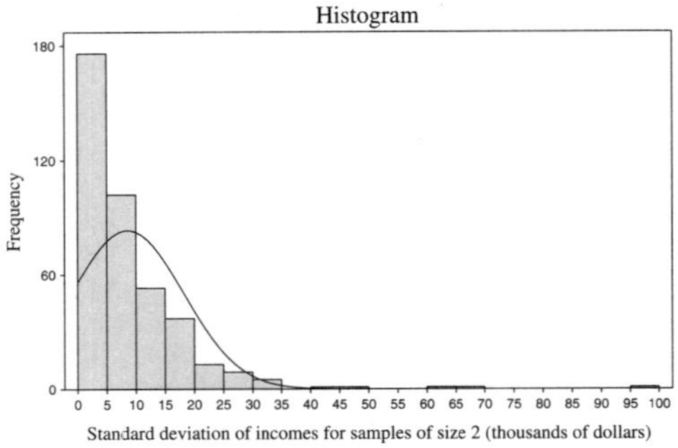

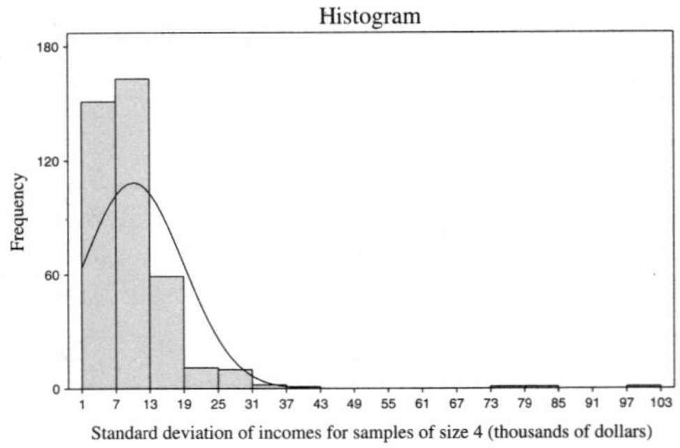

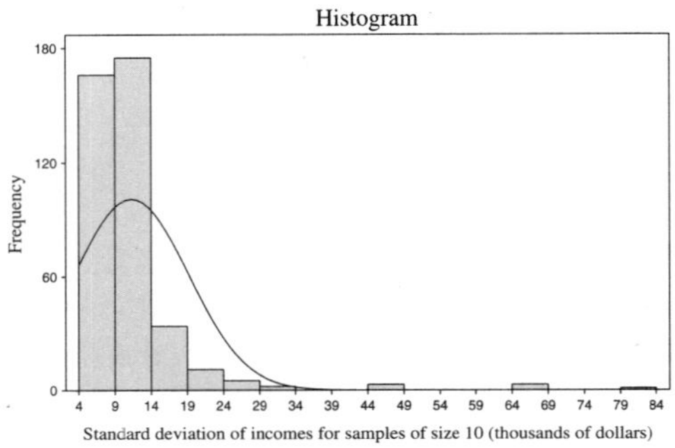

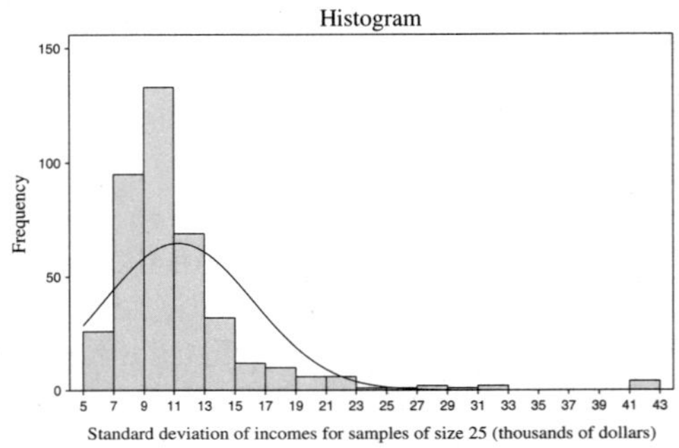

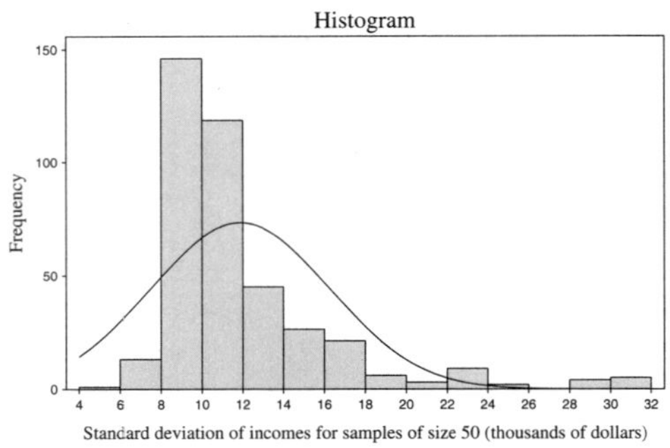

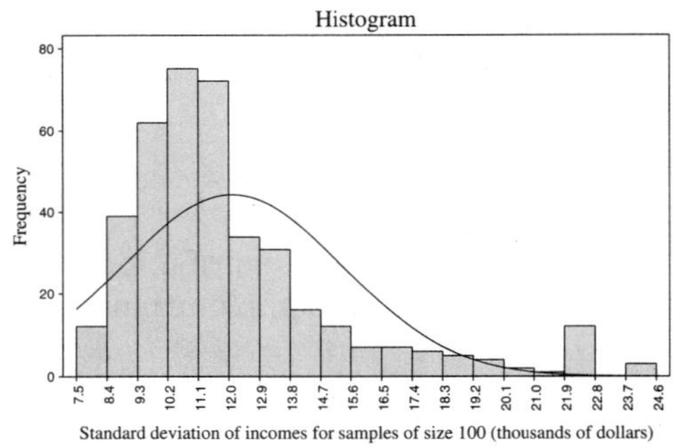

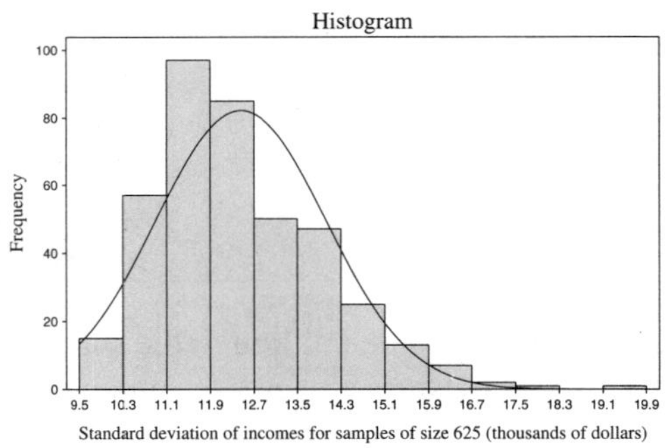

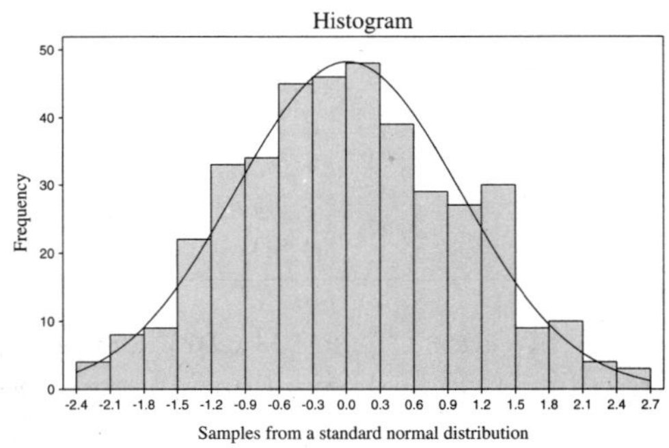

The last histogram is for a sample of size 400 from a standard normal distribution. This is included to see how a histogram for normal data varies about the normal curve.

We can make some observations:

(1) The distributions are very right–skewed, even for large sample sizes.

(2) As the sample size increases, the distributions become closer to the corresponding normal distribution in shape. In doing so, the skewness apparent in the population becomes progressively less noticeable.

(3) The variation of the sample standard deviations progressively decreases as the size of the sample increases. This can be seen from the range of values on the horizontal axis. This trend is slower than for the sample mean and sample median.

(4) At no point for the chosen sample sizes is the distribution well approximated by a normal distribution.

Summary

In this case we conduct an experiment on the computer to explore how applicable the Central Limit Theorem is for the sample standard deviation from income data from the 1990 census.

The Central Limit Theorem appears to work more slowly on the sample standard deviation than for the sample mean and sample median. As the sample size increases, the sampling distributions converge quite slowly to the normal shape. The variation in the sample standard deviation also decreases more slowly than for the sample mean. and sample median.

Additional technical note

Many text books show that, as the sample size (n) increases, the sampling distribution of the standard deviation will approach a normal distribution with standard deviation $\sigma/\sqrt{2n}$ and mean σ. This experiment shows that not only is the convergence to the normal shape slow, but that the formula for the standard error of the standard deviation itself can be inaccurate for moderate sample sizes. The reason for this is the non–normality of the population distribution.

Racial imbalance in Nassau County public schools

Topics covered: Binomial random variable. Central Limit Theorem. Extra–Binomial variation. Standardizing sample proportions.

Key words: Binomial random variable. Histogram. Normal plot. Sample proportions. Scatter plot. Side–by–side boxplots. z–score.

Data file: `lischool.dat`

On May 17, 1954, the United States Supreme Court issued its landmark *Brown vs. Board of Education of Topeka, Kansas* ruling, which declared that racially segregated public schools were inherently unequal. This ruling invalidated the "separate but equal" justification for segregated schools.

As a result of the ruling, massive programs designed to integrate public schools were begun throughout the United States. Despite this, there is a common public perception that public schools in the 1990's are still strongly segregated by race. Is this, in fact, true? Does the racial distribution in public schools reflect the general distribution in the community at large, or are there differences (reflecting *de facto* racial imbalance)?

The data examined here represent the proportion of white enrollment in the 56 school districts in Nassau County (Long Island, New York), for the 1992–1993 school year, as reported in the May 20, 1994, issue of *Newsday*. Given the total number of students enrolled in a given school district (n_i , for school district i), the number of white students (w_i) can be viewed as a **Binomial random variable**; that is,

$$w_i \sim \text{Binomial}(n_i, p_i),$$

where p_i is the true probability of a randomly selected student in school district i being white.

If school districts were perfectly integrated over the entire county, p_i would be the same for all school districts (say p_0). In 1992–1993, there were 174,556 students enrolled in Nassau County public schools, of whom 130,349 were white. Thus, a reasonable value for p_0 would be the overall ratio of white students, or

$$p_0 = \frac{130,349}{174,556} = .7467.$$

Note that since p_0 is based on the actual numbers for the entire county, it is not an estimate of the overall probability of a student being white, but rather the actual value itself.

If p_i did equal p_0 for all i , we would expect that the observed proportion of white students for each district ($\overline{p}_i = w_i/n_i$) would be reasonably close to p_0 . Thus, the distribution of values of $\overline{p}_i$ is informative about whether integration is suggested in the county. We must recognize, of course, that even if the actual probability of a randomly selected student being white *was* constant, we would expect some variation in the values of $\overline{p}_i$, because of random fluctuation.

Here is a histogram of the observed values of $\overline{p}_i$ for the 56 Nassau school districts:

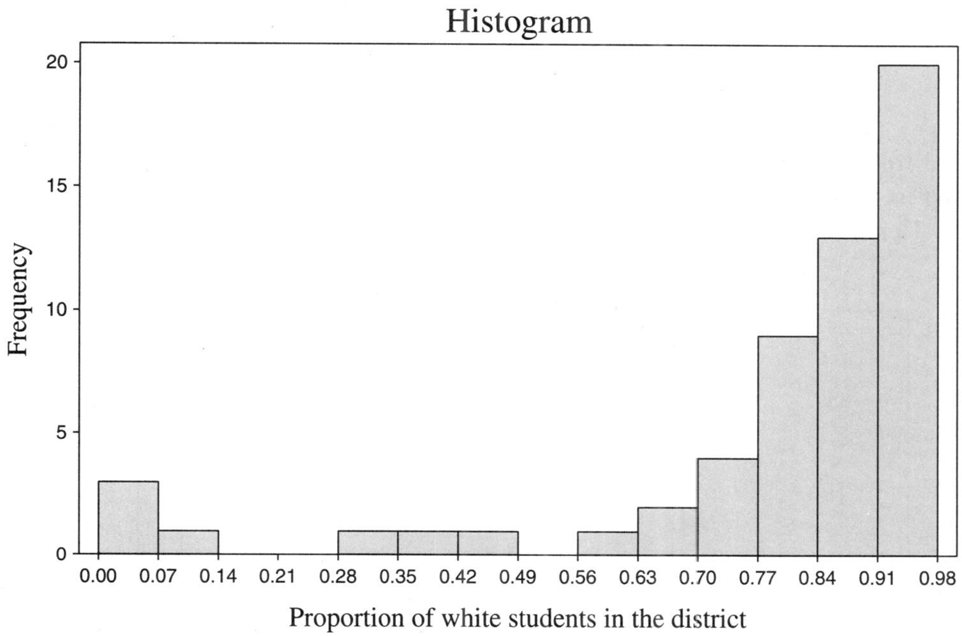

This is not the kind of picture we would expect to see under conditions of racial balance. The distribution is not at all centered on p_0, as we would expect, and there is a pronounced left tail, with a small mode at a very small proportion of white students (these are the Hempstead, Roosevelt, Uniondale and Westbury school districts).

It might be thought that this histogram alone is enough to decide that racial imbalance is occurring in Nassau County, but that is not quite true. The problem is that we have not taken into account the inherent variability of the estimates $\overline{p}_i$. School districts with small enrollments will have values of $\overline{p}_i$ that have much greater variability than school districts with large enrollments. For example, if the school districts whose values fall in the left tail all had very small enrollments, the large difference between $\overline{p}_i$ and p_0 could be just because of random fluctuation. Let's make this more explicit. Since $w_i \sim \text{Binomial}(n_i, p_i)$, we know that

$$E(w_i) = n_i p_i$$

and

$$Var(w_i) = n_i p_i (1 - p_i).$$

This implies that

$$E(\overline{p}_i) = p_i \tag{1}$$

and

$$Var(\overline{p}_i) = p_i(1 - p_i)/n_i. \tag{2}$$

Thus, the smaller n_i is, the larger the variance of $\overline{p}_i$, meaning that it would be less unusual that it is farther away from p_0 (if p_i equaled p_0 for all i).

We can say even more about the distribution of $\overline{p}_i$. By the Central Limit Theorem, not only does $\overline{p}_i$ have the mean and variance given in equations (1) and (2), respectively, it also (roughly) follows a normal (Gaussian) distribution. Since the values of n_i are in the thousands for these data, we would expect that the normal approximation would be quite accurate. Thus, if racial balance did hold, the values

$$z_i = \frac{\overline{p}_i - p_0}{\sqrt{p_0(1 - p_0)/n_i}} \tag{3}$$

would follow a standard normal distribution. Thus, examination of these z**–scores** allows us to see if racial balance seems reasonable for these data.

Here is a histogram of the z–scores:

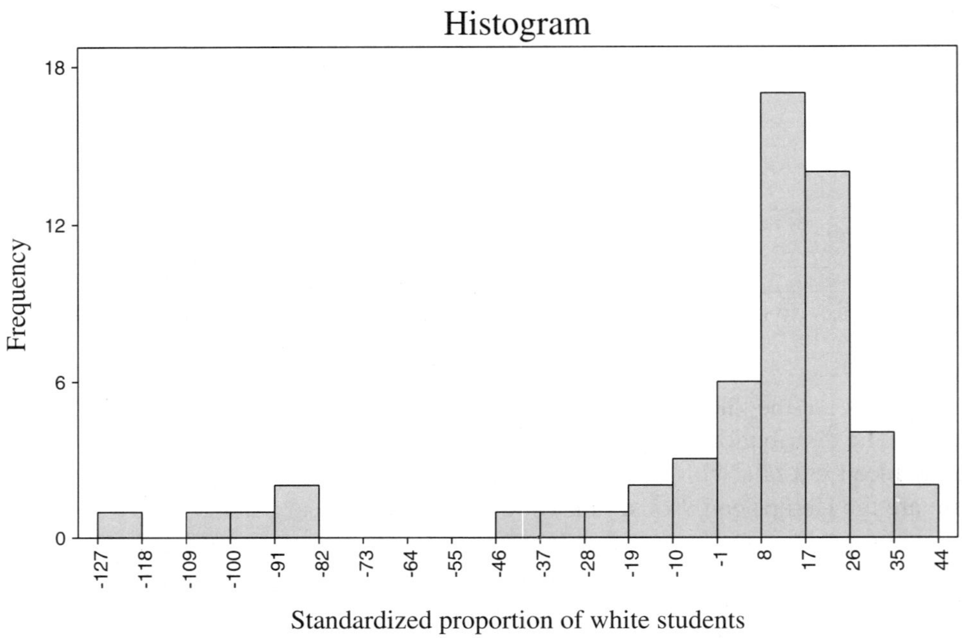

This picture gives strong evidence of racial imbalance, for two reasons. First, if the z_i followed a standard normal distribution, we would expect them to fall roughly in the interval $(-2.5, 2.5)$. This is not even remotely true here, with the observed values as high as 43, and as low as -125. In addition, the appearance of the histogram is quite non–normal, with a pronounced left tail, corresponding to school districts with unusually low white student percentages.

A very useful plot that can be used to identify potential non–normality even more clearly is the **normal plot** (sometimes also called a *rankits* plot). In this plot, the ordered values of the variable being examined are plotted along the vertical axis, with the expected values of these values (assuming that they came from a normal distribution) being plotted along the horizontal axis. If the data values are close to normally distributed, the plot will roughly follow a straight line.

On the next page is the normal plot for the z–scores.

The non–normality of the z–scores is quite apparent. The long left tail of their distribution shows up in the plot as the "bend" in the straight line at about -.7 (which corresponds to observed values being more negative than would have been expected under normality).

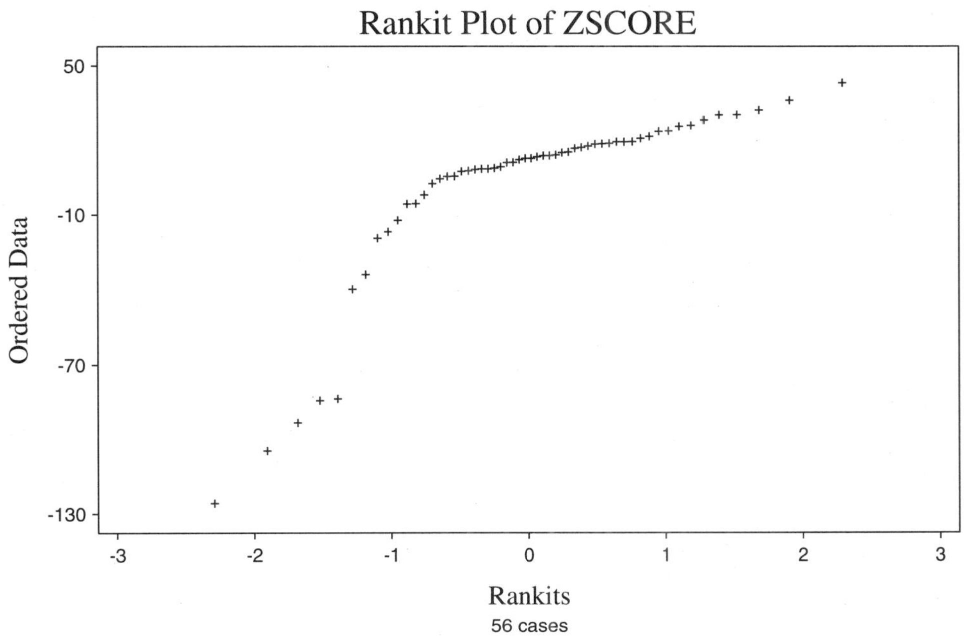

The long tails of the z–score distribution show that the actual variability of the z–scores is considerably greater than what would be expected based on the (normal approximation to the) Binomial distributions with constant p_i (each school district's count has a different n_i, of course). This greater–than–expected variation is often called **extra–Binomial variation**. One common cause of extra–Binomial variation is when a set of data is modeled as being based on Binomial distributions with constant p (as is being done here) when it actually represents a mixture of several different Binomial distributions (with different values of p).

We also need to recognize an implication of the large enrollments for the school districts, which is a result of equation (3). Consider the Herricks school district, which has an enrollment of 3432. Based on the form of the z–score, this large sample size implies that even small differences from p_0 in practical terms could lead to large z–scores. The observed proportion of white students in Herricks is 70.4%, which is not very far from the overall proportion of 74.7%; however, the z–score for Herricks is approximately

$$z = \frac{.704 - .747}{\sqrt{.747 \times .253/3432}} = -5.79,$$

which would be very unlikely under a normal distribution. For this reason, we need to decide if the observed racial imbalance is important from a practical point of view, even though we are already convinced (based on the observed z–scores) that it exists.

We can do this by examining the differences between $\overline{p}_i$ and p_0 for the 56 school districts (called `PDIFF` here), and the absolute values of those differences (called `ABSDIFF`).

Here are some summary statistics:

DESCRIPTIVE STATISTICS

	PDIFF	ABSDIFF
MEAN	0.0316	0.1875
SD	0.2495	0.1657
MINIMUM	-0.7440	0.0149
1ST QUARTILE	0.0237	0.0894
MEDIAN	0.1208	0.1545
3RD QUARTILE	0.1810	0.2027
MAXIMUM	0.2292	0.7440

It is apparent that the observed imbalance is of practical importance. Observed proportions of white students range from .744 less than the overall proportion to .2292 greater than the overall proportion. Even more worrying is that about 75% of the districts have an absolute difference from the overall proportion of at least .09, indicating strong racial imbalance.

From where could this imbalance be coming? The most natural explanation is that Nassau County itself is not characterized by a uniform racial distribution, but rather by individual neighborhoods that are predominantly white or nonwhite. If regions of racial homogeneity could be identified, then the extra–Binomial variation could be at least partially accounted for by estimating p_i by the proportion of white students for the appropriate region, rather than for the county as a whole (p_0).

A first attempt at identifying such regions can be motivated by a recent court case. On June 30, 1993, Federal District Court Judge Arthur D. Spatt determined that the structure of the Nassau County government was illegal, due to violation of the principle of "one person, one vote." He ordered that the government be reorganized away from the current Board of Supervisors system. A Commission on Government Revision was set up to oversee this reorganization. On March 29, 1994, the Commission presented its proposal to form a 19–district County Legislature, which it subsequently formally recommended on May 24, 1994. The districts for this legislature were created to be consistent with the Voting Rights Act, which requires that minorities have a reasonable opportunity to elect representatives of their choice.

These 19 districts can be used as a classification of the districts into (perhaps) more racially homogeneous regions. If that is the case, then the z–scores that are calculated based on equation (3), using the proportion of white students for the entire district as the true p_i (in place of p_0), should be more closely normally distributed. This would then confirm the geographical source of the racial imbalance.

Is this a reasonable hypothesis to pursue? Here are side–by–side boxplots of the proportion of white students for the 56 districts, separated by the proposed legislative district number. Note that most proposed legislative districts have only 2 – 4 school districts in them, so we can only get general impressions from the boxplots:

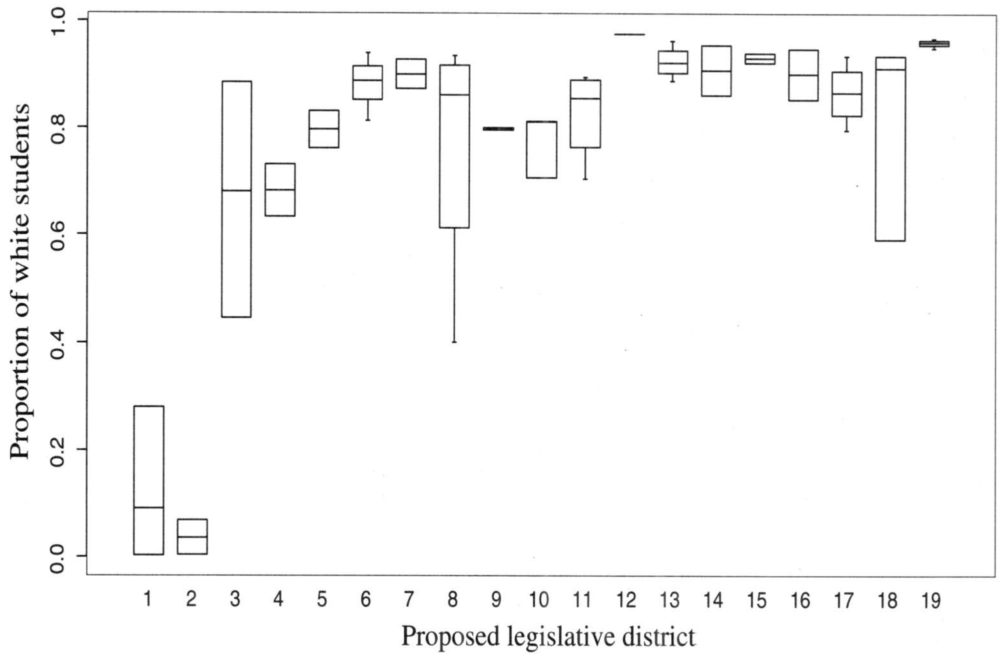

Obviously there is a good deal of variation between districts. The school districts in proposed legislative districts 1 and 2 apparently have 0 – 20% of the students being white; those in proposed legislative districts 3, 4 and 5 have roughly 60 – 80% students being white; while the other proposed legislative districts often have over 80% of the students being white.

Descriptive statistics separated by proposed legislative district make this more explicit:

Proposed legislative district	Total enrollment	White enrollment	Proportion white
1	13885	2158	0.1554
2	8345	222	0.0266
3	11266	7210	0.6400
4	7829	5319	0.6794
5	7828	6183	0.7899
6	11092	9790	0.8826
7	8857	8094	0.9139
8	8522	6696	0.7857
9	3535	2818	0.7972
10	11527	8969	0.7781
11	8907	7112	0.7985
12	6678	6517	0.9759
13	13261	12087	0.9115
14	8220	7342	0.8932
15	8686	8125	0.9354
16	9415	8433	0.8957
17	10278	8763	0.8526
18	6830	5333	0.7808
19	9595	9178	0.9565

The entries given under "Proportion white" in the table above can now be used as the true p_i for school districts in the given proposed legislative district.

On the next page are a histogram and normal plot of the new z–scores for the school districts.

The values look much more normally distributed, but the scale is still too wide for a standard normal (by a factor of about ten or so). Once again, this could possibly reflect only that the sample sizes are so large that even small differences between a school district's proportion of white students and the proposed legislative district's proportion lead to large z–scores. Descriptive statistics can help determine that:

```
DESCRIPTIVE STATISTICS

                     PDIFF     ABSDIFF
MEAN              4.510E-03     0.0618
SD                   0.0937     0.0701
MINIMUM             -0.3851     0.0000
1ST QUARTILE        -0.0300     0.0164
MEDIAN            4.936E-03     0.0405
3RD QUARTILE         0.0489     0.0796
MAXIMUM              0.2446     0.3851
```

The observed differences are considerably smaller than before, but many are still too large. The school district's proportion is as much as .3851 less than the proposed legislative district's proportion, and as much as .2446 greater. Still, about 75% of the differences are less than .08 in absolute value, as compared with about 75% being *greater* than .09 previously. Thus, this regionalization of the districts has accounted for much of the racial imbalance, but not all of it.

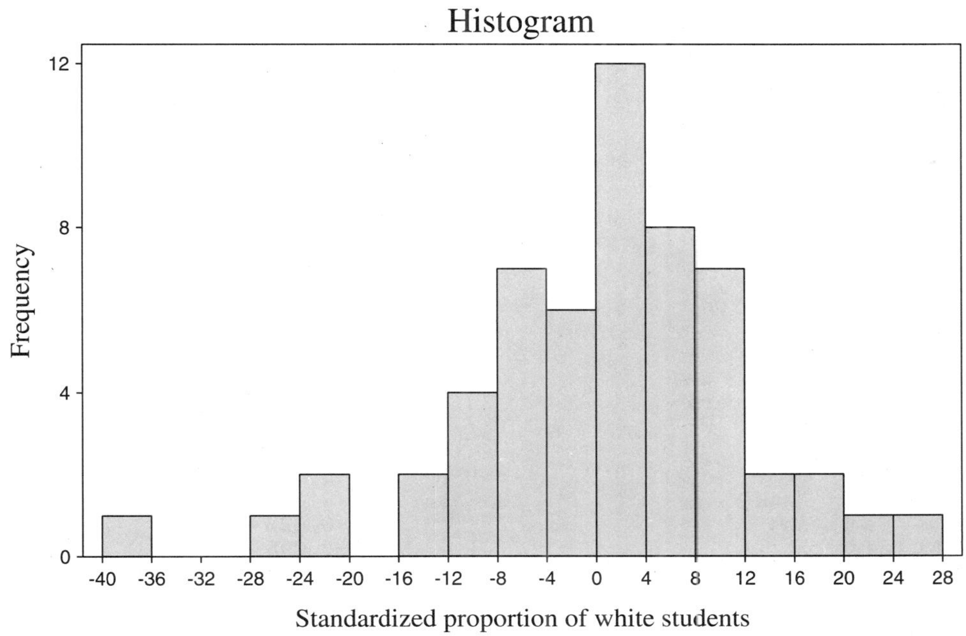
Histogram
Frequency
12
8
4
0
-40 -36 -32 -28 -24 -20 -16 -12 -8 -4 0 4 8 12 16 20 24 28
Standardized proportion of white students

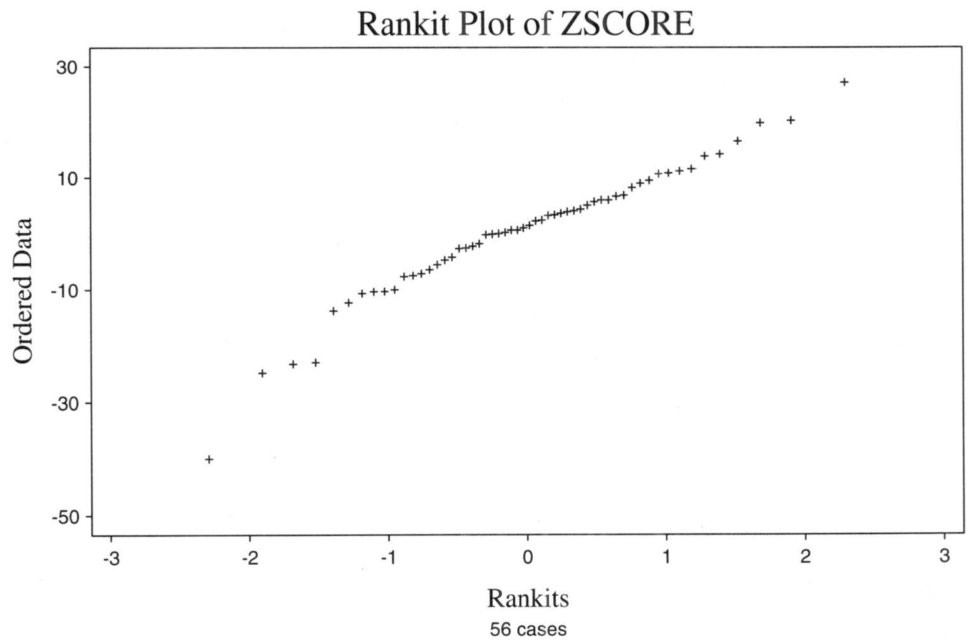
Rankit Plot of ZSCORE
Ordered Data
30
10
-10
-30
-50
-3 -2 -1 0 1 2 3
Rankits
56 cases

Where does this geographical model fail? Here are the PDIFF values for the 56 school districts, along with the associated proposed legislative district (LEGISLAT):

CASE	DISTRICT	LEGISLAT	PDIFF
1	Baldwin	5	-0.028
2	Bellmore	19	0.003
3	Bellmore - Merrick	19	-0.007
4	Bethpage	17	0.082
5	Carle Place	11	0.086
6	East Meadow	13	-0.023
7	East Rockaway	6	0.057
8	East Williston	11	0.096
9	Elmont	3	-0.194
10	Farmingdale	14	-0.032
11	Floral Park	8	0.112
12	Franklin Square	3	0.245
13	Freeport	1	0.125
14	Garden City	8	0.149
15	Glen Cove	18	-0.190
16	Great Neck	10	0.034
17	Hempstead	2	-0.023
18	Herricks	11	-0.095
19	Hewlett Woodmere	7	-0.014
20	Hicksville	17	-0.057
21	Island Park	7	-0.042
22	Island Trees	15	-0.014
23	Jericho	17	0.026
24	Lawrence	4	0.052
25	Levittown	15	0.004
26	Locust Valley	18	0.131
27	Long Beach	4	-0.046
28	Lynbrook	6	0.032
29	Malverne	8	-0.385
30	Manhasset	9	-0.001
31	Massapequa	12	0.000
32	Merrick	19	0.011
33	Mineola	11	0.026
34	New Hyde Park	9	0.002
35	North Bellmore	13	0.015
36	North Merrick	13	0.005
37	North Shore	18	0.154
38	Oceanside	7	0.014
39	Oyster Bay - East Norwich	17	-0.001
40	Plainedge	14	0.062
41	Plainview - Old Bethpage	16	0.052
42	Port Washington	10	-0.071
43	Rockville Centre	5	0.041
44	Roosevelt	1	-0.153
45	Roslyn	10	0.033
46	Seaford	19	0.004
47	Sewanhaka	3	0.040
48	Syosset	16	-0.043
49	Uniondale	1	-0.065
50	Valley Stream 30	6	-0.070
51	Valley Stream 13	6	0.029
52	Valley Stream 24	6	-0.019
53	Valley Stream Central	6	-0.031
54	Wantagh	13	0.051
55	West Hempstead	8	0.038
56	Westbury	2	0.042

Three school districts with particularly large (absolute) values of PDIFF are Elmont (unusually low proportion of white students), Franklin Square (unusually high proportion of white students) and Malverne (unusually low proportion of white students). Each of these anomalies can be explained geographically. Elmont is a community on the county border with New York City, near neighborhoods with relatively low white populations. Franklin Square is in proposed legislative district 3, but borders district 8, which has considerably higher proportion of white students. Malverne is in proposed legislative district 8, but is wedged between districts 2 and 3, which have considerably lower proportions of white students. Glen Cove, in proposed legislative district 18, also has a large (absolute) value of PDIFF, which is probably because of the large size of the district (it is roughly 10 miles wide from east to west, when the entire county is only about 15 miles wide from east to west), and the resultant heterogeneity. Thus, if the proposed legislative districts had been drawn slightly differently, the observed deviations from the mixture of Binomials model would be even less.

Summary

Examination of the distribution of the proportion of white students in the 56 school districts of Nassau County indicates strong racial imbalance. This is supported by observed extra–Binomial variation, which is often the result of trying to model a set of sample proportions as being from Binomial random variables with constant probability of success, rather than as a mixture of Binomials with different probabilities of success. This possibility is investigated for these data, based on separating the districts according to proposed county legislative districts. This geographical effect accounts for much of the observed extra–Binomial variation, but not all of it. Some of the largest observed errors based on the effect can be accounted for by the specific ways the proposed legislative districts were drawn, since slight changes in the lines would reduce the errors.

Technical terms

Binomial random variable: a discrete random variable that is often used to model data that come as a set of binary trials. If one of the two outcomes is arbitrarily termed a "success," then the random variable is appropriate as a representation of the number of successes in a given number of trials if the outcomes are independent from trial–to–trial, and the probability of success is constant from trial–to–trial.

Extra–Binomial variation: an observed pattern where counts that could be reasonably modeled as coming from a Binomial random variable (or a set of Binomial random variables with constant probability of success) exhibit more variation than would be expected under such a model. One common source of extra–Binomial variation is when the counts come from a mixture of several different Binomial random variables (with different probabilities of success), rather than Binomials with the same probability of success.

Normal plot: a graphical procedure for examining the normality of a variable, it is also known as a **rankits plot**. It also can be used for identifying outliers. The plot is obtained by plotting the ordered values of the variable against their expected values if they came from a normal population. The plot should appear to be more or less a straight line. Outliers are identified as points far removed from this approximate straight line, particularly at the extremes. Long tails can be identified by a pattern of points deviating off a straight line at either end.

The return on stocks in the Over the Counter market

Topics covered: Confidence interval. Prediction interval.

Key words: Confidence interval. Histogram. Mean. Normal plot. Prediction interval.

Data file: `nyseotc.dat`

Any investor in the stock market is interested in the financial performance of the stocks. Typically, as is usually the case for financial data, proportional returns are more meaningful than absolute returns. The following data come from a random sample of 30 stocks listed on the Over the Counter (NASDAQ) market, and represent the proportional change in price of the stock for the week May 9 – May 13, 1994 (as a proportion of starting price). What was the market like that week? This question can be answered by determining an estimate for a "typical" return.

First, let's take a look at the data. Here's a histogram:

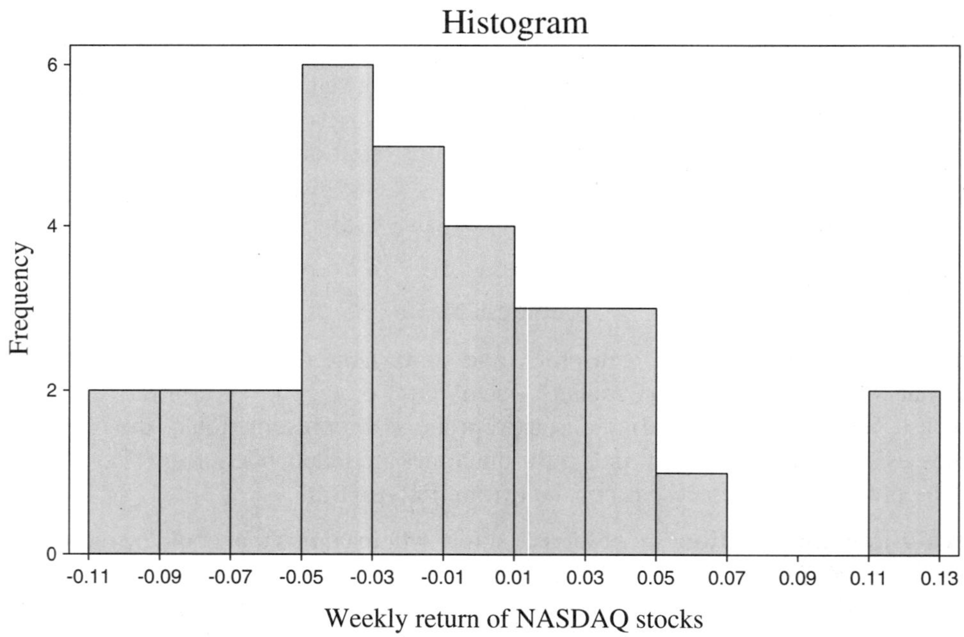

It is important to try to assess whether the sample appears to have been drawn from a normally distributed (Gaussian) population. The reason is that various methods of determining the accuracy of location estimates, and for making predictions of future values, are based on this assumption. Based on the histogram, the data look reasonably normally distributed. A normal (rankits) plot can be even more effective in assessing normality. For these data, it suggests little difficulty with the normality assumption:

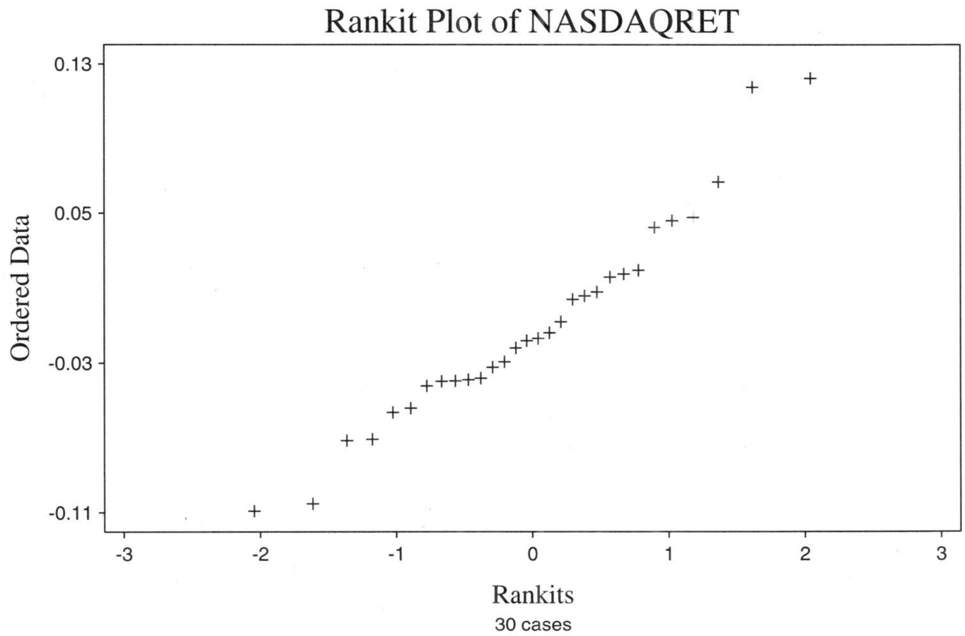

Let's look at descriptive statistics now:

```
                  NASDAQRET
N                        30
MEAN             -9.824E-03
SD                   0.0551
MINIMUM             -0.1092
1ST QUARTILE        -0.0404
MEDIAN              -0.0176
3RD QUARTILE         0.0177
MAXIMUM              0.1224
```

The average return is -0.98%. The median return is -1.76%. Apparently the market was down a bit during this week. There are two kinds of questions that we might want to answer from these data. One is as follows: what is our best guess for the general performance of the entire NASDAQ market for this week? Clearly, the most sensible estimate of the true (population) mean return for all stocks in the market is the sample mean, or -0.98%. This number alone is not enough, however, since it does not give us any indication about the *accuracy* of this estimate. For that, we need to construct a **confidence interval** for the true average return that week. The formula for a 95% confidence interval is as follows:

$$\overline{X} \pm t^{29}_{.025} \frac{s}{\sqrt{n}}, \tag{1}$$

where s is the sample standard deviation and $t^{29}_{.025}$ represents the critical value for a t–distribution on 29 degrees of freedom, and (one–sided) significance level of .025. For these data, a 95% confidence interval has the form

$$\begin{aligned} -.009824 \pm (2.05)(.0551)/\sqrt{30} &= -.009824 \pm .020623 \\ &= (-.030447, .010799). \end{aligned}$$

We need to be very clear about what this interval does, and does not, mean. A 95% confidence interval is defined as an interval estimate, such that in repeated sampling from the given population of NASDAQ stocks, 95% of the time the interval constructed from a given sample using (1) will, in fact, contain the true average return of NASDAQ stocks. Thus, a statement like "The true mean return is in the interval $(-.030447, .010799)$ with probability .95" is **not** correct. The true mean either is, or is not, in the interval, and we have no way of knowing which situation is true here.

Of what use, then, is this interval? Well, first we must be willing to accept the following principle: since 95% of all intervals constructed this way do, in fact, contain the true average return, we would have to be unlucky to be in the one out of 20 intervals that does not. Chances are the interval $(-.030447, .010799)$ does contain the true average return, so we will act as if it does. If we accept this principle, then the constructed confidence interval can tell us a lot. For example, for this week, the interval includes zero, indicating that while the return was a bit negative for the week, it was a "breakeven" kind of week for NASDAQ stocks, in general.

A very different question we might want to answer is as follows: what is our best guess for the performance of a randomly chosen NASDAQ stock for this week? Once again, the most sensible estimate for the expected performance of an individual stock is the sample mean, -0.98%. But how accurate is this estimate? The previously derived confidence interval is **not** the appropriate interval now, since it is an estimate for the *average* return over *all* stocks. Rather, what we want now is a **prediction interval**, which is an interval for the behavior of any one member of the population. The prediction interval has the form

$$\overline{X} \pm t^{29}_{.025} s \sqrt{1 + \frac{1}{n}}. \tag{2}$$

For these data, a 95% prediction interval has the form

$$\begin{aligned} -.009824 \pm (2.05)(.0551)\sqrt{1 + \frac{1}{30}} &= -.009824 \pm .114822 \\ &= (-.12465, .10500). \end{aligned}$$

Note that this interval is considerably wider than the confidence interval derived earlier. This is to be expected, since the prediction interval reflects two sources of variability (the variability of $\overline{X}$ as an estimate of the population mean μ, and the variability of the population around μ), while the confidence interval only reflects one source of variability (the variability of $\overline{X}$ as an estimate of μ). A way to think of this difference is as follows. Let's say we knew the actual average return for all NASDAQ stocks for this week exactly. Then, the confidence interval for the true mean would, of course, have width zero (since we know the value exactly). The prediction interval would still have nonzero width, however, since even if we know the mean return exactly, we don't know the return of any given stock exactly.

How can we use the prediction interval? It gives us an idea of the inherent variability of returns in the NASDAQ market. So, the observed interval $(-.12465, .10500)$ tells us that returns exhibit a good deal of variability, and that the return of any one particular stock can easily cover a (perhaps) surprisingly wide range of a loss for the week of about 12.5% to a gain for the week of about 10.5%, even in a "breakeven" week. This is the reason that most financial advisors tell investors that they should never invest money in the stock market that they cannot afford to lose, since there is a good deal of random variability in performance from stock to stock. More exactly, a 95% prediction interval has the property that approximately 95% of the stocks in the population can be expected to fall in the given interval.

Both the confidence interval and prediction interval constructed here require that the data represent a random sample from a normally distributed population. Based on the histogram and normal plot, that assumption seems reasonable. In fact, the intervals are relatively insensitive to the normality assumption, unless observed non–normality is quite strong (especially for large samples). Thus, we can feel confident about the intervals derived here.

Summary

Weekly returns for 30 stocks listed in the Over the Counter (NASDAQ) stock exchange for the week May 9 – May 13, 1994 were examined. A 95% confidence interval for the true average return of all NASDAQ stocks for this week was $(-.030447, .010799)$, indicating that, on average, the return for this week was not statistically significantly different from zero. A 95% prediction interval for the return of any one randomly selected stock was $(-.12465, .10500)$, reflecting the great variability in performance of individual stocks.

Technical terms

Confidence interval: an interval that provides guidance about the position of the true value of some unknown parameter. A 95% confidence interval (for example) has the following interpretation: in repeated sampling from the given population, approximately 95% of all confidence intervals constructed from those samples would, in fact, contain the true unknown parameter.

Prediction interval: an interval that provides a guidance about the value for one randomly chosen case from a population. A 95% prediction interval (for example) has the property that approximately 95% of the values in the population can be expected to fall in the given interval.

Volume and weight from a vineyard harvest

Topics covered: Forming hypotheses and models. Prediction intervals. Spatial–Temporal data. Variation.

Key words: Confidence interval. Logarithmic transformation. Mean. Prediction interval. Ratios. Side–by–side boxplots. Skewness. Standard deviation. Stem–and–leaf display.

Data Files: `vine1.dat`, `vine2.dat`, `vine3.dat`

South Bass Island, popularly called Put–in–Bay, is located 3 miles off the Northern Ohio coast of Lake Erie. The island is known for the 352 foot tall Perry's Victory and International Peace Memorial, a bustling summer season and an isolated, deserted winter. A 6 acre vineyard is located on the tip of the north–east peninsula of the island. The vineyard is comprised of 30 rows of vines (each 120 yards long) and 22 shorter rows (each about 100 yards long) arranged in the following way:

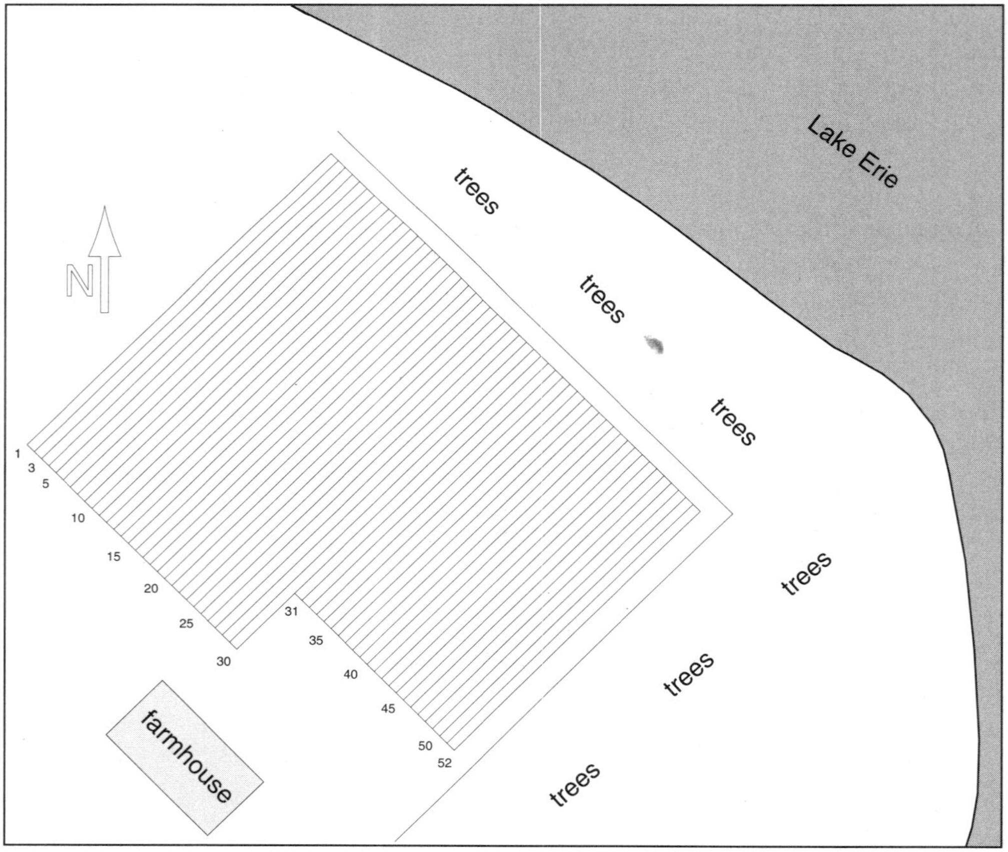

The area surrounding the vineyard is low and flat, with thick trees providing a barrier to the winds that sweep the island. The hardy vines produce the Catawba grapes that are used in the local winery.

The vineyard is run by Mark and Barbi Barnhill, who also run a bed–and–breakfast out of a 125 year old farmhouse. The vineyard has been redeveloped by them after falling into disrepair during the early 1970's. All of the data presented here are from their detailed diaries starting from 1976.

Over the years vines have been added where there were gaps in the rows. In early spring the vines are pruned and manicured. The Barnhills choose how many buds to leave and thus determine the number of bunches that will appear that year. Wouldn't leaving more buds lead to more bunches and hence increased yield? Not necessarily; the more bunches there are, the smaller their average size, as the vine divides its resources. The relationship between the number of buds and total yield is complex.

The harvest begins in late September or early October, depending on the maturity of the grapes. The bunches of grapes are harvested by clipping each bunch from the vine and placing it in a hand–held basket. When the basket is full it is dumped into a 1.5' × 3' × 6" basket called a "lug." The empty lugs are dropped at regular intervals down a row off the back of a tractor. The pickers move down each row dumping into the closest non–filled lug. After the row is completely picked the filled lugs are left between the rows and later counted while being collected by a tractor. The lugs are then emptied into plastic lined dumpsters. When the dumpsters are filled they are transported to the winery where their total contents are weighed.

The lug counts are easy to measure (e.g., "This vine produced 3 lugs!"). The Barnhills need the detailed information from the lug counts to plan their pruning and fertilizing for the next year. However, the important feature is the weight of grapes produced, not the lug counts themselves. The weight is only determined after the harvest is complete and then only for the total harvest. The weight of the harvest is often called the "yield." Understanding the relationship between the lug counts and the yield is thus essential.

Since 1983 the Barnhills have kept records of the lug counts from each row for each harvest. In total there are 52 rows (numbered 1 through 52) and 9 years (1983, ..., 1991) forming a 52×9 table of values.

The data–table looks like this:

ROW	1983	1984	1985	1986	1987	1988	1989	1990	1991
1	5	5	2	4.5	1.5	1.5	1	5	9.5
2	10	5	5	9	4	4	3	8	17.5
3	11	11	4	9	6	6	3	11	18
4	11.25	10	5	9.5	7	5	3	9	20
5	11	7	4	10	7	6	5	9.5	20.5
6	11	13	3	7	10	5	3.5	10	20
7	9	10.5	2.5	7.5	7	5	3	12	19
8	9	10	2	7	7	6	4.5	10	21
9	9	13	2	8	10	8	5	11.5	19
10	9	12	2	7	7	5	5	10	26
11	6	10	3	7	6	8	6	11	20
12	7	9	2.5	5	7	7	5	10	23
13	5	11	2	5	9	10	4	10.5	22
14	7	7	3	4	9	9	6	10	20
15	10	8	3	3	8	9	5	11	20
16	13	10.5	3	4	12	9	6.5	11	24
17	5.5	10	4	4	13	7	4	11.5	23
18	10.5	11.5	5	4	15	7	6	12	23
19	9	9	6	4.5	17	7	3	12.5	20
20	6	11	8	3.5	20	7	6	13	21
21	8	13	7	6	16	8	5	11	23
22	15	15	11	5	17	9	8	11	23
23	13	14	9	6	15	7	4.5	11	22
24	14	17	10	6	12	8	4	13	22
25	15	16	10	5	13	8	4	11	21
26	14.5	14	11	7	13	8	6.5	13	20
27	14	11	14	6	18	8	4	12	18
28	14	16	15	5	20	8	5	14	21
29	15	12	12	3	16	7	3	10	16
30	16	11.5	10	5	16	7	5.5	13	16
31	13	11	12	4	16	10	3	8	15
32	12	13	13	4.5	16	8	2	7.5	16
33	9	12	9.5	4	15	7	2	8	15
34	11.5	12	8	4	11	7	2	8	15
35	9.5	13	8	4	11	11	2	7	17
36	12	9	9	4	11	11	3	10	17
37	13	15	11	6	8.5	12	2	7	21
38	13.5	12	13	6	10	10	2	10	17
39	15	7	16	4	6	11	2	8	20
40	15.5	7	18	3	10	11	2	10	16
41	14	10	12	3.5	8	11	2	9	20
42	14	5	15	8	8	8.5	1.5	10	15
43	10	3.5	14	6	6	10	1.5	8	17
44	13	5	20	4.5	8	9	1.5	11	17
45	11	14	14	6	6	11	1	10	20
46	8.5	13	7	4.5	6	7	1	7	15
47	7.5	19	7.5	4	7	5	1	5	14
48	6	11.5	4	3	3	3	1	7	12
49	5	10	5	3	3	3	0.5	8	11
50	4	6	3	2	2	1.5	0.5	6	11
51	3	6.5	1.5	1	2	1	0	7.5	8.5
52	2	5	0.5	0.5	2	0	0	2.5	2.5

To convert these row lug counts to pounds, we can use a proportionate adjustment, that is,

$$\text{Yield of row } i \text{ in year } j \;=\; \text{Average weight per lug in year } j \times \text{Number of lugs of row } i \text{ in year } j.$$

This allows the average weight of a lug to differ from year to year, but assumes that it is the same from row to row in a given year. This approximation to the true yield from each row is reasonable, especially as we have no information on the differences between the rows in a given year.

How does the yield change by year and by row? First let's look at these factors separately. Here are side–by–side boxplots of the yields separately for each harvest year:

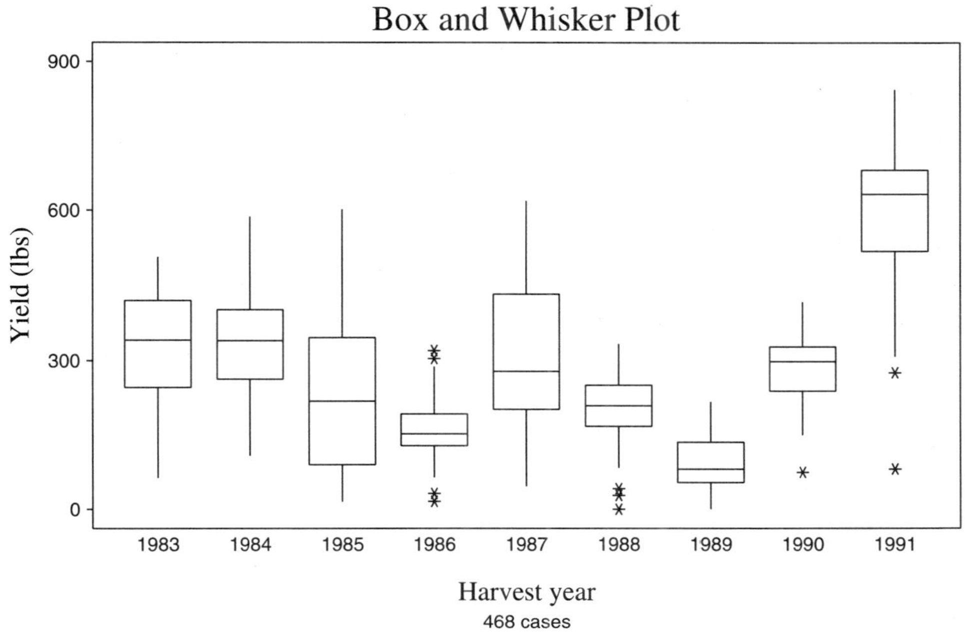

The yields certainly are not constant from year to year. The median yield per row is a little above 300 pounds in 1983 and 1984 and then drops in 1985 and 1986. There is a recovery in 1987 and then a decline to a low in 1989. The harvest yields in 1990 and 1991 were increasing and 1991 had the highest yield recorded. One measure of the variability of the row yields is the inter–quartile range (graphically, the length of the box). The variability also changes over time, although not in a consistent way. The harvests in 1985 and 1987 had the most variable row yields. As the row yields are roughly proportional to counts of lugs we would expect the variability to increase with the median number of counts. There is some evidence of this as the smaller yields tend to be less variable. However, the harvests that were most variable were 1985 and 1987, which had moderate yields.

Here are side–by–side boxplots of the yields separately for each row. We have separated the long rows (1–30) from the short rows (31–52).

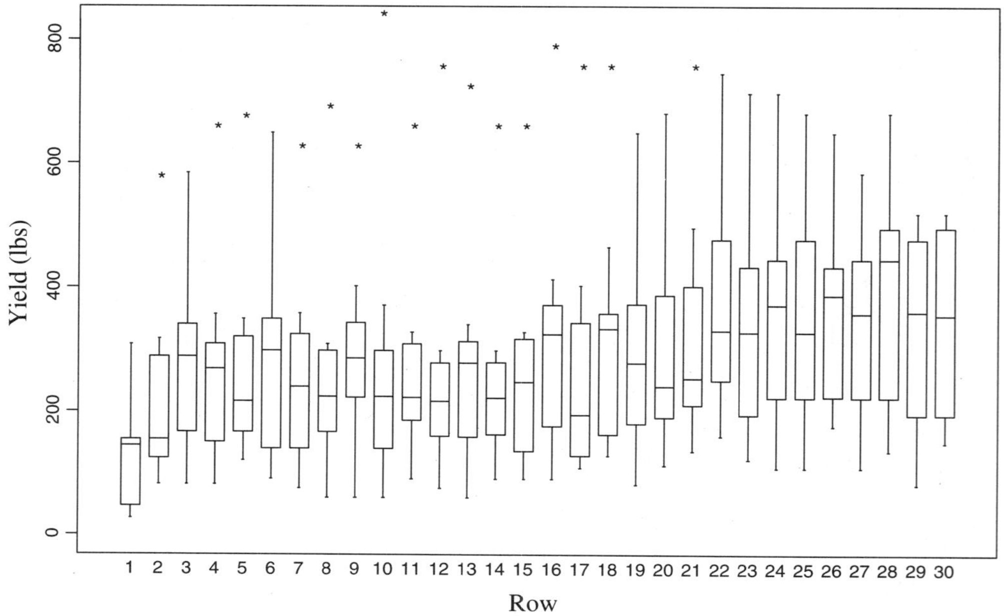

Boxplots for long rows

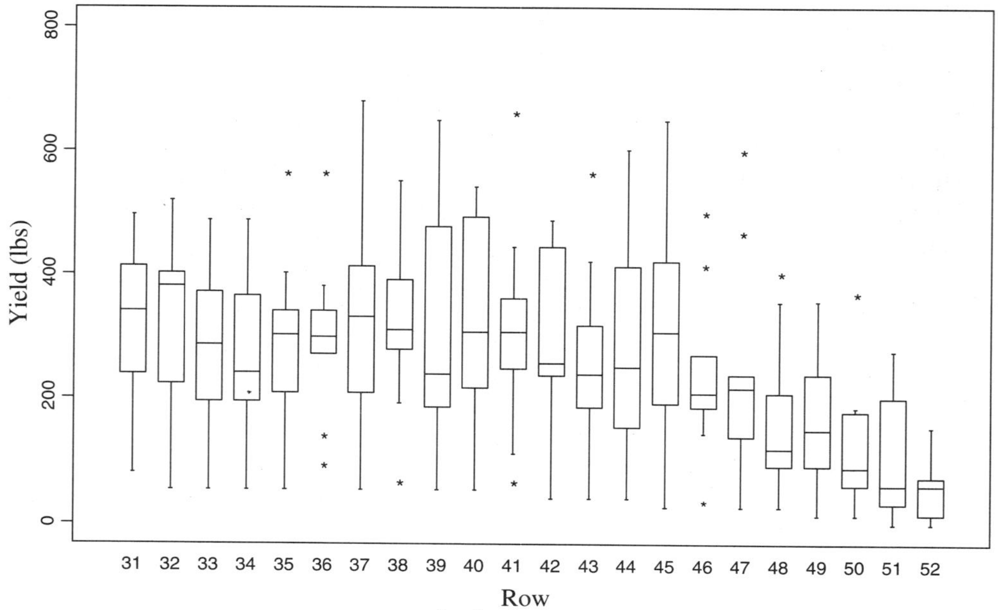

Boxplots for short rows

Each of the boxplots is based on only nine points, so we must take some care in their interpretation. The most striking feature of these plots is the consistency of pattern from row to row. Let's look at the long rows first. The first row has substantially lower yield than the subsequent rows. Indeed the second row also seems to be lower. Their medians are close, with the first row being skewed left and the second skewed right. These two rows are on the outside of the vineyard and are exposed to invasion by raccoons as the grapes ripen. In addition they are more exposed to winds than the interior rows. As we move eastward (increasing row number), the level of the yield hovers at about 220 pounds per row until about row 15. After that there is a slow increase settling at about 300 pounds per row. For the short rows the yield is reasonably stable at about 300 pounds per row for rows 30 through 41. After that there is a slow decline in yield to the end. These outside rows are also subject to raccoon damage and, in addition, bird damage. The flocks of birds attack from the safety of the trees on the eastern side. They attack the nearby rows more often. In addition, the number of surviving vines on the outside row is much less than those in the middle rows. Over time, the greater exposure of the outside rows has caused a greater mortality in the vines.

To get more detail, let's look at the plots of the row yields for each year on the same plot. Each of the three plots covers three successive years (as noted in the legend).

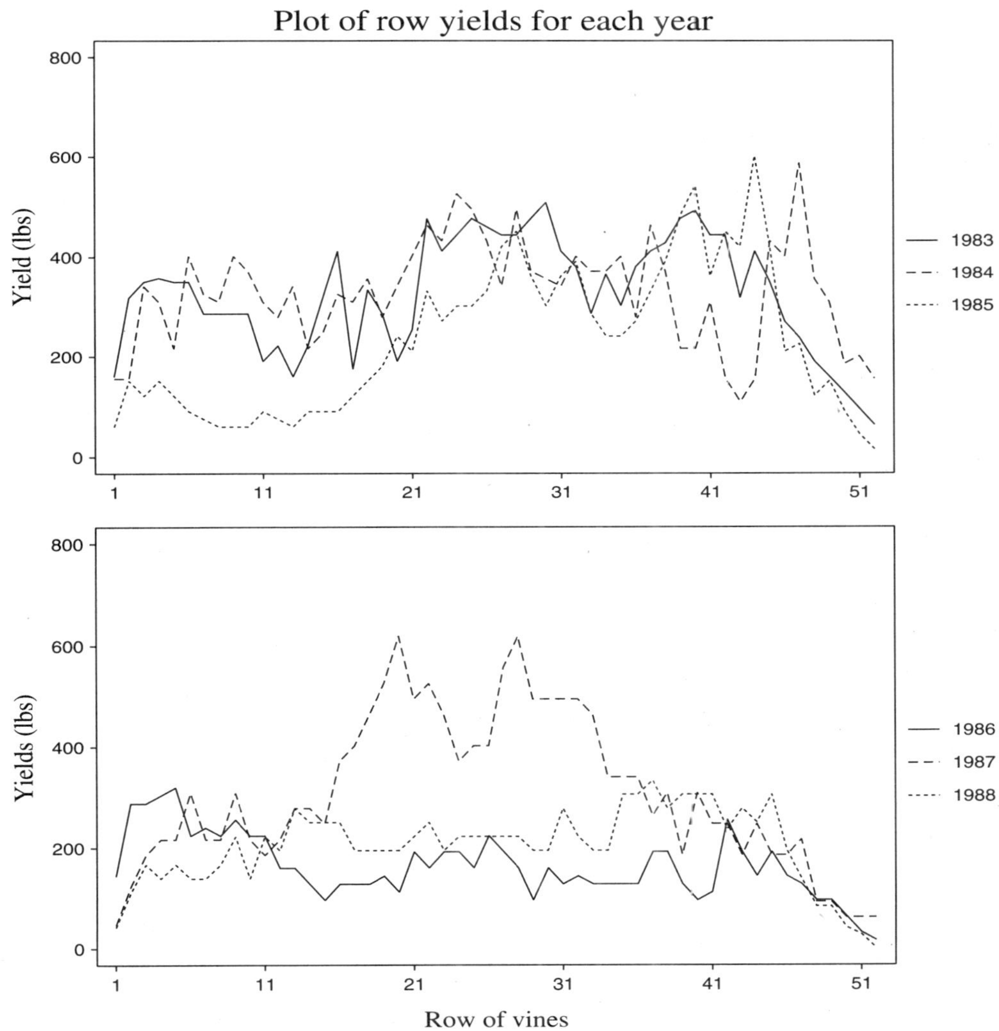

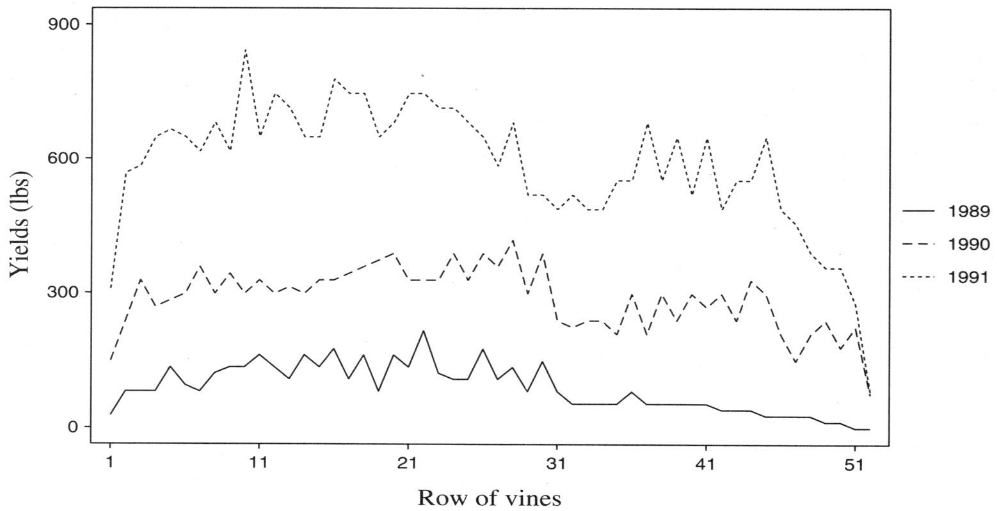

There are definite patterns of differences from year to year. Over the years the Barnhills have experimented with the agricultural management of the vineyard. They developed fertilization and insecticide schemes, varied the number of buds on the vines, developed traps for the raccoons and built scarecrows for the birds. Each of these factors will alter the yield in a complicated interacting way. The weather for the year preceding the harvest is also a big factor in determining the total yield.

The harvests of 1983 and 1984 were similar. In 1985 the first 26 rows were substantially lower than in the previous years while the latter rows were comparable. It was this deviation that produced the high variance for 1985 apparent in the earlier figure. Interestingly, 1986 and 1988 were similar, with 1988 a little higher. In 1987 rows 14 through 36 were greatly elevated relative to 1986 and 1988 and the other rows were similar. According to the Barnhills' diaries, it happened to rain hours after an insecticide was sprayed on many of the rows. It is possible that the insecticide did not have a full effect on these rows. Again the greater variation apparent for 1987 in the earlier plot was induced by these relatively elevated levels. Comparisons with the other years make it clear that the 1987 levels in these rows were unusually high, rather than the other years being unusually low. The harvests in 1989, 1990 and 1991 have a similar pattern with the general level increasing dramatically from year to year.

In each of these plots the decline in yield for the outer rows is apparent.

How does the total yield of the harvest depend on the number of lugs? Let's look first at the total number of filled lugs and the total yield of the harvested grapes for each year. Thinking about it, we can always express their relationship as:

$$\text{Total yield} = \text{Average weight per lug} \times \text{Number of lugs.} \quad (1)$$

How does the average weight of the lugs change from year to year? If the average weight per lug is constant (from year to year or row to row, say) then the yield would be proportional to the lug count, and the latter could be used as a convenient measure of the yield. The Barnhills have recorded these totals since 1976. Let's calculate the average weight per lug (`AVEWEIGHT`) and then look at these data:

CASE	YEAR	LUGS	YIELD	AVEWEIGHT
1	1976	11	313	28.455
2	1977	476	13299	27.939
3	1978	60	1848	30.800
4	1979	102	2894	28.373
5	1980	71	1792	25.239
6	1981	203	5483	27.010
7	1982	127	3490	27.480
8	1983	534.75	16970	31.734
9	1984	552.50	17083	30.919
10	1985	401	12067.5	30.094
11	1986	266	8519	32.026
12	1987	514	15912	30.957
13	1988	377.50	10507	27.833
14	1989	170.50	4618	27.085
15	1990	502	14975	29.831
16	1991	940.50	30521	32.452

Here are some descriptive statistics:

	LUGS	YIELD	AVEWEIGHT
N	16	16	16
MEAN	331.80	10018.2	29.264
SD	249.90	8029.4	2.119
MINIMUM	11.00	313.0	25.239
1ST QUARTILE	108.25	3043.0	27.569
MEDIAN	321.75	9513.0	29.143
3RD QUARTILE	511.00	15677.8	30.948
MAXIMUM	940.50	30521.0	32.452

The harvest in the first year was quite meager; 11 lugs and only 313 pounds of grapes. In contrast, the last harvest had 940.5 lugs and 30,521 pounds (over 15 tons).

Let's look at a histogram of the harvest yields:

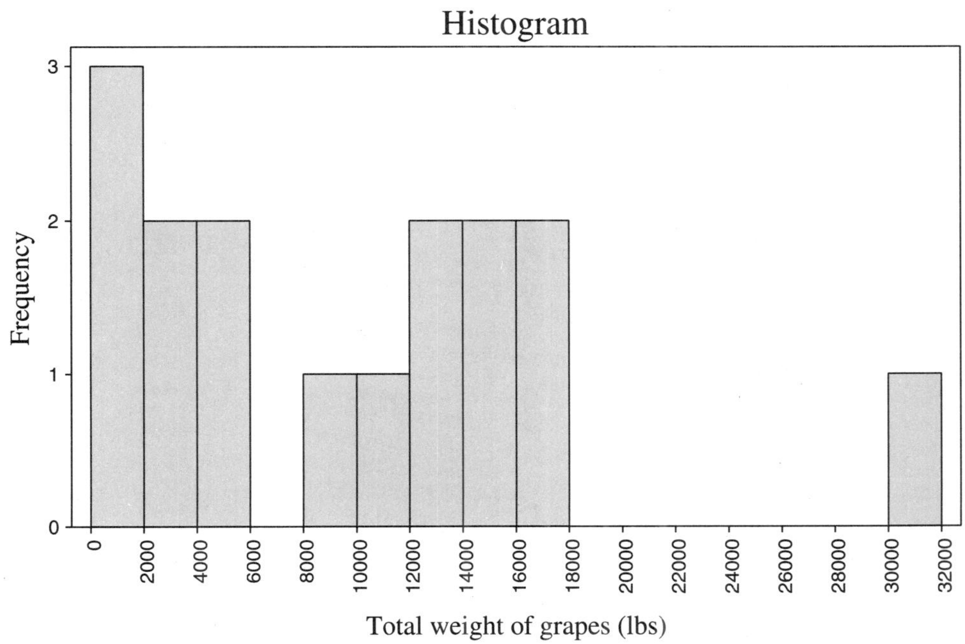

Now, for the number of lugs:

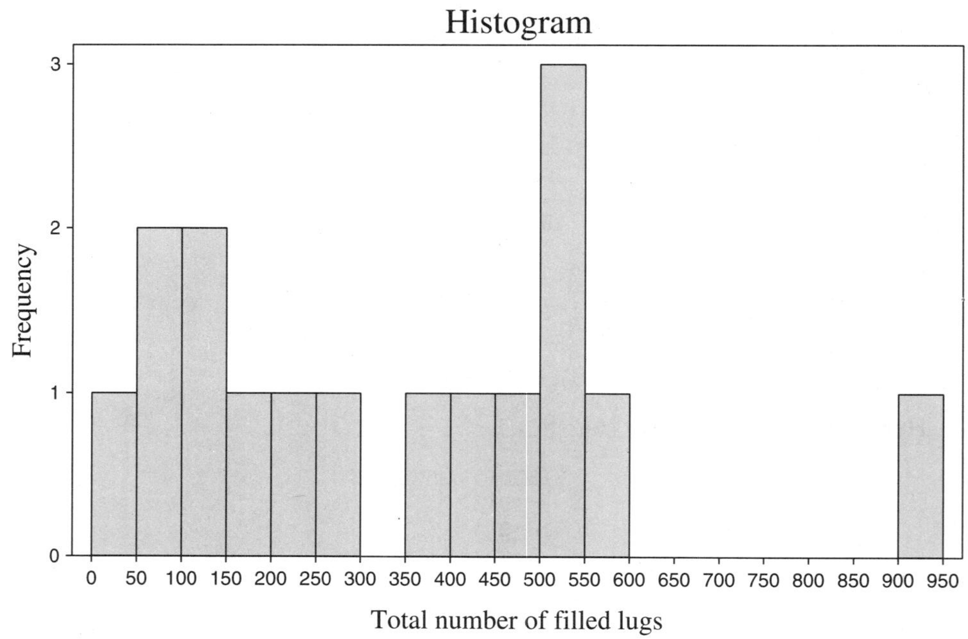

With 16 observations, the distributions are poorly defined. There is no clear skewness, and the distributions appear vaguely similar in shape. The 1991 value shows up as extreme in each plot. We will have to look further here.

What can we say about the average weight per lug (AVEWEIGHT)? It has a mean of 29.26 pounds per lug and a standard deviation of 2.12 pounds per lug. The standard deviation tells us that the values are somewhat spread out. Let's look at how the values are distributed:

```
STEM AND LEAF PLOT OF AVEWEIGHT
   Average weight per lug (pounds per lug)

   LEAF DIGIT UNIT = 0.1
   1  2  REPRESENTS 1.2

          STEM  LEAVES
        1   25  2
        1   26
        6   27  00489
        8   28  34
        8   29  8
        7   30  0899
        3   31  7
        2   32  04

16 CASES INCLUDED      0 MISSING CASES
```

There appear to be concentrations at about 27 pounds per lug and about 30 pounds per lug. The median and mean are close, so there is little evidence of skewness. The value for 1980 seems unusually low.

The average weight per lug does vary from year to year. What are some possible causes of this variation?

Well, the decision about when a lug is filled is a subjective one. Over–filled lugs are avoided as they tend to crush the bottom bunches. The pickers consciously try to fill each lug the same so that their count can be used to monitor the yield of each row (the lugs are not weighed individually). However, other factors might affect their unconscious choice. For a harvest where there are few bunches on the vines, the pickers might dump the baskets in the closest lug and leave some partially filled lugs as they move down the row. Lugs close to vines with few bunches tend to be less filled, as the pickers are less likely to walk back with their baskets to fill them, preferring to dump in a closer lug. Perhaps the picker in the same circumstance next year will bypass the lug they filled this year. These choices are essentially impossible to predict, even for the pickers themselves. They depend on factors that, realistically, cannot be measured.

What will be their effect? The average weight of the filled lugs will naturally vary with these choices. Thus even if, on average, the weight of the lugs is the same from year to year, we would still expect to see some variation in these values. If we knew this (assumed) common value of the average lug weight, we could then predict the total yield of the harvest from the total lug count using (1). We can express our uncertainty about the average lug weight in a given year using a prediction interval. Our point estimate of the average weight is the mean over the $n = 16$ years, $\overline{\texttt{AVEWEIGHT}} = 29.264$ pounds per lug. The sample standard deviation of the weights is SD(AVEWEIGHT) = 2.119 pounds per lug. Thus a 95 % prediction interval for the average weight of filled lugs for another individual harvest is

$$\overline{\texttt{AVEWEIGHT}} \pm t^{15}_{.025} \times \text{SD}(\texttt{AVEWEIGHT})\sqrt{1 + \frac{1}{n}};$$

$$\text{that is,} \quad 29.264 \pm 2.131 \times 2.119\sqrt{1 + \frac{1}{16}}$$

$$\text{or,} \quad 29.264 \pm 4.655 \text{ pounds per lug.}$$

Here the t–multiplier for 95% confidence is $t^{15}_{.025} = 2.131$. In words, based on these data we are 95% confident that the average weight per filled lug for another individual harvest will be between 24.6 and 33.9 pounds per lug.

Using (1) we can translate this uncertainty about the average weight into a prediction interval for the total yield given a lug count for a harvest. For example, the total lug count is now known to have been 1057 in 1992. A 95 % prediction interval for the yield of the harvest is

$$1057 \text{ lugs} \times \left(29.264 \pm 4.655 \text{ pounds per lug}\right)$$

$$\text{that is,} \quad 30,932 \pm 4920 \text{ pounds.}$$

In words, based on these data we are 95% confident that the total harvest yield will be between 26,012 and 35,852 pounds (given that 1057 lugs were picked). Don't peek at the *Postscript*!

There is a practical drawback of this procedure; we need to know the total lug count to use it. Thus the harvest must be taken in before we can predict the total yield.

It would be useful for the Barnhills to know the total yield of the harvest as early as possible. They pick the first few rows early because these rows are variably affected by bird and raccoon damage close to the harvest time. The bulk of the harvest must be taken in over one week, or else the crop will rot. Sometimes they do not have enough people–power to do this locally and they ship in some family members and friends to help. If they can predict how large the bulk of the crop will be, based on the lug counts from the first few rows, they can call in commensurate people to harvest it. We want the model on which the prediction is based to be simple for two reasons. First, it must be easy to apply. Second, and more important, a more complex model might not predict future harvest yields as well as a simple model.

To start let's look at how well we can predict the total yield of the harvest from the total lug counts from rows 3 through 6. The first two rows are often heavily reduced by the damage, and so are likely to be less useful in prediction. We could weigh the lugs also, and this might lead to a better model; however it requires additional effort beyond counting the lugs, so we will keep it simple.

Let's define a new variable:

$$\texttt{INITIAL} = \texttt{ROW3} + \texttt{ROW4} + \texttt{ROW5} + \texttt{ROW6}$$

and

$$\texttt{INRATIO} = \texttt{YIELD}/\texttt{INITIAL},$$

the ratio of the total harvest yield to the number of lugs in the initial rows. Here are some descriptive statistics:

	INITIAL	INRATIO
N	9	9
MEAN	35.694	432.08
SD	19.382	146.88
MINIMUM	14.500	239.97
1ST QUARTILE	19.000	348.80
MEDIAN	35.500	388.80
3RD QUARTILE	42.625	504.00
MAXIMUM	78.500	754.22

There are only 9 values because 1983 was the first year in which a record was kept of the lug counts by row. If the total yield is proportional to the weight of these initial rows, then, on average, the values of INRATIO will be the same from year to year. Let's look at how the values are distributed:

```
STEM AND LEAF PLOT OF INRATIO
   Ratio of yield to initial lug count

   LEAF DIGIT UNIT = 10
   1  2  REPRESENTS 120.

          STEM  LEAVES
        1    2  3
      (4)    3  1788
        4    4  17
        2    5  3
        1    6
        1    7  5

9 CASES INCLUDED
```

The high outlying value is 754 pounds per lug in 1985. The other values appear to be concentrated at a little under 400 pounds per lug. As is common for variables that are formed as a ratio, the distribution appears to be positively skewed. We can reduce the skewness by transforming the ratio by the logarithmic function; let

$$\texttt{LINRATIO} \quad = \quad \texttt{LOG(INRATIO)}.$$

Here are some descriptive statistics:

```
DESCRIPTIVE STATISTICS

                 LINRATIO
N                       9
MEAN               2.6152
SD                 0.1395
MINIMUM            2.3802
1ST QUARTILE       2.5409
MEDIAN             2.5897
3RD QUARTILE       2.7018
MAXIMUM            2.8775
```

A stem–and–leaf display gives a representation of the distribution:

```
STEM AND LEAF PLOT OF LINRATIO
   Log of the ratio of yield to initial row counts

   LEAF DIGIT UNIT = 0.01
   1  2  REPRESENTS 0.12
          STEM  LEAVES
        1   23   8
        1   24
      (4)   25   0788
        4   26   17
        2   27   2
        1   28   7
9 CASES INCLUDED
```

As for the weight per lug, we can calculate a prediction interval for the logarithm of the ratio for a single harvest based on these data. Our point estimate of the logarithm of the ratio is the mean over the $n = 9$ years, $\overline{\texttt{LINRATIO}} = 2.615$ log–pounds per lug. The sample standard deviation of the ratio is SD(LINRATIO) $= 0.1395$ log–pounds per lug. Thus a 95% prediction interval for the logarithm of the ratio for another harvest is

$$\overline{\texttt{LINRATIO}} \pm t^{8}_{.025} \times \text{SD}(\texttt{LINRATIO})\sqrt{1 + \frac{1}{n}};$$

$$\text{that is,} \quad 2.615 \pm 2.306 \times 0.1395\sqrt{1 + \frac{1}{9}}$$

$$\text{or,} \quad 2.615 \pm 0.339 \text{ log–pounds per lug.}$$

Here the t–multiplier for 95% confidence is $t^{8}_{.025} = 2.306$. In words, based on these data we are 95% confident that the logarithm of the ratio for another individual harvest is between 2.276 and 2.954 log–pounds per lug. We can translate this back to an interval statement about the ratio itself by taking the values in this interval to the power of 10 (i.e., the inverse of the logarithm). That is, we are 95% confident that the ratio for another individual harvest is between $10^{2.276} = 188.8$ and $10^{2.954} = 899.5$ pounds per lug.

Finally, as for the average weight, we can translate this uncertainty about the ratio into a prediction interval for the total yield given an initial lug count for a harvest. In 1992, the lug counts from rows 3 through 6 were 19, 20, 28 and 24, for a total of 91 lugs. A 95 % prediction interval for the yield of the harvest is

$$91 \text{ lugs} \times \left(188.8, \quad 899.5\right) \text{ pounds per lug;}$$

$$\text{that is,} \qquad (17,181 \text{ pounds}, \quad 81,855 \text{ pounds}).$$

In words, based on these data we predict with 95% confidence that the harvest yield is between 17,181 and 81,855 pounds (given that 91 lugs were picked from rows 3 through 6). The interval is very large compared with the interval based on the model for the average weight of all lugs. This reflects that the uncertainty of the prediction based on the lug counts from a few rows is large.

Summary

There is considerable variation in the yields in the vineyard, both from row to row and year to year. The yields of rows close together are positively related to each other. The outer rows have reduced yields compared with the interior rows because of weather and animal damage. In addition, the pattern of the row yields changes from year to year.

The Barnhills need a simple model that will enable them to predict the total yield of the vineyard. The yield of the vines is measured by the total weight of the harvested grape bunches. As only the total harvest is weighed, we must use counts of the filled lugs as a surrogate for the yield.

If the average weight of the filed lugs is the same from year to year, the total weight of the yield can be predicted from the total number of lugs.

To improve the timeliness of the prediction, a similar model can be built using the total lug count of rows 3 through 6, rather than the total lug count. Not surprisingly, this prediction is noticeably less accurate than one based on all rows.

Postscript

These data were collected during the 1992 harvest and before the total weight of the harvest was known. Subsequently, the harvest yield was found to be 31,229 pounds. Thus, the actual value falls inside the 95% prediction intervals for each of the two models considered. If the models are correct, we expect this to happen for 95% of such intervals (so it is not too surprising!).

The model using the total lug count of 1057 predicts, with 50% confidence, that the total yield will be between 29,339 and 32,525 pounds. The point estimate is 30,932 pounds. The model using the initial lug count of 91 predicts, with 50% confidence, that the total yield will be between 29,488 and 47,692 pounds. Thus the actual value fell in the 50% prediction interval for both models. If the models were correct, we would expect it to fall inside about half of the 50% prediction intervals. Thus the predictions from both of the models appear to be reasonable.

The performance of stock mutual funds compared to the market as a whole

Topics covered: Comparison of location to a standard. Confidence interval. Hypothesis testing. Prediction interval.

Data File: `funds.dat`

Recall our earlier investigation of stock mutual funds in the case "The performance of stock mutual funds." For both long– and short–term performance measures, there were differences based on the type of fund. This fact does not address a perhaps more fundamental question, however — do any of these types of managed funds beat a generally diversified portfolio? If not, then perhaps a more appropriate investment strategy would be to save investment fees and just invest in a low–cost index fund (that is, a fund that is designed to track the performance of the market as a whole).

Investigate this question using the stock mutual fund data. A measure that is often used to represent the return of a fully diversified portfolio is Standard and Poor's (S & P) 500 index. The five–year performance figure for the S & P 500 index is $15,440, while its 1992 return figure is 8%. For each of these measures, answer the following questions:

(1) Treating the entire set of funds as a sample from a homogeneous population, is the average measure significantly different from the measure for the S & P 500 Index? Do the assumptions that you make in your analysis appear to be valid here?

(2) Now separate the data set into the seven groups defined by the fund type. Treating each of these subsets as a sample from a homogeneous population, is the average measure significantly different from the measure for the S & P 500 Index? Do the assumptions that you make in your analysis appear to be valid here?

(3) Based on your answers to (1) and (2), what advice (if any) would you give to someone looking to invest in stock mutual funds?

Comments to the Instructor

A 95% confidence interval for the mean `FIVEYR` performance is (14550, 15320), while one for the mean of `RETURN92` is (6.21, 8.49). The former interval is below the S & P 500 figure, so (taken as a whole) the average performance of the managed funds is significantly below that of the market as a whole. In contrast, the confidence interval for the mean of `RETURN92` does contain the S & P value, so the average 1992 return of the managed funds is not significantly different from that of the market as a whole.

The assumptions appear reasonable. The distribution of `FIVEYR` shows a slight long right tail, while one case (Skyline Equity) is a large outlier with respect to `RETURN92`. Removing this one fund does not change the form of the confidence interval appreciably.

We do know, however, that there are differences between types of funds here, so it seems more sensible to separate funds by their type. The resultant 95% confidence intervals for the means of the two variables are as follows:

Type of fund	FIVEYR	RETURN92
Balanced	(13730, 14410)	(5.92, 8.95)
Equity–Income	(13550, 14770)	(7.62, 11.20)
Growth–and–Income	(14030, 15260)	(7.02, 10.23)
Growth	(14650, 16420)	(7.91, 13.13)
Aggressive–Growth	(15970, 18700)	(4.07, 12.23)
Small–Company	(16560, 17880)	(7.87, 13.97)
International–and–Global	(10900, 11860)	(−5.09, −1.26)

Thus, the average five year performances for Balanced, Equity–Income, Growth–and–Income and International–and–Global funds were significantly below that of the market as a whole, while those for Aggressive–Growth and Small–Company funds were significantly above the market as a whole. The assumptions seem reasonable, except for the outliers noted in the case "The performance of stock mutual funds." Omitting these outliers does not change the results appreciably.

In contrast, all of the intervals for the mean 1992 return contain the S & P mean value of 8%, except for the International–and–Global funds, which are extremely poor performers. The only apparent problem with assumptions is the outlier mentioned above, which does not materially affect the results.

These data provide a good example of how long–term performance can be very different from short–term performance. It also supports the well–known phrase that "historical earnings do not necessarily indicate future performance"!

To maintain consistency with the unannotated edition,
the next case begins on the following page.

The return on stocks in the New York Stock Exchange

Topics covered: Confidence interval. Prediction interval.

Data file: `nyseotc.dat`

The case "The return on stocks in the Over the Counter market" examined the performance of 30 NASDAQ stocks for the week May 9 – May 13, 1994. Interval estimates were constructed for both the true average return of all stocks in the NASDAQ market, and the return of a particular randomly selected NASDAQ stock.

The given data set also provides return figures (for the same week) for 30 stocks listed on the New York Stock Exchange (NYSE). What can you say about the typical performance of all NYSE stocks? How about an individual randomly selected NYSE stock?

Comments to the Instructor

The typical performance of all NYSE stocks can be assessed using a confidence interval for the mean. The sample average return for the week is .0039; a 95% confidence interval for the average return for the week is $(-.0098, .0177)$. Thus, the average weekly return is not significantly different from zero.

The performance of an individual randomly selected NYSE stock can be examined by constructing a prediction interval; a 95% prediction interval has the form $.0039 \pm .0769$, or $(-.073, .0808)$. Since this interval is narrower than the corresponding one for NASDAQ stocks, apparently the variability of the Over the Counter market was greater than that of the New York Stock Exchange for this week.

Baseball free agency: do teams get what they pay for?

Topics covered: Confidence interval. Nonparametric tests. Paired samples. Tests for differences in location.

Key words: Histogram. Long–tailed data. Normal plot. Outliers. Paired samples t–test. p–value. Sign test. Wilcoxon signed rank test.

Data File: `free.dat`

Starting in 1976, major league baseball players achieved true free agency; with few restrictions or conditions, a player whose contract had expired could sell his services to the highest bidder. This has led to a veritable orgy of spending — sometimes rational spending, sometimes not so rational.

Clearly teams are interested in knowing whether the investment is worth it. Management theorists provide two possible answers to this question:

(1) *equity theory*: equity theory says that the reason that players seek free agency is that they feel unappreciated with their current team. This bad feeling causes them to slack off in their last year with the old club, and then play much better the next year for the new team, since they are happier (that is, people who feel that they are underpaid work less efficiently, while people who feel that they are being paid fairly work more efficiently). Thus, equity theory would imply that free agents who change teams would have better performance in the year after free agency than in the year before it.

(2) *expectancy theory*: expectancy theory says that since there is a strong connection between performance and success of a team, players will "look ahead" to becoming a free agent, and will push up their performance in order to get the big bucks. Thus, expectancy theory would imply that free agents who change teams would have worse performance in the year after free agency than in the year before it.

Note, by the way, that teams signing free agents would certainly prefer to see that equity theory is true than the expectancy theory.

The question of which theory seems to fit the data best can be investigated using **paired sample** tests. Consider, for example, studying the batting average statistic. Batting averages would be obtained for free agents before and after the player changes teams. Let μ_B be the average batting average for a player before becoming a free agent, with μ_A the average batting average for a player after becoming a free agent. The null hypothesis, which is that there is no difference between the means of the before and after years, is

$$H_0 : \mu_B = \mu_A.$$

The alternative hypothesis is

$$H_a : \mu_B \neq \mu_A.$$

In particular, equity theory would be consistent with $\mu_B < \mu_A$, while expectancy theory would be consistent with $\mu_B > \mu_A$. If the null hypothesis is not rejected, that says that there is no statistically significant difference in typical performance between the free agent year and the next year. The data appear in the form of a paired sample, with values for each individual player corresponding to the years before and after free agency.

The 162 players (not including pitchers) who changed teams via free agency in the years 1976 through 1989 are the subjects of this data analysis. There was a strike during the 1981 season that resulted in, on average, 55 games being cancelled per team (i.e., about one–third of the games of the season). In total, 11 players had 1981 as their year before becoming a free agent and 18 had 1981 as their year after becoming a free agent. As these results are likely to be different from those of most of the other free agents, we will exclude the affected data and study the remaining 133 players. We'll be focusing on the usual measures of hitting success: batting average (hits per at bat), with variable names `BAFA` and `BANEXT` representing batting average in the free agent year, and batting average in the next year, respectively; home runs (`HRFA` and `HRNEXT`); and runs batted in (`RBIFA` and `RBINEXT`). The data were collected by David Ahlstrom.

First we take a look at the data:

```
DESCRIPTIVE STATISTICS

                 BAFA     BANEXT      HRFA     HRNEXT      RBIFA    RBINEXT
N                 133        133       133        133        133        133
MEAN           0.2580     0.2471    8.9474     8.4737     42.188     39.511
SD             0.0337     0.0469    8.6824     9.2414     27.224     29.383
MINIMUM        0.1670     0.0000    0.0000     0.0000     1.0000     0.0000
1ST QUARTILE   0.2400     0.2185    2.0000     1.0000     22.000     12.500
MEDIAN         0.2610     0.2540    6.0000     5.0000     37.000     36.000
3RD QUARTILE   0.2790     0.2800    12.500     13.500     60.500     62.500
MAXIMUM        0.3530     0.3330    35.000     49.000     119.00     137.00
```

Informal examination of the numbers shows that batting averages, home runs and runs batted in all seem to drop after free agency. Overall, expectancy theory looks more likely to be true than equity theory. However, we would like to take a look at effects of free agency for each player, i.e. the change in performance for *each* player, rather than the change in the *average* player. To do this we will focus on the change in performance between the free agency year and the next year. We could look at other measures, such as the ratio, although the difference has a clearer interpretation in terms of total production. A difference of zero would mean that there was no change between the two years. We can do this using a **paired sample t–test**, where the mean of the differences in performance is compared to its standard error. Note that this is equivalent to a **one–sample t–test** based on whether the change in performance has mean zero. Let's take a look at the comparison for batting average first:

```
PAIRED T TEST FOR BAFA - BANEXT

MEAN           0.0109
STD ERROR    4.28E-03
T                2.54
DF                132
P              0.0121

CASES INCLUDED 133
```

A 95% confidence interval for the average change is

$$\begin{aligned}&\overline{\mathtt{BAFA}-\mathtt{BANEXT}} \pm t^{n-1}_{.025} \times \widehat{\mathrm{SE}}(\overline{\mathtt{BAFA}-\mathtt{BANEXT}})\\ &= 0.0109 \pm t^{132}_{.025} \times 0.00428\\ &= (0.0025,\ \ 0.0193),\end{aligned}$$

where $n = 133$ is the number of changes, $t^{132}_{.025} \approx z_{.025} = 1.96$ is the t–table value for a two–sided 95% confidence interval on 132 degrees of freedom, and $\widehat{\text{SE}}(\overline{\texttt{BAFA} - \texttt{BANEXT}})$ is the estimated standard error of $\overline{\texttt{BAFA} - \texttt{BANEXT}}$. As is usual for samples this large, $t^{132}_{.025}$ is approximated by $z_{.025}$, the value for a two–sided 95% confidence interval from the normal table. The values for $\overline{\texttt{BAFA} - \texttt{BANEXT}}$ and $\widehat{\text{SE}}(\overline{\texttt{BAFA} - \texttt{BANEXT}})$ can be read from the numerical summary given above. Thus, while the drop in batting average of about .0109 hits per at bat is statistically significant, it is modest in size. The p–value for the test (.012) indicates that a confidence interval with degree of confidence 98.8% would barely not contain zero.

We can check the assumption of the t–test that the distribution of the differences is Gaussian by examining a histogram of the difference between free agent batting average and the next year's batting average (called DIFBA). It looks acceptable, except for one possible high value:

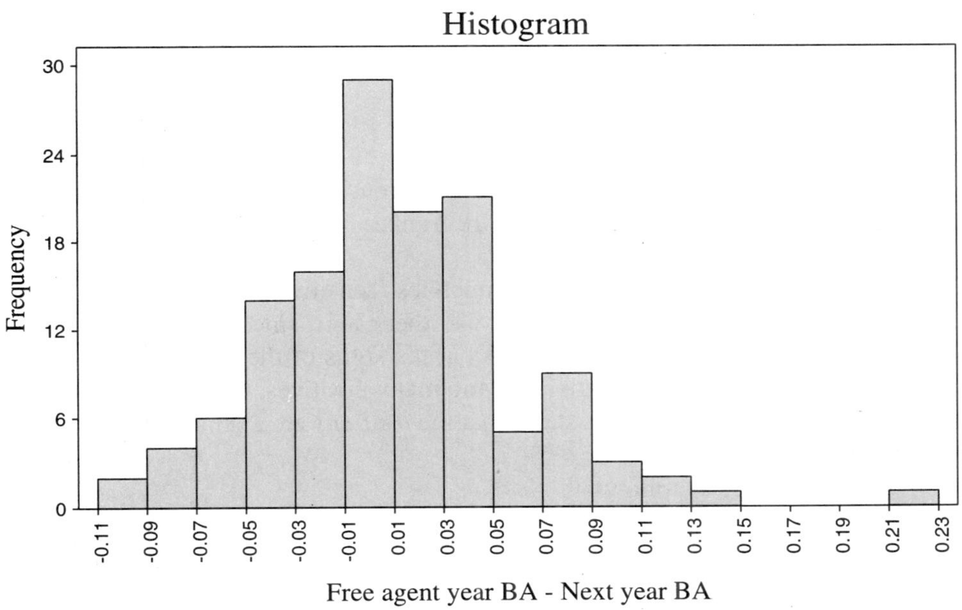

We also can see it in a normal (rankits) plot:

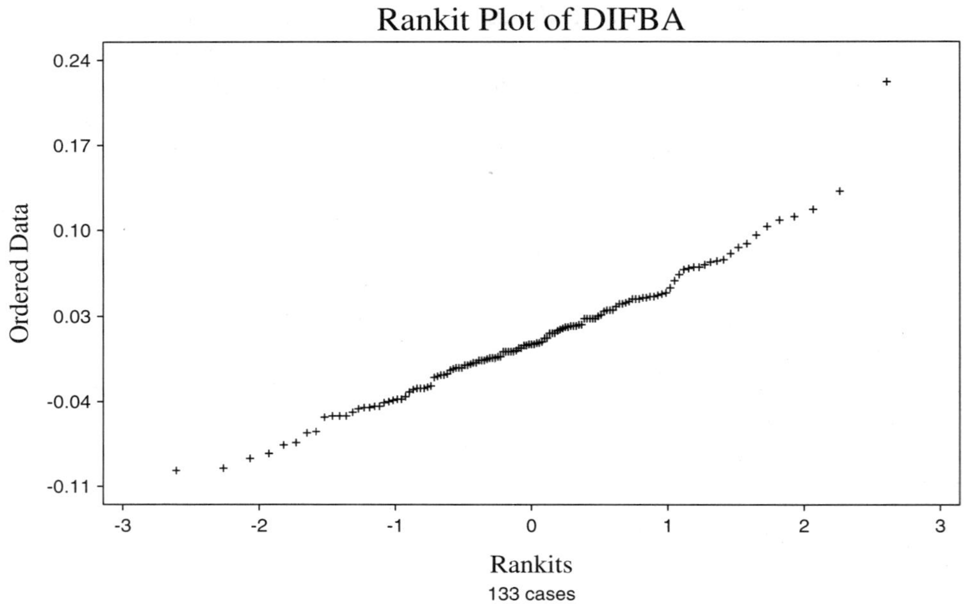

The outlier is Ivan DeJesus (Case 61). He went from a .222 batting average in 1985 for St. Louis to .000 for the New York Yankees in 1986 (he was only up four times, with no hits).

There exists a class of tests that are much less sensitive to unusual observations, known as **nonparametric** tests. We will use them here since we have an unusual observation in our data. The **sign test** looks at the signs of the differences in batting average; if there are too many negatives, or too many positives, that indicates rejection of the null (now the null hypothesis states that the *medians* are unchanged from year to year, rather than the means).

The result here is as follows:

```
SIGN TEST FOR BAFA - BANEXT

NUMBER OF NEGATIVE DIFFERENCES          55
NUMBER OF POSITIVE DIFFERENCES          78
NUMBER OF ZERO DIFFERENCES (IGNORED)     0

TWO TAILED PROBABILITY OF A RESULT AS
OR MORE EXTREME THAN OBSERVED       0.0560

A VALUE IS COUNTED AS A ZERO IF ITS
ABSOLUTE VALUE IS LESS THAN 0.00001

CASES INCLUDED 133    MISSING CASES 0
```

Note that the difference is marginally statistically significant, with a drop in batting average being indicated (since there are more positive values of `BAFA - BANEXT` than negative ones). A more sensitive nonparametric test is the **Wilcoxon signed rank test**, which substitutes the ranks of the observations for their values, and (roughly speaking) does a t–test based on the ranks (actually, it's a test based on a normal approximation). The result still indicates a drop in batting average, at about the same level of significance:

```
WILCOXON SIGNED RANK TEST FOR BAFA - BANEXT

SUM OF NEGATIVE RANKS                                     -3459.0
SUM OF POSITIVE RANKS                                      5452.0

NORMAL APPROXIMATION WITH CONTINUITY CORRECTION             2.237
TWO TAILED P-VALUE FOR NORMAL APPROXIMATION                0.0253

TOTAL NUMBER OF VALUES THAT WERE TIED      97
NUMBER OF ZERO DIFFERENCES DROPPED          0
MAX. DIFF. ALLOWED BETWEEN TIES       0.00001

CASES INCLUDED 133    MISSING CASES 0
```

Another approach is to remove the unusual observation, being sure to note its presence (specifically, that players who barely play in the year after free agency can be unusually different from the other players), and redo a paired t–test:

```
PAIRED T TEST FOR BAFA - BANEXT

MEAN         9.29E-03
STD ERROR    4.00E-03
T                2.32
DF                131
P              0.0218

CASES INCLUDED 132    MISSING CASES 0
```

Here a 95% confidence interval for the average change is

$$\begin{aligned}&\overline{\mathrm{BAFA} - \mathrm{BANEXT}} \pm t^{n-1}_{.025} \times \widehat{\mathrm{SE}}(\overline{\mathrm{BAFA} - \mathrm{BANEXT}})\\&= 0.00929 \pm t^{131}_{.025} \times 0.00400\\&= (0.00145,\ 0.01713),\end{aligned}$$

where $t^{131}_{.025} \approx z_{.025} = 1.96$.

If the 95% confidence interval does not include zero, then the difference is significantly different from zero at a $.05$ level. In this case, the interval does not include zero. Thus, free agents (on average, dropping one unusual player) appear to have a somewhat lower batting average in the year after free agency, and the drop is statistically significant at a $.05$ level according to the t–test.

Now, let's move on to home runs. Here is the t–test:

```
PAIRED T TEST FOR HRFA - HRNEXT

MEAN          0.4737
STD ERROR     0.5926
T               0.80
DF               132
P             0.4255

CASES INCLUDED 133    MISSING CASES 0
```

Here a 95% confidence interval for the average change is

$$\begin{aligned}&\overline{\mathrm{HRFA} - \mathrm{HRNEXT}} \pm t^{n-1}_{.025} \times \widehat{\mathrm{SE}}(\overline{\mathrm{HRFA} - \mathrm{HRNEXT}})\\&= 0.4737 \pm t^{132}_{.025} \times 0.5926\\&= (-0.6878,\ 1.6352).\end{aligned}$$

We can see that the average number of home runs dropped by about one–half of a home run, which is not statistically significant.

Once again, we need to verify that the differences are normally distributed:

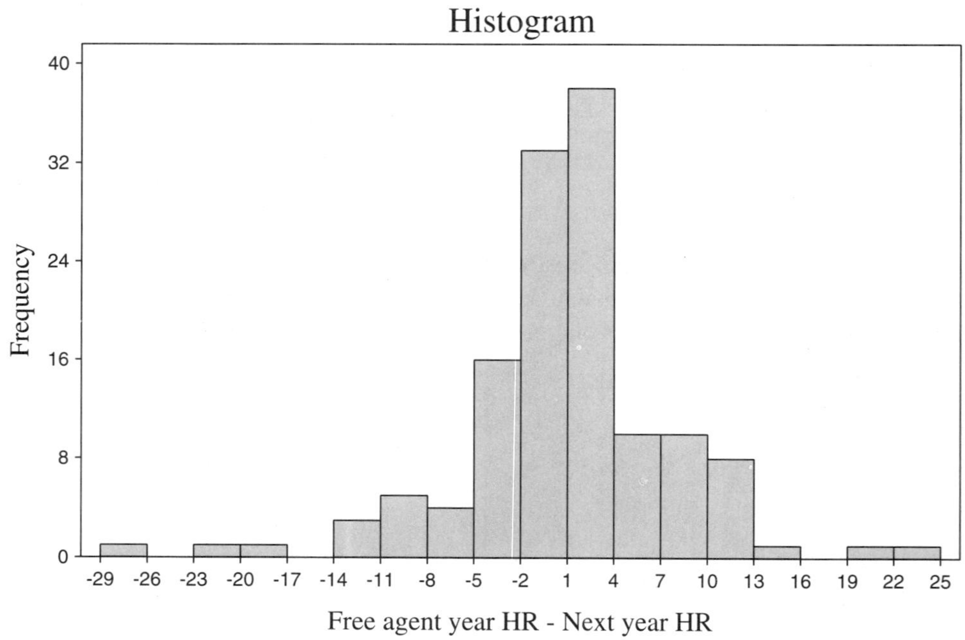

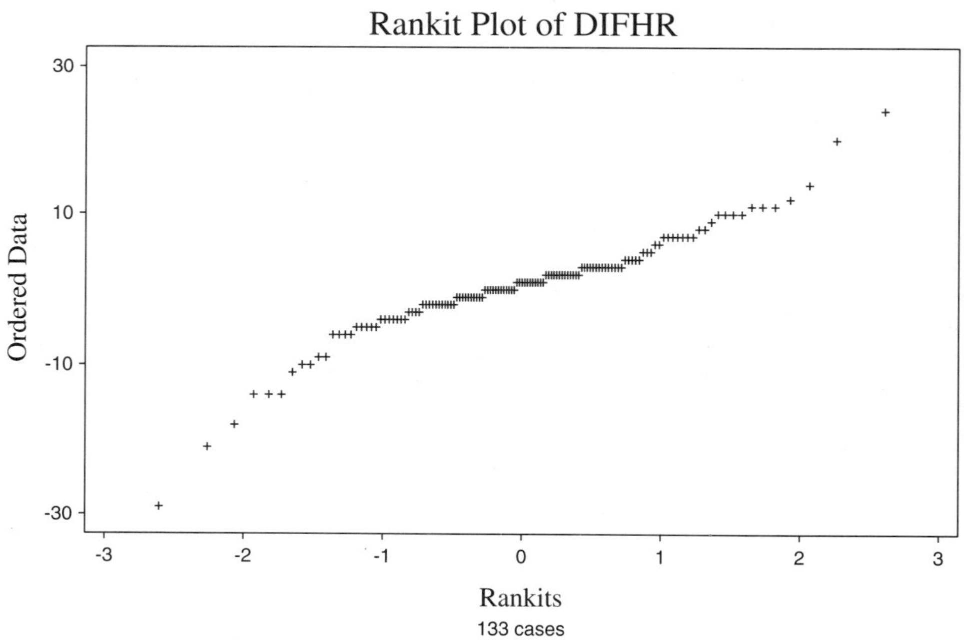

These plots suggest a problem — the distribution of difference in home runs is very long–tailed, in both the left and right tails. This can lead to misleading results from the t–test (this is less likely for large sample sizes, but it is still possible), but it does not affect the nonparametric tests, which can still be trusted.

Here are those results:

```
SIGN TEST FOR HRFA - HRNEXT

NUMBER OF NEGATIVE DIFFERENCES           52
NUMBER OF POSITIVE DIFFERENCES           69
NUMBER OF ZERO DIFFERENCES (IGNORED)     11

TWO TAILED PROBABILITY OF A RESULT AS
OR MORE EXTREME THAN OBSERVED        0.1455

A VALUE IS COUNTED AS A ZERO IF ITS
ABSOLUTE VALUE IS LESS THAN 0.00001

CASES INCLUDED 132    MISSING CASES 0


WILCOXON SIGNED RANK TEST FOR HRFA - HRNEXT

SUM OF NEGATIVE RANKS                                 -3145.5
SUM OF POSITIVE RANKS                                  4235.5

NORMAL APPROXIMATION WITH CONTINUITY CORRECTION         1.408
TWO TAILED P-VALUE FOR NORMAL APPROXIMATION            0.1590

TOTAL NUMBER OF VALUES THAT WERE TIED    115
NUMBER OF ZERO DIFFERENCES DROPPED        11
MAX. DIFF. ALLOWED BETWEEN TIES      0.00001

CASES INCLUDED 121    MISSING CASES 11
```

Each of the nonparametric tests provides only mild evidence against the null hypothesis that the median number of home runs is unchanged from the free agent year to the next year. While the summary statistics given earlier show that the drop in the median values is one home run, these tests indicate that the drop is not highly statistically significant.

Finally, analysis of runs batted in once again gives evidence for a drop from the free agent year to the next year:

```
PAIRED T TEST FOR RBIFA - RBINEXT

MEAN          2.6767
STD ERROR     1.6077
T               1.66
DF               132
P             0.0983


SIGN TEST FOR RBIFA - RBINEXT

NUMBER OF NEGATIVE DIFFERENCES           50
NUMBER OF POSITIVE DIFFERENCES           80
NUMBER OF ZERO DIFFERENCES (IGNORED)      3

TWO TAILED PROBABILITY OF A RESULT AS
OR MORE EXTREME THAN OBSERVED        0.0107

A VALUE IS COUNTED AS A ZERO IF ITS
ABSOLUTE VALUE IS LESS THAN 0.00001

CASES INCLUDED 133    MISSING CASES 0
```

```
WILCOXON SIGNED RANK TEST FOR RBIFA - RBINEXT

SUM OF NEGATIVE RANKS                                  -3317.5
SUM OF POSITIVE RANKS                                   5197.5

NORMAL APPROXIMATION WITH CONTINUITY CORRECTION          2.183
TWO TAILED P-VALUE FOR NORMAL APPROXIMATION             0.0290

TOTAL NUMBER OF VALUES THAT WERE TIED     117
NUMBER OF ZERO DIFFERENCES DROPPED          3
MAX. DIFF. ALLOWED BETWEEN TIES       0.00001

CASES INCLUDED 130    MISSING CASES 3
```

A histogram and normal plot indicate that the distribution of differences in runs batted in is slightly left–skewed (with one possible unusually high value). The paired t–test gives a somewhat higher significance level than the two non–parametric tests.

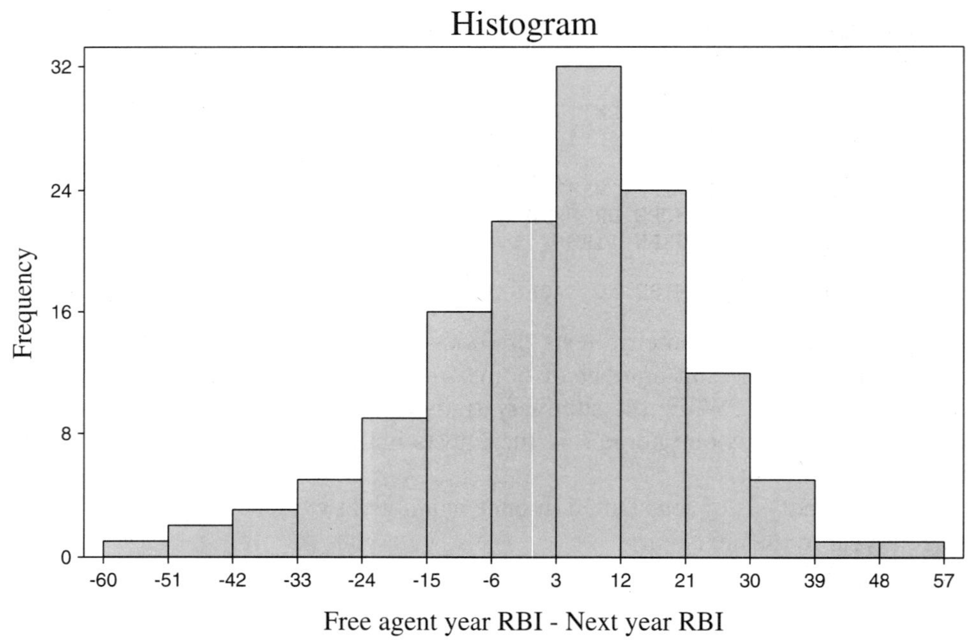

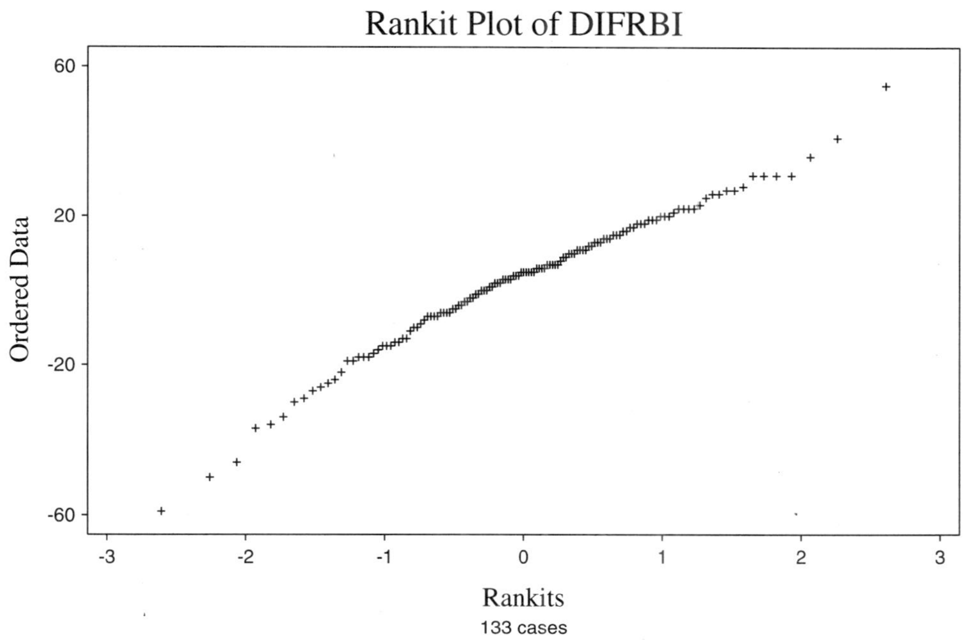

The number of times at bat for the players provides another interesting comparison for these data. Let's look at the distribution of difference between free agent at bats and the next year's at bats:

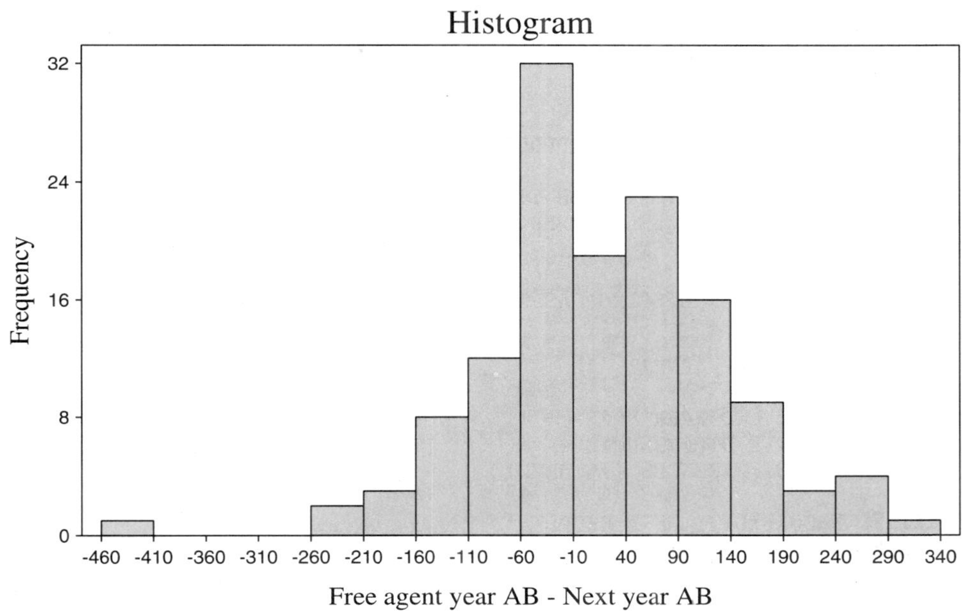

It looks reasonably normal, except for one very low value. The outlier is Lenny Randle (Case 107) who went from 39 at bats in 1979 to 489 in 1980, for a huge difference of -450. If we exclude Lenny Randle, the paired t–test based on the number of times at bat for the players is as follows:

```
PAIRED T TEST FOR ABFA - ABNEXT

MEAN          25.386
STD ERROR     9.4269
T               2.69
DF               131
P             0.0080

CASES INCLUDED 132    MISSING CASES 0
```

A 95% confidence interval for the average change is

$$\begin{aligned}&\overline{\mathtt{ABFA} - \mathtt{ABNEXT}} \pm t^{n-1}_{.025} \times \widehat{\mathrm{SE}}(\overline{\mathtt{ABFA} - \mathtt{ABNEXT}})\\ &= 25.386 \pm t^{131}_{.025} \times 9.4269\\ &= (6.909,\ \ 43.863).\end{aligned}$$

There is a statistically significant drop of over 25 in the average number of times at bat after free agency. We don't know if this is because of injury or some other factor, but it is interesting to note that teams don't get as full a season out of their free agents as they would expect. This could account for the significant drop in runs batted in, as tests based on home runs per at bat and runs batted in per at bat do not show a significant difference between years.

The Wilcoxon signed rank test also suggests that the drop in number of at bats is significant, while the sign test has a much higher p–value than the paired t–test.

```
WILCOXON SIGNED RANK TEST FOR ABFA - ABNEXT

SUM OF NEGATIVE RANKS                                     -3385.0
SUM OF POSITIVE RANKS                                      5526.0

NORMAL APPROXIMATION WITH CONTINUITY CORRECTION             2.403
TWO TAILED P-VALUE FOR NORMAL APPROXIMATION                0.0163

TOTAL NUMBER OF VALUES THAT WERE TIED        67
NUMBER OF ZERO DIFFERENCES DROPPED            0
MAX. DIFF. ALLOWED BETWEEN TIES         0.00001

CASES INCLUDED 133    MISSING CASES 0

SIGN TEST FOR ABFA - ABNEXT

NUMBER OF NEGATIVE DIFFERENCES              60
NUMBER OF POSITIVE DIFFERENCES              73
NUMBER OF ZERO DIFFERENCES (IGNORED)         0

TWO TAILED PROBABILITY OF A RESULT AS
OR MORE EXTREME THAN OBSERVED           0.2981

A VALUE IS COUNTED AS A ZERO IF ITS
ABSOLUTE VALUE IS LESS THAN 0.00001

CASES INCLUDED 133    MISSING CASES 0
```

Summary

Analysis of measures of batting quality for free agent baseball players indicates that batting average, number of home runs and number of runs batted in all drop from the free agent year to the following year, a result consistent with expectancy theory. Although the drops are not very large in an absolute sense (about ten percentage points in batting average, one to three runs batted in, and one home run), and might be of limited practical importance, the first two are statistically significant. The drop in home runs and runs batted in appears to be due mostly to fewer times at bat for the players in the year after free agency. It would seem that teams should not pay prospective free agents based on their free agent year, as they might very well be disappointed.

Technical terms

Alternative hypothesis: the hypothesis that will only be believed if the observed evidence is strong enough to warrant rejection of the null hypothesis.

Null hypothesis: the hypothesis that will be used unless there is sufficient evidence to warrant its rejection. That is, the null hypothesis gets "the benefit of the doubt," unless the evidence against it is strong enough.

p**–value**: the probability of observing a test statistic as extreme as, or more extreme than, the observed value, assuming that the null hypothesis is true. It is also termed a *tail probability*, and is typically used as a measure of the strength of evidence against the null hypothesis that is observed in the data.

Paired t**–test**: a method used to assess the statistical significance of the difference in means between two paired samples. The test is only appropriate when the variables have a natural pairing of observations, such as in a "before / after" data set. The test statistic is equivalent to a **one–sample** t**–test** assessing whether the mean of the differences is significantly different from zero. The test requires that the paired differences represent a random sample from a (roughly) normal distribution.

Sign test: a nonparametric test used to assess the statistical significance of the difference in medians between two paired samples. The test compares the number of positive differences to the number of negative differences, and determines if those numbers are significantly different from each other. The test does not require that the paired differences represent a random sample from a (roughly) normal distribution (rather, the distribution should be roughly symmetric), but is considerably less sensitive than the t–test if the differences actually are normally distributed.

Wilcoxon signed rank test: a nonparametric test used to assess the statistical significance of the difference in medians between two paired samples. The method involves ranking the absolute differences in the pair values, and then summing the ranks corresponding to positive differences, and the ranks corresponding to negative differences; the test then determines whether these sums are significantly different from each other. The test requires that the paired differences represent a random sample from a (roughly) symmetric, but not necessarily normal, distribution. Although it is not as sensitive as the t–test under normality, it is comparatively efficient then as well.

Baseball free agency: another look at player effectiveness

Topics covered: Nonparametric tests. Paired samples. Tests for differences in location.

Data File: `free.dat`

The performance of free agents after they have changed teams was pretty depressing — at least for the teams that have signed them (see "Baseball free agency: do teams get what they pay for?"). Perhaps the reason for this is the performance measures that were used in the earlier case.

Perform similar analyses to assess whether the equity or expectancy theories are apparently appropriate for other performance measures.

Good candidates would be total performance measures (such as home runs, runs scored or runs batted in) adjusted by the number of at bats. These measures represent the performance rate rather than total production. For example, you could look at the home runs per at bat.

Do the measures adjusted by the number of at bats appear to be on a more illuminating scale than the original values? Note that the scaling chosen can have a large impact on the interpretability of the results.

Another way of evaluating players, which attempts to allow comparison of different types of players, is to measure performance by the *total number of runs produced*,

$$\text{Total runs produced} = \text{Runs batted in} + \text{Runs scored} - \text{Home runs}$$

(the number of home runs is subtracted at the end since otherwise they would be double–counted as adding both a run batted in and a run scored). This measure also could be standardized by the number of at bats.

Do the observed patterns differ depending on the performance measure used?

Comments to the Instructor

The results for other performance measures parallel those of the ones previously examined. As was true for batting average, there is a small, but significant, drop in slugging average from before declaring free agency to the following year. Similarly, the overall production measures of runs scored and total runs produced drop. The runs scored drop by an average of 2.6 runs which is not significant at the 0.1 level of significance. The total runs produced drop by 4.8 runs, which is significant. There are a couple of outliers in terms of total runs produced that make the t–test less compelling, although there are no serious violations of the assumptions of the t–test for the other performance measures.

If home runs, runs batted in, runs scored and total runs produced are standardized by times at bat (to reflect production rate, rather than total production), the differences are no longer statistically significant. Histograms illustrate the usefulness of this scaling relative to the original scale. Thus, as was noted in the case "Baseball free agency: do teams get what they pay for?", it appears that the most important effect is that the average times at bat drops for free agents, thereby leading to a drop in total production. All of the tests agree, and the only model violation noted is an outlier with respect to runs batted in rate (case 33, John Grubb, increased his production from 17 runs batted in in 176 times at bat to 51 runs batted in in 210 times at bat).

To maintain consistency with the unannotated edition,
the next case begins on the following page.

Reporting of sexual partners by men and women

Topics covered: Confidence interval. Independent samples. Nonparametric tests. Tests for differences in distributional shape. Tests for differences in location.

Key words: Central Limit Theorem. Histogram. Long–tailed data. p–value. Two sample t–test. Wilcoxon rank sum test.

Data File: `sex.dat`

Sexually transmitted diseases, once thought to have been vanquished by the "magic bullet" penicillin, have come back with a vengeance. HIV, the virus that causes AIDS, is the most dangerous and well–known example, but the last 20 years have witnessed the emergence of other sexually transmitted viruses, such as herpes and hepatitis–B, as well as penicillin resistant mutations of the original bacterial agents responsible for syphilis and gonorrhea.

For now, the medical options for treatment are often nonexistent, and progress in vaccine development has been much slower than anticipated. Prevention is the only option. This has made knowledge and education about sexual behavior the primary defense against epidemic spread.

Accurate knowledge of the sexual behavior of the population is essential to guide public policy efforts. Without such data, the most basic questions regarding the epidemiology of the disease cannot be answered. What is the epidemic potential of HIV? That depends on how many sex partners people have. How likely is HIV to spread from the initially infected groups to the general population? That depends on who is having sex with whom, and how much contact there is among different groups in the population.

Collecting accurate information on sexual practices is not easy. This is sensitive information that people may be reluctant to disclose, and politicians are reluctant to support openly scientific studies. Indeed, a report of the National Academy of Sciences' Institute of Medicine released July 27, 1994, concluded that Congress should authorize federal funding of extensive research on American sexual and drug use habits, which had not been allowed because of federal and congressional restrictions. At present the only regular random survey that asks questions about sexual behavior is the General Social Survey (GSS). The GSS is an annual cross–sectional survey, using face–to–face interviews and a stratified probability sample designed to be representative of the U.S. population 18 years of age and older. The survey has been carried out every year for the past 20 years, and response rates are 71–80%, among the highest in the field. The questions on sexual behavior were added in 1989, designed to provide a small amount of reliable representative data on current patterns of adult sexual behavior in the general population while another more comprehensive survey was embroiled in funding controversy. Respondents were asked to report the number of male and female sexual partners they had had since they were 18 years of age (referred to here as "lifetime" partners). These items are distributed as a self–administered questionnaire in order to minimize the potential discomfort and embarrassment of the respondent. Non–response on the questionnaire for these questions is less than 10% overall, and less than 5% for those under 55 years old. Data on the number of male and female sex partners the respondent has had since they were 18 years of age were collected in

1989, 1990 and 1991. The data used here are a simple random sample from the male and female U.S. population 18 years of age and older. There are 1682 men and 1850 women in the sample. These sizes were chosen to reflect the true proportion of males to females in the U.S. population 18 years of age and older (which is 47.62% male).

A puzzling anomaly often noted in even the most carefully conducted surveys of sexual behavior is that the number of female sexual partners reported by men far exceeds the number of male partners reported by women. By simple mathematical identity, for the entire population these two numbers must equal each other. This discrepancy is sometimes interpreted as evidence that, as a result of the controversial nature of the subject, sexual behavior surveys produce unreliable data. Without a vaccine or cure for HIV infection, however, behavioral surveys provide crucial information for managing the current and long–term spread of the disease. It is thus important to evaluate whether the male–female reporting discrepancy in these surveys reflects a systematic lack of reliability in the data, or a relatively minor aberration.

Do men and women, on average, report the same number of opposite–sex partners?

We will try to answer this question based on the GSS data. The variable `MALE` is the number of female partners reported by the male respondents, while the variable `FEMALE` is the number of male partners reported by the female respondents. First let's get some descriptive statistics on the samples:

```
DESCRIPTIVE STATISTICS

                      MALE          FEMALE
N                     1682            1850
SUM                  19711            6146
MEAN                11.719           3.322
SD                  24.479           6.054
MINIMUM              0.000           0.000
1ST QUARTILE         1.000           1.000
MEDIAN               4.000           1.000
3RD QUARTILE        10.000           3.000
MAXIMUM            253.000         100.000
```

Thus the total number of female partners reported by men was 19,711, while the total number of male partners reported by women was 6146. As this is a sample of the population, and not the population itself, we would expect the sums to differ. However, men report well over three times as many lifetime partnerships with women as women report with men, on average. (The median values of partners per respondent, which are less affected by unusually large values, still reflect this discrepancy, with the median number of female partners reported by men being four times the median number of male partners reported by women.) This creates a massive discrepancy, with more than two–thirds of all contacts reported by men ((19711-6146)/19711 = 68.8%) unaccounted for. Could this discrepancy be due solely to chance? That is, is the observed discrepancy consistent with the hypothesis that:

Expected total number of female partners reported by males =
Expected total number of male partners reported by females.

In the context of our sample, we can restate this hypothesis in terms of the means of each variable:

$$1682 \times \text{Expected average number of female partners reported by males} =$$
$$1850 \times \text{Expected average number of male partners reported by females,}$$

or,

$$\frac{1682}{1850} \times \text{Expected average number of female partners reported by males} =$$
$$\text{Expected average number of male partners reported by females.}$$

To test this hypothesis we can create a new variable BMALE:

$$\texttt{BMALE} \;=\; 1682 \times \texttt{MALE}/1850.$$

As there are slightly more females than males, we would expect the average number of female partners reported by males to be slightly greater than the average number of male partners reported by females. We can interpret BMALE as the number of female partners that the male would have had if the number of males and females in the population had been equal.

The null hypothesis, which is that there is no difference between the average number of partners reported by men and women, can be stated as

$$H_0 : \mu_{\texttt{BMALE}} = \mu_{\texttt{FEMALE}}.$$

The alternative hypothesis is

$$H_a : \mu_{\texttt{BMALE}} \neq \mu_{\texttt{FEMALE}}.$$

It is important to note that there is no ambiguity that the hypothesis is correct for the U.S. population. The purpose of this test is to see if the responses in the survey sample are representative of the true values in the U.S. population. It is the characteristics of the population that the survey sample actually represents, rather than those of the U.S. population, that are being tested.

Let's look again at descriptive statistics for the two groups:

```
DESCRIPTIVE STATISTICS

                      BMALE       FEMALE
N                      1682         1850
SUM                   17921         6146
LO 95% CI             9.590        3.046
MEAN                 10.655        3.322
UP 95% CI            11.719        3.598
SD                   22.256        6.053
MINIMUM               0.000        0.000
1ST QUARTILE          0.909        1.000
MEDIAN                3.637        1.000
3RD QUARTILE          9.092        3.000
MAXIMUM             230.025       100.00
```

The mean number of partners for men (10.655) is substantially greater than that for women (3.322). A 95% confidence interval for $\mu_{\texttt{BMALE}}$ is from 9.590 to 11.719

partners. A 95% confidence interval for $\mu_{\texttt{FEMALE}}$ is from 3.046 to 3.598 partners. With such large sample sizes, the uncertainty in the expected values is small. These intervals suggest that the average number of female partners reported by the males is higher than the average number of male partners reported by the females. We can address the hypothesis testing question formally by conducting a **two–sample (unpaired) t–test.** The idea of the test is to see if the estimate of the difference in the means is statistically significantly different from zero.

The difference in the averages is

$$\overline{X}_{\texttt{BMALE}} - \overline{X}_{\texttt{FEMALE}} = 10.655 - 3.3222 = 7.333 \text{ partners.}$$

This is the natural estimate of

$$\mu_{\texttt{BMALE}} - \mu_{\texttt{FEMALE}}.$$

We can interpret this difference as the difference in total lifetime partners per woman (e.g., a value of 2 would mean that each woman in the population would have had to report 2 more partners to make the reported totals equal).

How uncertain is this estimate? There are two ways to answer this question and hence two versions of the two–sample (unpaired) t–test. The first version assumes that the variance of the number of partners is the same for males and females. From the descriptive statistics, the sample standard deviation for the males is 22.256 partners and for the females it is 6.053 partners. Thus the variation of the males appears to be much greater than the variation of the females. That the difference is not a result of chance variation is confirmed by an F–statistic for the hypothesis test of equal variances (that is, the null hypothesis for this F–test is that the variance of BMALE equals the variance of FEMALE):

```
                          F      NUM DF    DEN DF       P
TESTS FOR EQUALITY     -------   ------    ------    ------
      OF VARIANCES      13.52     1681      1849     0.0000

CASES INCLUDED 3532
```

The second version of the test estimates the variances separately for each group (but is less powerful when the two variances are identical). For this test the estimated standard error of the difference $\overline{X}_{\texttt{BMALE}} - \overline{X}_{\texttt{FEMALE}}$ is

$$\begin{aligned}\widehat{\text{SE}}(\overline{X}_{\texttt{BMALE}} - \overline{X}_{\texttt{FEMALE}}) &= \sqrt{\frac{s^2_{\texttt{BMALE}}}{n_{\texttt{BMALE}}} + \frac{s^2_{\texttt{FEMALE}}}{n_{\texttt{FEMALE}}}} \\ &= \sqrt{\frac{22.256^2}{1682} + \frac{6.053^2}{1850}} \\ &= 0.5606 \text{ partners.}\end{aligned}$$

The t–statistic for the two–sample (unpaired) t–test with different group variances is then

$$t = \frac{\overline{X}_{\texttt{BMALE}} - \overline{X}_{\texttt{FEMALE}}}{\widehat{\text{SE}}(\overline{X}_{\texttt{BMALE}} - \overline{X}_{\texttt{FEMALE}})} = \frac{7.333}{0.5606} = 13.08.$$

Note that the difference is over 13 standard errors above zero. Hence there is strong evidence that the means are different and that the mean number of partners for men is greater than the mean number for women. Note, however, that with 3532 cases we should be able to detect even small differences between the means. It is a separate question whether the difference can help our ultimate understanding of the epidemic.

To get an idea of the size of the difference in the averages, let's calculate a 95% confidence interval. A 95% confidence interval for the true difference is

$$\overline{X}_{\mathtt{BMALE}} - \overline{X}_{\mathtt{FEMALE}} \pm 1.96 \times \widehat{\mathrm{SE}}(\overline{X}_{\mathtt{BMALE}} - \overline{X}_{\mathtt{FEMALE}});$$
$$\text{that is, } 7.333 \pm (1.96)(0.5606)$$
$$= (6.234,\ 8.432) \text{ partners.}$$

Thus we are 95% confident that the true difference in the mean number of reported partners between men and women is between 6.234 and 8.432 lifetime partners. Wow, that is a big difference! Another way to state this is that each woman in the population would have had to report between 6.234 and 8.432 more lifetime partners to make the reported totals equal. As the mean number of partners per woman is 3.222, this does not seem likely.

Note that the 95% confidence interval does not contain zero, so the hypothesis of equality of the two means can be rejected at the 5% level of significance. This test is, of course, identical to the one based on the above t–statistic.

We can check the assumption of the t–test (and the confidence interval) that the distributions are normally distributed by examining the histograms of the number of partners for each group:

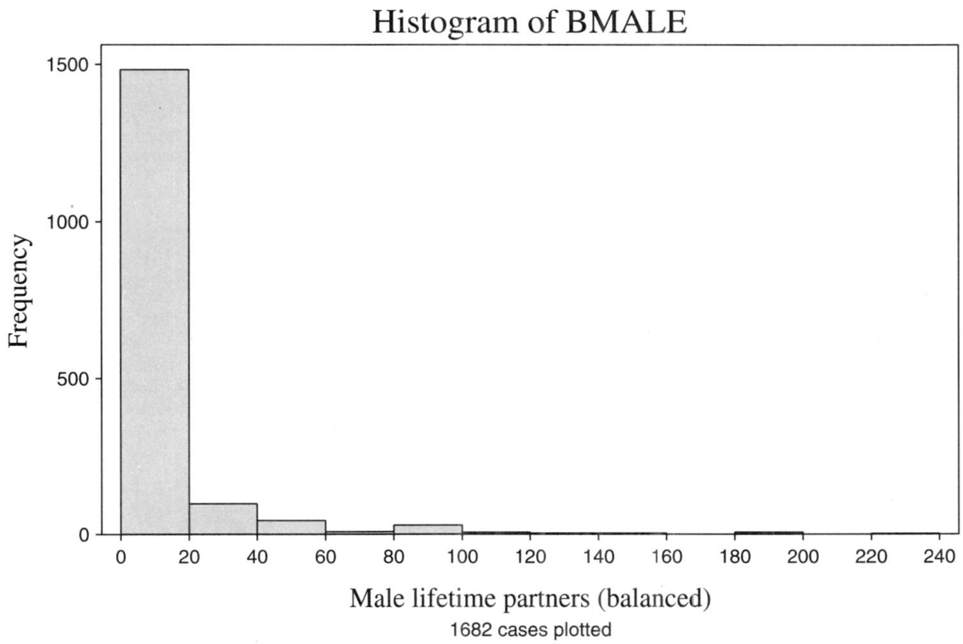

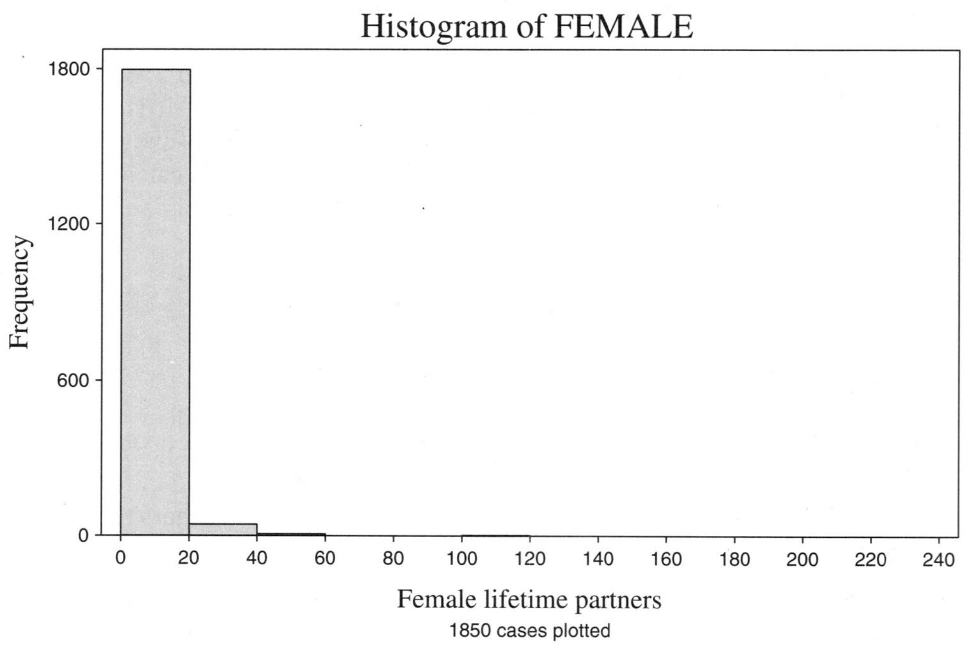

We see that the distributions are very right–skewed; so much so that these figures do not show the true shape. Let's look at the logarithm of the total lifetime partners to overcome the skewness in the data. This transformation makes sense for the current problem because the process of acquiring partners might be more closely multiplicative than additive. Here are the histograms:

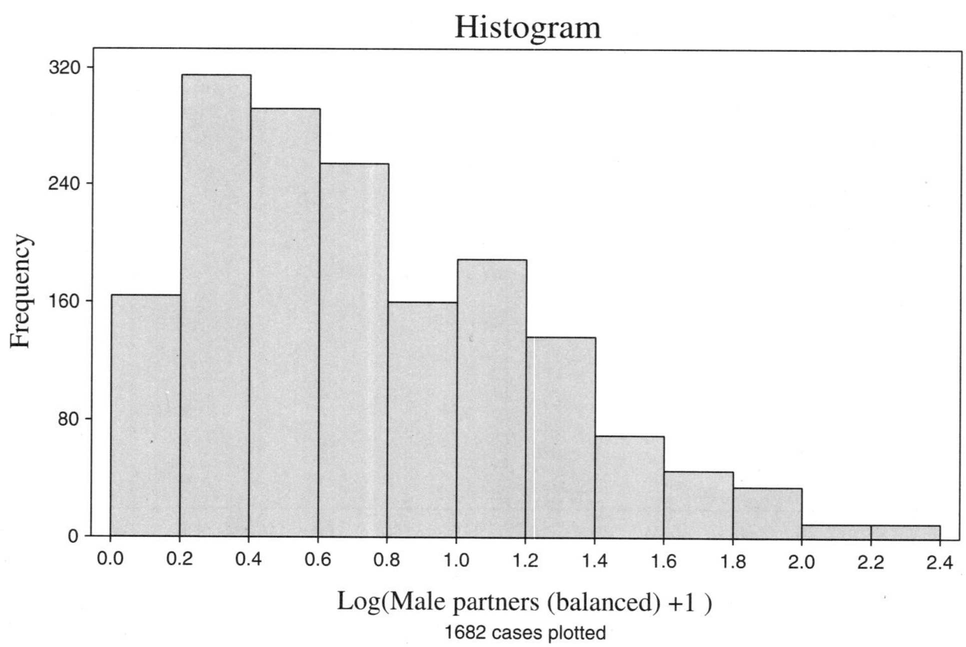

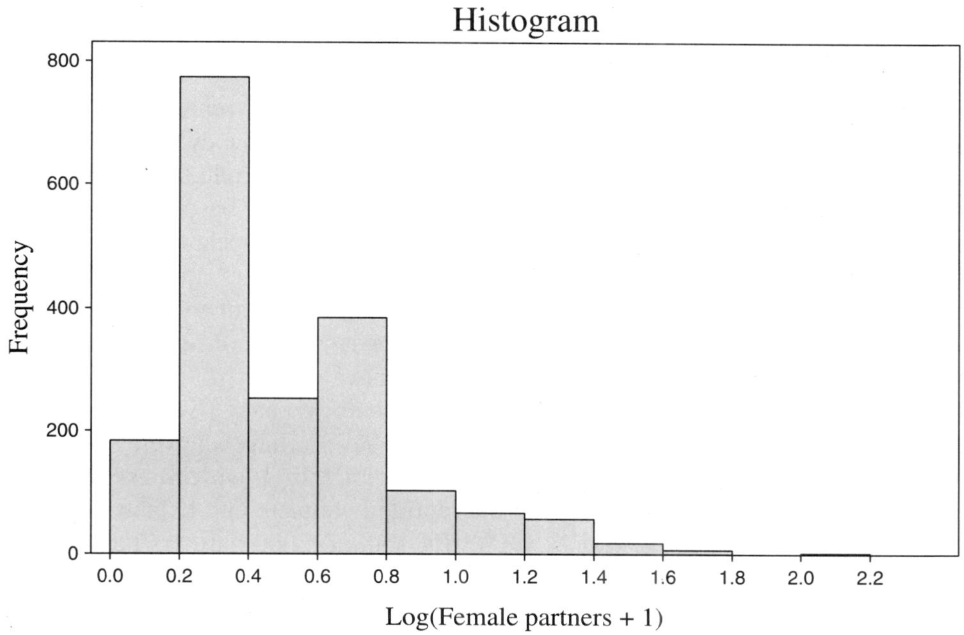

Both BMALE and FEMALE take on zero values, so the logarithm has been applied after adding one to each value. The distributions are still somewhat right–skewed, even after the transformation. We can see that many more of the women report between 1 and 3 partners than of the men. In contrast, a greater proportion of men report values in the upper tail (at least 10 partners).

The original distributions are very non–normal in shape. What does this mean for the validity of the t–test and the confidence intervals? In some circumstances, it can invalidate the results, but here it will have relatively little impact. This is because the tests are based on a normal distribution for the sample averages (i.e., $\overline{X}_{\texttt{BMALE}}$ and $\overline{X}_{\texttt{FEMALE}}$) rather than a normal distribution of the values themselves. These averages are based on very large sample sizes (1682 and 1850, respectively). As a result, even though the distribution of the values is quite non–normal, the distribution of the sample averages will be quite close to normal. This is the effect described by the Central Limit Theorem.

Given that the data confirm the discrepancy between the number of lifetime partnerships reported by women and the number reported by men, let's think about what could cause it. There need not be a single cause; many factors could be involved. Keep in mind that we are speculating now; we probably cannot use the data we have to confirm our hypotheses.

One cause of the difference can be that the data do not represent a random sample of the population. This is called "sample bias." Sample bias could explain the discrepancy if important parts of the population are systematically excluded from the survey. It has been suggested that patterns of age–matching in forming relationships, with men usually slightly older than their female partners, may mean that men in the sample are reporting contacts with women who fall below the sampled age range for women. For the GSS data, however, this can be shown to imply that over 60% of the lifetime sexual partners reported by all men are with women who are currently under

18 years old, a highly unlikely scenario, and one that also neglects the distribution of contacts for under–18 year old men.

Sample bias also could affect the results if highly active women were less common, and more likely to be excluded from the sample either by chance or for other reasons. The classic explanation of this type ascribes the discrepancy to the exclusion of prostitutes (commercial sex workers, or CSWs) from the sample. There are good reasons to think that this explanation is not sufficient here. First, based on other questions asked in the GSS, the fraction of men who visit CSWs appears to be relatively low: 1% report a CSW contact in the last year, 3% in the last 5 years, and the prevalence of first intercourse with a CSW is substantially less likely for recent cohorts of men. If, inflated for lifetime estimates, 15% of the men in the sample were reporting contacts with non–sampled CSWs, each man would have to have roughly 45 (different) CSW partners apiece in their lifetime in order to explain the discrepancy. Even if half of the men were reporting CSW contacts, each man would have to have an average of about 16 different CSW partners. Working from the other direction, we can estimate how many CSWs would be necessary to bring the reported contacts into balance. For this sample, it would require about 7 to 14 CSWs (at 1000 to 2000 partners per lifetime) to compensate for the discrepancy. At that rate, past or current CSWs would comprise about one–third to three-quarters of a percent of the female population. While little is known about the prevalence of CSWs, the most reliable estimates suggest the prevalence of current (active) CSWs is about 30–70 per 100,000 women, well below the level necessary to explain the discrepancy here.

Another cause of the observed discrepancy could be that the respondents are not giving accurate (or honest!) estimates of the number of partners they have had. This is called "reporting bias" (also called "response bias"). Reporting bias could explain the discrepancy if under– and over–reporting patterns are gender–dependent. Standard role expectations for men and women make this a reasonable hypothesis, and a commonly cited explanation. Having a large number of sexual partners can enhance a man's reputation (up to a point), but similar patterns among women are generally discouraged. Thus one might expect a tendency for men to over–report and/or women to under–report their lifetime partner totals. To explain the observed discrepancy here, however, one would have to assume a substantial amount of reporting bias. Splitting the bias equally between men and women by taking the mean of their reports as the true number of contacts would imply that all men are over–reporting by 65% *and* all women are simultaneously reporting values that are less than a third of their true numbers.

How could these survey biases manifest themselves? One possibility is that the value reported by the men is proportionately inflated from its true value (e.g., men reporting 12 rather than 10 and 120 rather than 100 partners). Is it then possible that the distributions are the same, except for an inflation factor? We can test this by proportionately deflating the men's values so that they have the same mean as the women's values, and conducting a **Wilcoxon (Mann–Whitney) rank sum test** against the female's values. This test checks to see if the two samples are drawn from the same distribution. Based on the descriptive statistics, we can define the adjusted men's values:

$$\texttt{AMALE} = 3.322 \times \texttt{MALE}/11.719.$$

The results of the test are then:

```
RANK SUM TWO-SAMPLE (MANN-WHITNEY) TEST FOR AMALE VS FEMALE

                            SAMPLE
VARIABLE      RANK SUM      SIZE      U STAT      MEAN RANK
---------    ---------     ------   ---------    ---------
AMALE        2.72E+06      1682     1.31E+06      1618.3
FEMALE       3.52E+06      1850     1.81E+06      1901.2
TOTAL        6.24E+06      3532

NORMAL APPROXIMATION WITH CONTINUITY CORRECTION        8.235
TWO-TAILED P-VALUE FOR NORMAL APPROXIMATION           0.0000

TOTAL NUMBER OF VALUES THAT WERE TIED        3519
MAXIMUM DIFFERENCE ALLOWED BETWEEN TIES 0.00001

CASES INCLUDED 3532
```

The average rank of the men's values is 1618.3, while that of the women's values is 1901.2. Thus, on average, the adjusted men's values are somewhat less than the women's values. This difference is statistically significant ($p = 0.0000$). Thus, even adjusting for possible inflation, the sample of men's values come from a different distribution than that of the women's values.

The histograms given earlier showed that the difference in shape of the two distributions comes from the many large values reported in the original men's sample compared with the women's sample. Recent commentaries have suggested that reporting bias is more specifically located in the upper tail of the reported partnership distribution, and this explanation appears to be more tenable based on these data. From the histograms, we see that about 82% of the respondents report less than 10 partners in their lifetime, 90% report less than 20 and 96% report less than 40.

Is it possible that the distributions are the same except for bias in the upper tail? That is, could both reports be consistent for, say, less than 10 partners being reported, and be less consistent for the larger reports? One way to test this is to omit the respondents in the sample who reported 10 or more partners and conduct a Wilcoxon rank sum test for the two truncated samples. The proportion of respondents that report less than 10 lifetime partners is $2888/3532 = 82\%$. Thus the test relates to most of the respondents.

The results of the test are as follows:

```
RANK SUM TWO-SAMPLE (MANN-WHITNEY) TEST FOR BMALE VS FEMALE

                            SAMPLE
VARIABLE      RANK SUM      SIZE      U STAT      MEAN RANK
---------    ---------     ------   ---------    ---------
BMALE        1.72E+06      1185     1.02E+06      1451.5
FEMALE       2.45E+06      1703     1.00E+06      1439.6
TOTAL        4.17E+06      2888

NORMAL APPROXIMATION WITH CONTINUITY CORRECTION        0.377
TWO-TAILED P-VALUE FOR NORMAL APPROXIMATION           0.7064

TOTAL NUMBER OF VALUES THAT WERE TIED        2888
MAXIMUM DIFFERENCE ALLOWED BETWEEN TIES 0.00001

CASES INCLUDED 2888
```

The average rank of the men's values is 1451.5, while that of the women's values is 1439.6. Thus, on average, the two samples are much closer. The difference between them is also not statistically significant ($p = 0.7064$). This is consistent with the similarities of the quartiles of the two distributions (the first quartiles are 0.909 and 1, respectively, for `BMALE` and `FEMALE`; the medians are 1.818 and 1, respectively; and the third quartiles are 3.637 and 3, respectively). Thus, after omitting the upper tail of both samples, the sample of men's values does not appear to come from a different population than that of the women's values.

Summary

A consistent anomaly observed in recent surveys of sexual behavior is that men report substantially more partnerships with women than women report with men. In data examined here, men report over 3 female partners for every 1 male partner reported by women. We are 95% confident that the true difference in the number of reported partners between men and women is between 6.234 and 8.432 lifetime partners per woman in the population. As the mean number of partners per woman is 3.222, the discrepancy appears to be of practical importance, as well as statistically significant.

Explanations for this discrepancy usually invoke two factors, sex–linked response bias, with men over–reporting and women under–reporting their true contact rates, and sampling bias, with typical survey sampling and field methods likely to miss the most highly active women ("prostitutes"). Neither of these explanations appears tenable in explaining the size of the observed discrepancy based on this data set.

The discrepancy does not appear to be explainable in terms of simple proportionate inflation in reported contacts by the men (or deflation by the women).

This discrepancy is sometimes interpreted as evidence that surveys apparently produce unreliable data because of sex–linked response and sampling bias. However, much of the discrepancy appears to be driven by the upper tail of the partnership distribution. The implication is that sexual behavior surveys apparently provide reliable data in the main, and that improvements that could increase precision in the upper tail would make these data more useful for modeling the spread of AIDS and other sexually transmitted diseases.

Technical terms

Unpaired (Two sample) t–test: a method used to assess the statistical significance of the difference in means between two independent samples. The test is only appropriate when the variables come from two independent samples or groups. The test statistic is analogous to a one–sample t–test assessing whether the mean of the differences is significantly different from zero. The test requires that each sample represent a random sample from a (roughly) normal distribution.

Wilcoxon (Mann–Whitney) rank sum test: a nonparametric test used to assess the statistical significance of the difference in distributions between two independent samples. The method involves comparing the average rank of the two samples in the sample formed by "pooling" both samples together; the test then determines whether the average ranks from the two samples are significantly different from each other. Although it is not as sensitive as the two–sample unpaired t–test under normality, it is comparatively efficient then as well, while still being appropriate for non–normal data.

The comparative volatility of stock exchanges

Topics covered: Comparison of volatility. Nonparametric tests. t–test.

Key words: Confidence interval. Histogram. Median test. Normal plot. Rank sum test. Two–sample t–test.

Data file: `nyseotc.dat`

The Over the Counter (NASDAQ) market advertises itself as being the market for exciting growth companies. Since it is a commonly accepted truism in investments that greater return is accompanied by greater risk, we might expect this to translate into more volatility in NASDAQ stocks compared with the more stable ones listed on the New York Stock Exchange (NYSE).

A reasonable measure of volatility (if the average return for the given time period is not too far from zero) is the absolute proportional change (return) in prices of stocks in the market. That is, a market where stocks average a weekly 10% change in price (whether it is positive or negative change) is more volatile than one that averages a 5% change.

We again consider the data previously examined in the cases "The return on stocks in the Over the Counter market" and "The return on stocks in the New York Stock Exchange." Recall that the given values were the returns for 30 stocks listed on each exchange for the week May 9 – May 13, 1994. The difference is that now we are considering the absolute value of the return, rather than the return itself. We will call these variables `ABSNASD` and `ABSNYSE`, respectively.

This is a two–sample hypothesis testing problem. The standard approach would be to use a two–sample t–test. Remember that there are a couple of key assumptions when using this test: that the two populations being sampled from are Gaussian, and that they have the same variance. Let's take a look at histograms first:

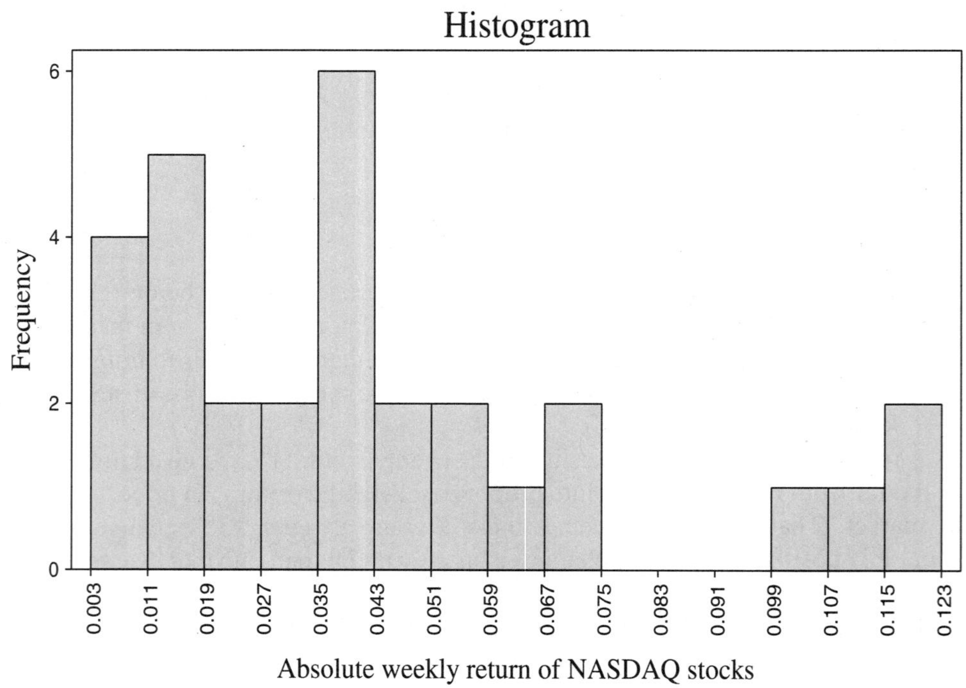

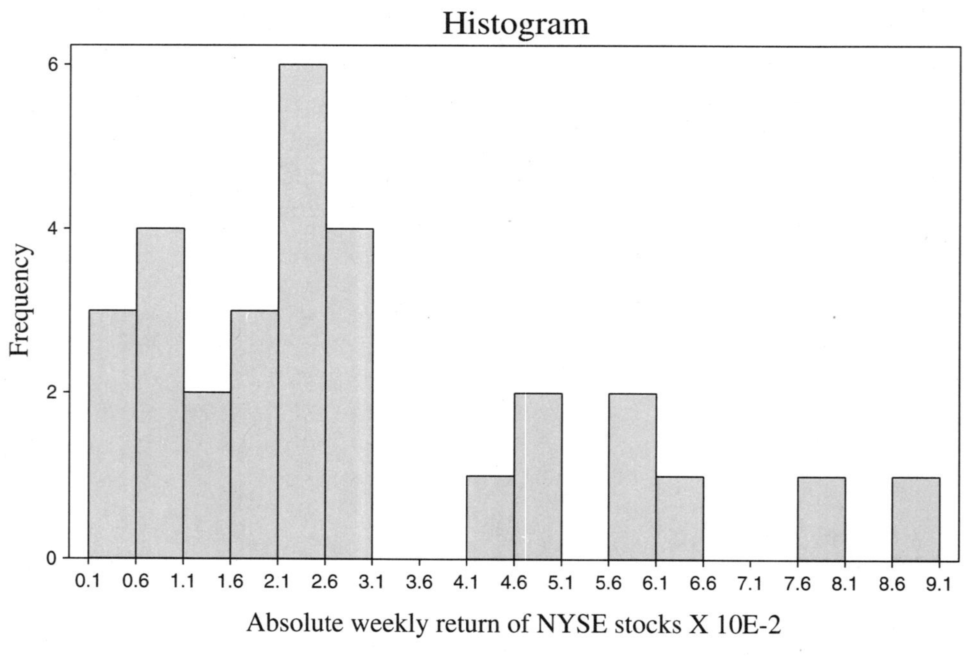

Normal plots also can be instructive:

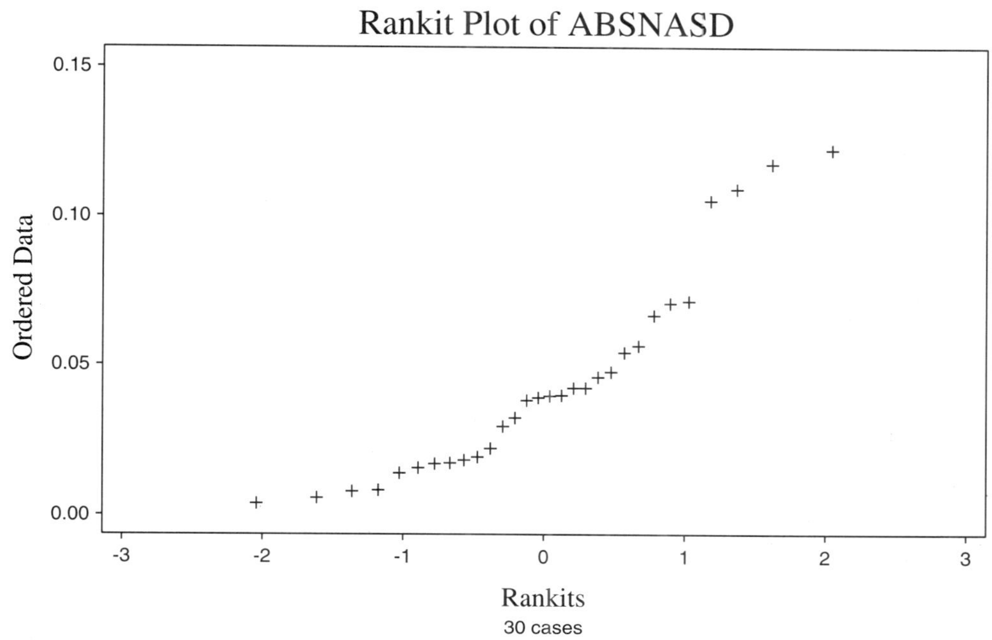

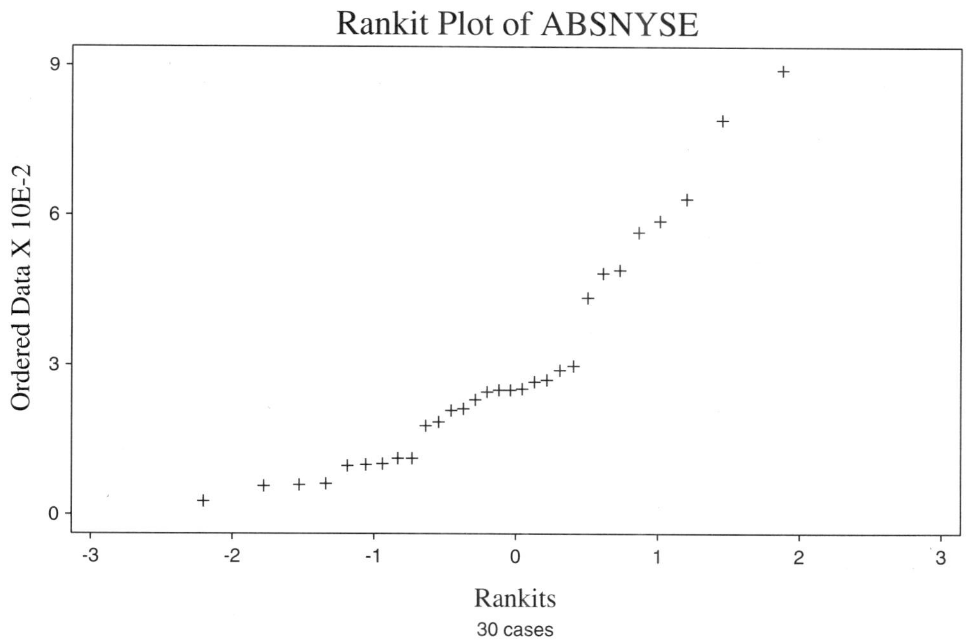

These plots suggest potential problems. The distributions of both NASDAQ and NYSE stocks look distinctly long–tailed. Let's look at some descriptive statistics.

```
                    ABSNASD      ABSNYSE
N                        30           30
MEAN                 0.0439       0.0291
SD                   0.0337       0.0223
MINIMUM           3.788E-03    2.625E-03
1ST QUARTILE         0.0172       0.0110
MEDIAN               0.0393       0.0248
3RD QUARTILE         0.0589       0.0447
MAXIMUM              0.1224       0.0891
```

Two things are apparent: first, the NASDAQ stocks *do* appear to be more variable than the NYSE stocks. In particular, the average proportional absolute change in price for the NASDAQ stocks is 4.39%, while that for the NYSE stocks is 2.91%. A 95% confidence interval for this difference has the form

$$\overline{X}_1 - \overline{X}_2 \pm t_{.025}^{n_1+n_2-2}\sqrt{\left(\frac{(n_1-1)s_1^2+(n_2-1)s_2^2}{n_1+n_2-2}\right)\left(\frac{1}{n_1}+\frac{1}{n_2}\right)},$$

or

$$.0439-.0291 \pm (2.00)\sqrt{\left(\frac{(29)(.0337^2)+(29)(.0223^2)}{58}\right)\left(\frac{1}{30}+\frac{1}{30}\right)}$$
$$= (.000044, .029556)$$

(this is a confidence interval based on assuming that the variances of absolute returns are equal for the two stock exchanges).

Thus, the observed difference in absolute return is a noticeable, but relatively small, 1.48%, with the absolute return being higher for the NASDAQ stocks. A second important point is that the assumption of constant variance might be in some doubt. Here is the two–sample t–test:

```
TWO-SAMPLE T TESTS FOR ABSNASD VS ABSNYSE

                              SAMPLE
VARIABLE        MEAN           SIZE        S.D.          S.E.
---------    ----------       ------    ---------     ---------
ABSNASD        0.0439           30        0.0337      6.15E-03
ABSNYSE        0.0291           30        0.0223      4.07E-03

                                 T          DF          P
                              ------      ------      ------
EQUAL VARIANCES                2.01         58        0.0492
UNEQUAL VARIANCES              2.01       50.3        0.0499

                                 F        NUM DF      DEN DF        P
TESTS FOR EQUALITY            -------     ------      ------      ------
      OF VARIANCES             2.28         29          29        0.0148

CASES INCLUDED 60
```

The test for equality of variances given above rejects the hypothesis of constant variance, so we can't really trust the standard t–test (given above on the line labeled `EQUAL VARIANCES`). The test that does not assume equal variances, and corrects the degrees of freedom to try to account for heteroscedasticity (given above on the line labeled `UNEQUAL VARIANCES`) still has a tail probability of .0499, however; that is, there **is** a significant difference (at the .05 level) in volatility between NASDAQ and NYSE stocks (as noted earlier, NASDAQ stocks are more variable).

We still haven't addressed the problem of apparent non–normality of the samples. In order to do this, we can appeal to tests that do not require the assumptions of normality and constant variance (nonparametric tests). The first of these tests is the **median test**. This test calculates the median of the joint sample of both groups. If there was no difference in location between the two groups, we would expect that about half of the values in each sample would be above the joint median, and about half would be below it. A pattern where one sample has most of its values above the joint median, while the other sample has most below the joint median, indicates a difference in location. A chi–squared test is used to assess the significance of the pattern. Note, by the way, that this test actually tests the hypothesis of whether the distribution of absolute returns for the NASDAQ stocks differs from that of the NYSE stocks (without specifying what that distribution might be). From a practical point of view, the test is best able to detect shifts (differences) in the medians of the two groups.

```
MEDIAN TEST FOR ABSNASD - ABSNYSE

                     ABSNASD      ABSNYSE       TOTAL
                    ---------    ---------    ---------
ABOVE MEDIAN            19           11           30
BELOW MEDIAN            11           19           30
TOTAL                   30           30           60
TIES WITH MEDIAN         0            0            0

MEDIAN VALUE        0.0268

CHI-SQUARE  4.27    DF 1   P-VALUE 0.0389
MAX. DIFF. ALLOWED BETWEEN A TIE   0.00001
CASES INCLUDED 60
```

Note that the NASDAQ stocks are mostly above the joint median, while the NYSE stocks are mostly below the joint median, indicating a significant difference with a tail probability of .0389. This result is consistent with the observed median absolute returns of 3.93% for the NASDAQ stocks and 2.48% for the NYSE stocks.

A second nonparametric test is the Wilcoxon (Mann–Whitney) rank sum test. Roughly speaking, this is similar to a t–test, except that instead of using the true data values, the data used are the ranks of the values in the joint sample.

```
RANK SUM TWO-SAMPLE (MANN-WHITNEY) TEST FOR ABSNASD VS ABSNYSE

                              SAMPLE
VARIABLE     RANK SUM   SIZE     U STAT    MEAN RANK
---------   ---------  ------  ---------   ---------
ABSNASD       1020.0     30      555.00       34.0
ABSNYSE       810.00     30      345.00       27.0
TOTAL         1830.0     60

NORMAL APPROXIMATION WITH CONTINUITY CORRECTION       1.545
TWO-TAILED P-VALUE FOR NORMAL APPROXIMATION          0.1224

TOTAL NUMBER OF VALUES THAT WERE TIED           6
MAXIMUM DIFFERENCE ALLOWED BETWEEN TIES 0.00001
CASES INCLUDED 60
```

The NASDAQ stocks have a higher sum of ranks in the joint sample, since they tend to be the larger values. However, the hypothesis of same distribution for both groups is *not* rejected here, with a tail probability of .1224 indicated.

So, the results are fairly clear — NASDAQ stocks are more volatile than NYSE stocks, although the difference in volatility for this week was small. So, if you're a gambler, the Over the Counter market is for you; but you'd better be willing to lose bigger too!

Summary

An examination of the absolute returns for 30 stocks listed in the Over the Counter market and New York Stock Exchange, respectively, shows that the NASDAQ stocks are more volatile than the NYSE stocks. This result is supported by both parametric and nonparametric tests. It is consistent with the generally held perception that the NASDAQ market focuses more on growth companies, which would presumably combine greater growth potential with greater risk.

Technical terms

Median test: a nonparametric test used to assess the difference in distributions between two independent samples. The test evaluates whether the groups have a statistically significantly different proportion above or below the joint median than 50%. The test does not require that the distributions of the variables be Gaussian, but it has relatively low efficiency if the distributions are Gaussian.

Validity of t – testing for the sexual partners reported by men and women

Topics covered: Central Limit Theorem. Computer experiment. Independent samples. Tests for differences in location.

Key words: Histogram. Long–tailed data. Normal plot. Two sample t–test.

Data File: `sex.dat`

In the case "Reporting of sexual partners by men and women," we found that the distributions of the numbers of opposite–sex sexual partners reported by men and women were very right–skewed. However, based on an appeal to the Central Limit Theorem, it was claimed that the t–tests and the confidence intervals based on those distributions were still quite accurate.

In this case we will conduct an experiment on a computer to check if that claim is correct.

The tests are based on a normal distribution for the sample averages (i.e., $\overline{X}_{\texttt{BMALE}}$ and $\overline{X}_{\texttt{FEMALE}}$) rather than the normal distribution of the values themselves. These averages are based on very large sample sizes (1682 and 1850, respectively). As a result, even though the distribution of the values is quite non–normal, the distribution of the sample averages could be quite close to normal. To check this we will conduct an experiment by randomly sampling from the distribution of samples and calculating the averages of those samples and look at the distribution of the sample means.

Draw a simple random sample with replacement of size 1682 from the population defined by `BMALE`. Calculate the average of this sample. This average is a single observation from the distribution of the averages of `BMALE`. Is it close to the average of `BMALE` (10.655)?

Repeat this process 100 times, each time calculating the average of the sample and storing the result in a variable called `AVEBMALE`. These 100 values are then a random sample from the distribution of the sample mean of the population defined by `BMALE`. The question then is how normally distributed these values are. We can use the same data analysis tools to analyze the results of our computer experiment as we used to analyze the original data.

Plot a histogram of `AVEBMALE`. Does it look normal in shape? Create a normal plot of `AVEBMALE`. Is it clearer from this more focused tool?

Calculate some descriptive statistics for `AVEBMALE`. Is the mean value close to the sample mean of the population defined by `BMALE`? Is the standard deviation close to the value of the standard error of the mean of `BMALE` in the original case ($22.256/\sqrt{1682} = 0.543$ partners)?

Do you trust the result of the t–test? Do you think that the confidence intervals calculated in the previous case are accurate?

Conduct a similar computer experiment for the number of male partners reported by the females (`FEMALE`). How valid are the results here?

Comments to the Instructor

Obviously, this case requires the availability of software that is capable of generating multiple samples with replacement from a given sample. It is fairly straightforward to write such code, if no software package is available to do it.

Considering the large sample sizes being considered here, it will undoubtedly be the case that the distribution of sample means will look quite Gaussian, with mean close to the average of the variable from which it was sampled, and with standard deviation close to the theoretical standard error of the mean.

Note, by the way, that this exercise differs from the cases "The Central Limit Theorem for census data," "The sampling distribution for the median" and "The sampling distribution for the standard deviation" in that we are using an observed sample to try to understand the sampling distribution of a statistic, rather than using the more obvious (but generally unavailable) population. This is an example of a use of the **bootstrap**, first introduced by Bradley Efron in 1979.

To maintain consistency with the unannotated edition,
the next case begins on the following page.

Subgroups in the "Old Faithful" geyser data

Topics covered: Comparison of means for two group data.

Data File: `geyser1.dat`

Our earlier investigation of the intereruption times of the "Old Faithful" geyser ("Eruptions of the 'Old Faithful' geyser") showed that there were two subgroups in the data. Further, these subgroups could be usefully defined by whether the duration of the previous eruption was less than or greater than three minutes.

Determine whether the average intereruption times are statistically significantly different from each other, based on whether the duration of the previous eruption was less than or greater than three minutes.

What does this say about the simple prediction rule proposed for prediction of intereruption times in the earlier case?

Would you say that the observed difference in average intereruption times between the two groups would be of practical importance to tourists (for example), irrespective of its statistical significance?

Comments to the Instructor

The use of the simple prediction rule proposed in the case "Eruptions of the 'Old Faithful' geyser" is strongly supported when investigating subgroups in the data. We separate the data by whether `DURATION` is greater than, or less than, 3 minutes. Then, two–sample t–tests that test whether the means of `INTERVAL` for the two groups are equal are $t = 24.12$ on 220 degrees of freedom (equal variances) and $t = 25.00$ on 136.3 degrees of freedom (unequal variances), respectively. It appears that the variance of `INTERVAL` is similar in the two groups. As was noted in the case "Eruptions of the 'Old Faithful' geyser," the mean intereruption time differs by about 25 minutes for the two groups, which is certainly of practical importance to tourists.

There are apparently a few outliers in the two groups, but the nonparametric tests still strongly reject the hypothesis that the distribution of `INTERVAL` is the same in the two groups. The median test takes on a chi–squared value of 86.95 on one degree of freedom, while the rank sum test has a z–score of 11.47.

Thus, it is clear that the intereruption time is significantly different based on the duration of the previous eruption.

Mortgage rates for different types of mortgages

Topics covered: Comparison of location for two groups.

Data File: `mort.dat`

The single most important property of a mortgage from the point of view of a potential home buyer is the interest rate being charged. Having said this, it is certainly true that the decision of what bank to apply to for a mortgage is more complicated than just looking at the interest rate being charged. In particular, there are two basic types of mortgages: a fixed rate mortgage, in which the interest rate is fixed for the entire length of the mortgage, and an adjustable rate mortgage, in which the interest rate can vary in the future (within certain prescribed limits on its movement). Given that an adjustable rate mortgage is a riskier choice for a home buyer, one might expect that the interest charged would be lower. Is this the case?

The data set for this case represents information about mortgages in the New York metropolitan area as reported in *Newsday* on March 27, 1993, as given below:

	Lender	Type	Rate
1	Anchor Savings	Fixed	7.875
2	Bank of the Hamptons	Fixed	7.125
3	Bay Ridge Federal	Adjustable	5.000
4	Brooklyn Federal	Adjustable	4.500
5	Chemical Bank	Fixed	7.500
6	Citibank	Fixed	7.125
7	Columbia Federal	Fixed	7.500
8	Countrywide Funding	Fixed	7.500
9	Dale Mortgage Bankers	Fixed	6.750
10	EAB	Adjustable	6.000
11	East NY Savings Bank	Adjustable	5.250
12	Fairmont Funding	Fixed	7.250
13	GMAC Mortgage	Adjustable	4.250
14	Home Federal Savings	Fixed	7.250
15	Maspeth Federal	Fixed	8.000
16	Ridgewood Savings	Adjustable	4.500
17	Roosevelt Savings	Fixed	8.000
18	Roslyn Savings Bank	Fixed	6.875
19	The Mortgage Corner	Fixed	7.375
20	Triumphe Financial	Fixed	6.875

Is there a significant difference in interest rate charged for the two types of mortgage? Can you think of any other factors that might be related to the interest rate charged?

Do you think that the observed differences would be considered important by a potential home buyer?

Comments to the Instructor

Fixed rate mortgages average an interest rate of 7.357%, while adjustable rate mortgages average 4.917%. Over the course of 15 to 30 years, this would translate into many thousands of dollars less in interest paid for adjustable rate mortgages. Of course, this is not really a fair comparison, since the actual interest paid for an adjustable rate mortgage could be higher or lower than is implied by the initial interest rate, since the rate can change over the life of the loan.

The observed difference is, in fact, statistically significant. Two–sample t–tests equal $t = 10.35$ on 18 degrees of freedom (equal variances) and $t = 8.57$ on 6 degrees of freedom (unequal variances). Unequal variances are weakly indicated for these data.

The nonparametric tests agree that the distributions of rate differ for fixed and adjustable rate mortgages. The median test equals a chi–squared value of 9.00 on one degree of freedom, while the z–score for the rank sum test is 3.42.

There are many other factors that would go into choosing a mortgage bank than merely the interest rate. Different lenders charge very different application fees and closing costs; some banks offer mortgages at different rates based on the extent to which they examine the credit history of the applicant; banks may offer discounts to their own customers; and so on.

To maintain consistency with the unannotated edition,
the next case begins on the following page.

Condom use and the prevention of AIDS

Topics covered: Comparison of Binomial proportions.

Key words: Confidence interval. z–test.

A controversial public health issue these days is whether governmental organizations should encourage the use of condoms in the general population, in order to reduce the spread of AIDS. This is clearly a political, religious and cultural issue, but it is also a statistical one — does the use of condoms in fact reduce the risk of HIV infection?

Two studies undertaken during 1991 through 1993 attempted to address this question. In each study, heterosexual couples, where one (and only one) partner was infected with the HIV virus, were followed during the period of the study. In the Saracco study group, in Italy, of 171 couples who always used condoms, 3 partners became infected with HIV, while of 55 couples who did not always use condoms, 8 partners became infected. In the European study group organized by Dr. Iona DeVincenzi, of 123 couples who always used condoms, no partners became infected with HIV, while of 122 couples who did not always use condoms, 12 partners became infected. What do these two studies say about condom usage and new infection of HIV? That is, within each study, is there a statistically significant difference between the probability of being infected if the couple always uses condoms versus if the couple does not?

This is a two–sample problem, referring to Binomial proportions. We want to examine if the observed proportions in the two samples can be considered equal. Let $\overline{p}$ be the sample frequency estimate of p, the probability of new infection of HIV. We know that

$$var(\overline{p}) \approx \overline{p}(1-\overline{p})/n,$$

where "$\approx$" is representing "can be estimated by." This is all that is needed to construct a **two–sample z–test** for a comparison of Binomial proportions. It is similar to the two–sample t–test for the comparison of means that assumes unequal variances, except that we assume that the degrees of freedom are large enough so that we can use a normal approximation. Suppose we have two Binomial samples, with the observed proportions of new infection being $\overline{p}_1$ and $\overline{p}_2$, respectively, based on samples of size n_1 and n_2, respectively. These sample proportions are estimates of the true probabilities of new infection, p_1 and p_2, respectively. We wish to test the hypotheses

$$H_0 : p_1 = p_2$$

versus

$$H_a : p_1 \neq p_2.$$

The z–statistic then has the form

$$z = \frac{\overline{p}_1 - \overline{p}_2}{\sqrt{\overline{p}_1(1-\overline{p}_1)/n_1 + \overline{p}_2(1-\overline{p}_2)/n_2}}.$$

How does this testing procedure apply to the studies? For the Saracco study, the test statistic is

$$z = \frac{.1455 - .0175}{\sqrt{.0022605 + .0001005}} = 2.63,$$

which has a tail probability of .0085, while for the DeVincenzi study,

$$z = \frac{.0984 - 0}{\sqrt{.00072719 + 0}} = 3.65,$$

which has a tail probability of .0003. Thus, each study provides highly statistically significant evidence that condom use is associated with lower risk of contracting HIV (note, by the way, that we cannot definitely say that condom use *causes* lower risk of contracting HIV, because the level of condom use has not been assigned randomly to couples; it is possible that not using condoms could be associated with other activities that would increase the likelihood of HIV infection).

An equivalent way to approach this question is by using confidence intervals. Since the standard error of $\overline{p}_1 - \overline{p}_2$ can be approximated by

$$\text{SE}(\overline{p}_1 - \overline{p}_2) \approx \sqrt{\overline{p}_1(1 - \overline{p}_1)/n_1 + \overline{p}_2(1 - \overline{p}_2)/n_2},$$

a 95% confidence interval for the true difference in probabilities $p_1 - p_2$ has the form

$$\overline{p}_1 - \overline{p}_2 \pm (1.96)\sqrt{\overline{p}_1(1 - \overline{p}_1)/n_1 + \overline{p}_2(1 - \overline{p}_2)/n_2}.$$

If the 95% confidence interval does not contain zero, the hypothesis of equality of the two proportions can be rejected at the 5% level of significance. The 95% confidence intervals for the Italian and European study are (0.0328, 0.2232) and (0.0455, 0.1513), respectively. These results confirm that the difference in proportions is deemed statistically significantly different from zero. As zero lies outside both the intervals, this indicates that the proportion of HIV infection is significantly different between the group always using condoms and the other group who did not always use condoms. The results of each study are consistent with the true difference in HIV infection probability being roughly between 3 to 4% and 15 to 20%.

On August 7, 1993, the Centers for Disease Control and Prevention, National Institute of Health, and Food and Drug Administration, in response to these results, issued a joint statement declaring that latex condoms, when properly used, are highly effective in blocking the spread of HIV.

Summary

Two European studies conducted during the period 1991–1993 provide evidence whether condom use prevents the transmission of the HIV virus. Both studies provide statistically significant evidence that couples who use condoms regularly have less of a chance of new HIV infection than couples that do not. This does not, however, address the question of whether lack of condom use causes increased chance of HIV infection, since other factors could be associated with both increased chance of infection and lack of condom use.

Technical terms

z**–test**: a hypothesis test, usually related to hypotheses concerning statements about location. This test is based on assuming that the statistic used to carry out the test has a normal distribution.

Air bags and types of automobiles

Topics covered: Comparing proportions. Contingency table. Tests of independence.

Key words: χ^2 test.

The presence of air bags in an automobile is often a key factor in the decision to purchase a particular model. The following data set refers to the 1993 model year and is based on data compiled by Robin Lock for the *Journal of Statistics Education,* volume 1 (1993), from data given in *Consumer Reports* and the *PACE New Car and Truck Buying Guide*. The data set was archived electronically on the `statlib` facility, accessible through the Internet. Automobiles can be classified into six categories, roughly corresponding to size: small, sporty, compact, midsize, large cars and vans. Is there a relationship between the type of vehicle and what level of air bag implementation is standard equipment?

Here is a cross–classification (**contingency table**) of these two variables for the 93 vehicles:

	Air bags			
Type	None	Driver only	Driver and passenger side	
Small	16	5	0	21
Sporty	3	8	3	14
Compact	5	9	2	16
Midsize	4	11	7	22
Large	0	7	4	11
Van	6	3	0	9
	34	43	16	93

If there was no relationship between the two factors, the observed counts would be fairly close to those expected assuming the factors were independent. This would mean that the probability distribution of airbag implementation does not depend on the type of vehicle. For example, the probability of a small car having no airbags would be the same as for a sporty car, or a van.

We can estimate the probabilities by the corresponding proportions and then compare them graphically. On the next page we have plotted the proportions of automobiles for each type of automobile that fall into the first two categories of airbag implementation. The proportion of each type with airbags on both sides is determined by the other two probabilities and the fact that they sum to one. If the two variables were independent we would expect the two lines to be flat.

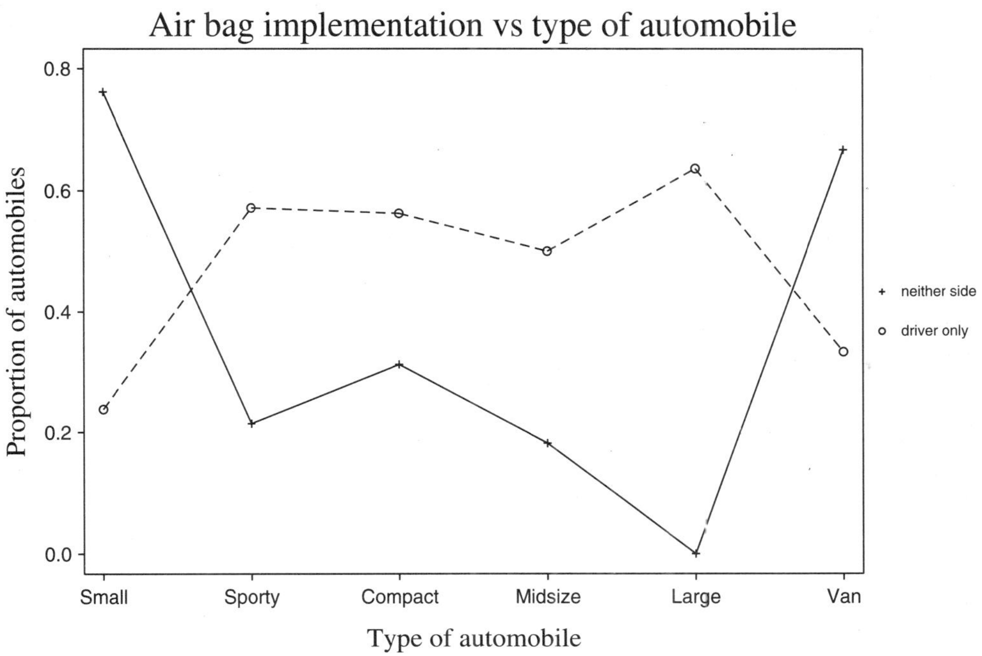

We see that the proportion of small cars and vans with no airbags is much larger than that for the other types. We also can see that as the type of car goes from small to large, the level of use of air bags as standard equipment generally becomes greater. This is not a safety issue (since larger cars would be expected to be safer in an accident), but rather a cost issue. Vans are different; they're fairly expensive (a bit less than midsize and large cars in general, but more than the smaller cars), but rarely have air bags as standard equipment (and none on the passenger side). Perhaps the different design of the vehicle itself (with people closer to the front of the vehicle) makes it more expensive to install air bags.

The expected counts under independence are as follows:

Air bags

Type	None	Driver only	Driver and passenger side	
Small	7.68	9.71	3.61	21
Sporty	5.12	6.47	2.41	14
Compact	5.85	7.40	2.75	16
Midsize	8.04	10.17	3.78	22
Large	4.02	5.09	1.89	11
Van	3.29	4.16	1.55	9
	34	43	16	93

We can see a pattern of deviation between the expected counts and the observed values in the table. How real are these patterns? We would expect to see variation from the expected values, regardless of whether or not there is a relationship. Are the deviations between the observed counts and the expected counts consistent with those that we would expect if the two factors were independent? To check this we can conduct a hypothesis test using either the Pearson or likelihood ratio χ^2 statistics. Both tests of independence are highly significant (the Pearson statistic $X^2 = 33.00$, $p = .0003$; the likelihood ratio statistic $G^2 = 39.49$, $p < .0001$). Thus, there is strong evidence that the association we can see in the table and the plot is real and not the result of chance variation.

Summary

The level of air bag protection provided as standard equipment in 1993 automobiles is directly related to the size of the car, with larger cars being more likely to provide air bags for both driver and passenger, and smaller cars being more likely not to offer air bags at all. This is apparently related to the cost of the car. Vans are different, in that they rarely have air bags as standard equipment, despite their relatively high cost.

Technical terms

χ^2 **test**: a test of the adequacy of the fit of a model to a given set of data, such tests are often called *goodness–of–fit* tests. The observed value, which could be determined using a Pearson statistic or a likelihood ratio statistic, is compared to a χ^2 distribution with the appropriate degrees of freedom.

Contingency table: a cross–classification of variables that are discrete (or categorical). The entries in a cell of the table correspond to the number of observations that share the characteristics defined by the given row and column.

A further look at the reporting of sexual partners by men and women

Topics covered: Comparing proportions. Contingency table. Independent samples. Tests of independence.

Key words: χ^2 test.

In the case "Reporting of sexual partners by men and women," we found that in data from the General Social Survey (GSS) men reported substantially more lifetime partnerships with women than women reported with men. Overall, men reported over 3 female partners for every 1 male partner reported by women. Our analysis suggested that much of the discrepancy could be a result of dramatic differences in the number and size of reports for those people with large reported numbers of lifetime partners. In this case we will explore the upper tail of the partnership distribution and the changes over time in the reports.

From the original case, we see that about 82% of the respondents report less than 10 partners in their lifetime, 90% report less than 20 and 97% report less than 40. To explore things further, let's partition the respondents into the people that report less than 10 lifetime partners and those that report 10 or more partners. This variable is a surrogate for the overall lifetime behavior. We would like to understand how this behavior relates to the discrepancy in reporting between men and women.

First, let's look at the total lifetime number of opposite–sex partners reported, cross–classified by gender and promiscuity:

	Respondent activity			
Gender	< 10	≥ 10	Total partners	Total respondents
Male	3240	16471	19711	1682
Female	3429	2717	6146	1850
Partners	6669	19188	25857	
Respondents	2888	644		3532

Note that the table entries are the total number of partnerships reported by each category of respondent, rather than the number of respondents itself. In that sense, the population being sampled from is that of *partners*, rather than *respondents*.

The highly active group corresponds to $644/3532 = 18.2\%$ of the respondents. However, as a group, they contributed $19188/25857 = 74.2\%$ of the reported partnerships. For all of the respondents, the ratio of the number of partnerships reported by men to that reported by women is

$$\frac{\text{Number of partnerships reported by men}}{\text{Number of partnerships reported by women}} = \frac{19711}{6146} = 3.21.$$

For the 81.8% of respondents in the less active group we have

$$\frac{\text{Number of partnerships reported by less active men}}{\text{Number of partnerships reported by less active women}} = \frac{3240}{3429} = 0.94.$$

For the 18.2% of respondents in the highly active group we have

$$\frac{\text{Number of partnerships reported by highly active men}}{\text{Number of partnerships reported by highly active women}} = \frac{16471}{2717} = 6.06.$$

Note that in these last two comparisons there is no reason for the ratio to be one in the population. People in the less active group can form partnerships with those in the highly active group, leading to a general balance. These ratios do, however, seem to suggest that there is little discrepancy in reporting in the less active group and a large amount in the highly active group.

We have a total of 25,857 partnerships reported in this sample. We know that 74.2% of the partnerships come from respondents in the highly active group. We also note that 23.8% of the partnerships were reported by females.

Is there a relationship between gender and activity level?

Does the gender of the respondent affect the likelihood of a partnership being reported by a highly active respondent rather than a less active respondent? For males, 83.6% of the partnerships come from respondents in the highly active group. For females, only 44.2% of the partnerships come from respondents in the highly active group. These proportions seem quite different, suggesting that the two factors are related.

Let's look at the counts we would expect to see in the table assuming that the two variables were independent of each other:

	Respondent activity			
Gender	< 10	≥ 10	Total partners	Total respondents
Male	5083.8	14627.2	19711	1682
Female	1585.2	4560.8	6146	1850
Partners	6669	19188	25857	
Respondents	2888	644		3532

There is a clear pattern of deviation from the counts observed in the sample. The number of partnerships reported by the highly active men is much more (16471) than expected if the variables were independent (14627.2).

How real are these patterns? We would expect to see variation from the expected values, regardless of whether or not there is a relationship. Are the deviations between the observed counts and the expected counts consistent with those we would expect if the two factors were independent? To check this we can conduct a hypothesis test using either the Pearson or likelihood ratio χ^2 statistics. Both tests of independence are highly significant (the Pearson statistic $X^2 = 3791.28$, $p <.0000$; the likelihood ratio statistic $G^2 = 3468.5$, $p <.0000$). Thus there is strong evidence that the association we can see in the table is real and not a result of chance variation.

As the GSS is conducted annually, there is a time factor to consider. Is the pattern of discrepancy the same from year to year? That is, are the differences we see common to each of the independently conducted surveys? Let's investigate this by looking at the cross–classification of the total number of partners by gender and year:

Gender	**Year** 1989	1990	1990	Total partners	Total respondents
Male	7675	6878	5158	19711	1682
Female	2125	1647	2374	6146	1850
Partners	9800	8525	7532	25857	
Respondents	1285	1092	1155		3532
Male/Female	3.612	4.176	2.173		

If the reported values were representative of the U.S. population in each year we would expect the ratio of male to female reported partners to be one. We see that the ratio in each year is substantially greater than one, although it does vary from year to year. This suggests that there is a relationship between the gender and year (i.e., the ratio changes from year to year). How real are the differences in these ratios? We would expect to see variation because they are based on samples, regardless of whether or not there is a relationship. Are the deviations between the ratios consistent with those we would expect if the two factors were independent? To check this we can conduct a hypothesis test using the χ^2 statistics. The tests of independence are highly significant; the Pearson statistic $X^2 = 366.32$ on 2 degrees of freedom ($p < .0000$), while the likelihood ratio statistic $G^2 = 355.57$ on 2 degrees of freedom ($p < .0000$). Thus there is strong evidence that the variation in the ratios is real and not a result of chance variation.

Our previous analysis suggests that much of the discrepancy is in the highly active group. Let's see if that is the case here also by breaking down the table further into less and highly active groups:

Respondent reporting less than 10 opposite–sex partners

Gender	**Year** 1989	1990	1990	Total partners	Total respondents
Male	1115	1025	1100	3240	1185
Female	1200	1012	1217	3429	1703
Partners	2315	2037	2317	6669	
Respondents	1033	884	971		2888
Male/Female	0.929	1.013	0.904		

The proportion of respondents that report less than 10 lifetime partners is $2888/3532 = 82\%$. Thus this test relates to most of the respondents. We see that the ratios for each year are quite close to one, and the variation is much less than for the full table. As before we can conduct a hypothesis test to see if these differences are real using the χ^2 statistics. The tests of independence are not significant; the Pearson statistic $X^2 = 3.76$ on 2 degrees of freedom ($p = 0.1527$), while the likelihood ratio statistic $G^2 = 3.76$ on 2 degrees of freedom ($p = 0.1527$). Thus there is only modest evidence that the variation in the ratios is not simply a result of random fluctuation.

Let's look at the highly active group:

Respondent reporting at least 10 opposite–sex partners

	Year				
Gender	1989	1990	1990	Total partners	Total respondents
Male	6560	5853	4058	16471	497
Female	925	635	1157	2717	147
Partners	7485	6488	5215	19188	
Respondents	252	208	184		644
Male/Female	7.092	9.217	3.507		

These ratios are quite variable, as confirmed by the χ^2 statistics ($X^2 = 398.43$ on 2 degrees of freedom, $p <.0000$; $G^2 = 375.69$ on 2 degrees of freedom, $p < .0000$). Thus the tests of independence are highly significant and there is a great deal of evidence that the variation in these ratios from year to year is not simply because of random fluctuation.

Summary

Further analysis of the data from a survey on sexual behavior better explains the discrepancy in the number of opposite–sex partners reported by men and women by examining the upper tail of the distribution of total lifetime partnerships. Among the 82% of respondents reporting less than 10 partners in their lifetime the ratio of male to female reports drops to 0.95:1, while for the most active 18%, the report ratio is 6.06:1 that men's partnerships are more likely to be from the highly active group than women's partnerships. A hypothesis test confirms this.

As the GSS data are collected each year we are able to compare the pattern of discrepancy for each of 1989, 1990 and 1991. In each year there was a large discrepancy, although the size of the discrepancy changes from year to year. A hypothesis test confirms this.

Finally, the data were decomposed into less and highly active groups. For the low activity groups the discrepancy was found to be small and not (statistically significantly) changing from year to year. In contrast, the high activity group retained a large discrepancy for each year, and the size of the discrepancy changed from year to year.

The discrepancy thus appears to be driven by the upper tail of the partnership distribution, suggesting errors and uncertainty in the reports by highly active people.

The implication is that sexual behavior surveys can produce reliable measures of sexual conduct for the great majority (80–85%) of survey respondents. Simple improvements in the data collection methods, however, could substantially enhance the usefulness and precision of the data: shorter retrospective time frames for more active respondents, and the collection of some life–cycle and network information. These improvements could increase reliability in measures of the upper tail of the contact distribution. The role of behavior and behavioral change in the spread of sexually transmitted diseases could then be systematically assessed.

Betting on professional football — can you beat the bookies?

Topics covered: Contingency table. Tests of independence.

Data File: `foot.dat`

Sports gambling is a (mostly illegal) multi–billion dollar industry in the United States. Despite pressure from sports leagues, newspapers routinely report odds, betting lines, over/unders, etc., before all major sporting events. Is it possible to beat the odds and make money by gambling on sports?

This analysis examines this question for professional (National Football League) football. For each game, oddsmakers in Las Vegas (where sports gambling is legal) determine a *point spread.* For example, in a game between the San Francisco 49ers and Kansas City Chiefs, the 49ers might be favored by 4 points. Most casual fans interpret this to mean that the oddsmakers are predicting that the 49ers will win by 4 points, but that's not quite what the point spread means; actually, this is the value such that the bets for the 49ers and the bets for the Chiefs will balance each other out (this is why the idea that "the bookies took a bath" when an upset occurs is fallacious — by balancing out the bets, the bookies can't possibly lose. It is only in very unusual circumstances, when a lot of money is bet at a disadvantageous point spread, that the bookies lose money.).

Anyway, back to the original question: is it possible to beat the bookies? Well, clearly, if you understand football better than the general betting public does, you can bet on games where you think the point spread has been set at an inappropriate value, and make money. But are there any patterns in past games that might help us develop some general rules for betting? For example, one bit of folk wisdom is that you should always bet on home underdogs (the idea being that a home team will do better than expected because of home field familiarity, friendly fans, etc., so a less–talented team will make the game closer than expected). Is that really a good idea?

The data come from all regular season National Football League games played in the 1989, 1990 and 1991 seasons (a total of 672 games). The following variables are examined:

```
BETTING  : Favorite covers (wins by more than point spread):  1
           "Push" (favorite wins by exactly point spread)   :  0
           Favorite doesn't cover (wins by less than the
           point spread, or loses)                          : -1
DAY      : Sunday afternoon (0), Sunday night (1),
           Monday night (2), Thursday (3), Saturday (4)
HOMEAWAY : Favored team is at home (1) or away (0)
WEEK     : 1, 2, ..., 17
YEAR     : 1989, 1990 or 1991
```

These data were compiled by Professor Hal Stern of the Statistics Department at Harvard University. The data set was submitted by Robin Lock and archived electronically on the `statlib` facility, accessible through the Internet.

These variables are categorical, and we are interested in the question of whether there is any association between them; thus, we are interested in tests of independence. Perform the appropriate tests of independence to see if the betting result (the variable `BETTING` above) is related to whether the favorite is at home or away, the year, the week that the game was played, or the day of the week that the game was played. What do the results say about the possibility of beating the oddsmakers using simple prediction rules such as these?

Comments to the Instructor

The news is not too good for bettors. There is no evidence of any association between the betting result and any of the other available (simple) predictors. In particular, here are Pearson χ^2 values, with tail probabilities, for the tests of independence of `BETTING` and the other variables:

Variable	X^2	p
DAY	6.51	.590
HOMEAWAY	1.41	.494
WEEK	24.44	.828
YEAR	2.92	.571

Apparently the point spreads are set in a pretty reasonable fashion.

Formal investigation of the perceptions of the New York City subway system

Topics covered: Tests of independence.

Data File: `subway.dat`

In the cases "Perceptions of the New York City subway system: safety and cleanliness" and "Perceptions of the New York City subway system: other issues," the associations between various pairs of variables were examined informally. In this case, you will make these comparisons in a formal way.

Perform the appropriate tests of independence to see if meaningful pairs of variables are related to each other.

Examine pairs of variables whose associations were noteworthy in earlier cases, such as the cleanliness and safety variables. Also examine any pairs of variables that (based on informal comparisons) seemed to be unrelated to each other.

Do the formal tests of independence support your earlier informal impressions? Are there any potential problems in applying the usual tests of independence to these data?

Comments to the Instructor

The Pearson χ^2 goodness–of–fit statistic can be used to formally assess whether the observed joint distributions are significantly different from those assuming independence. As would be expected, the tests confirm the patterns noted earlier. Here is a selection of pairs of variables, with associated values of X^2 and p–values:

Variables		X^2	p
Cleanliness of stations	Cleanliness of trains	90.45	.000
Cleanliness of stations	Safety in stations	39.04	.001
Cleanliness of stations	Crowdedness of stations	89.29	.000
Safety in stations	Police presence in stations	56.40	.000
Convenience of schedule	Frequency of trains	86.89	.000
Ease of token purchase	Safety on trains	17.42	.359
Cleanliness of trains	Speed of trains	7.49	.824
Crowdedness of trains	On–train announcements	11.76	.761

It should be noted, however, that each of these constructed tables yields several cells with small fitted counts under independence. This could cast some doubt on the tail probabilities derived based on the χ^2 approximations.

Emergency calls to the New York Auto Club

Topics covered: Prediction. Residual plots. Simple regression. Tests of hypotheses.

Key words: F–statistic. Fitted value. Normal plot. p–value. Prediction interval. R^2. Residual. Scatter plot. Type I error.

Data File: `ers.dat`

The Automobile Club of New York provides many services to its members, including travel planning, traffic safety classes and discounts on insurance. The service with the highest profile is its Emergency Road Service (ERS). If a club member's car breaks down, the member can call the Club to send out a tow truck for assistance. This service is especially useful in the winter months, when Club members can be stranded with frozen locks, dead batteries, weather induced accidents and spinning tires.

If the weather is very bad, the Club can be overwhelmed with calls. By tracking the weather conditions the Club can divert resources from other Club activities to the ERS for projected peak days. This will lead to better service for Club members and also greatly reduce stress on the Club staff.

Are the number of calls that the Club will receive in a day predictable from the weather forecast given on the previous day?

We will investigate this with data from the second–half of January in 1993 and 1994. The Club reports the number of ERS calls answered each day (`CALLS`). We have also recorded the lowest temperature for the previous night (`LOW`), the highest temperature for the day (`HIGH`), and forecasts of both of these from the previous day (`FLOW` and `FHIGH`, respectively).

Here is a **time series** plot of `CALLS` for these two periods:

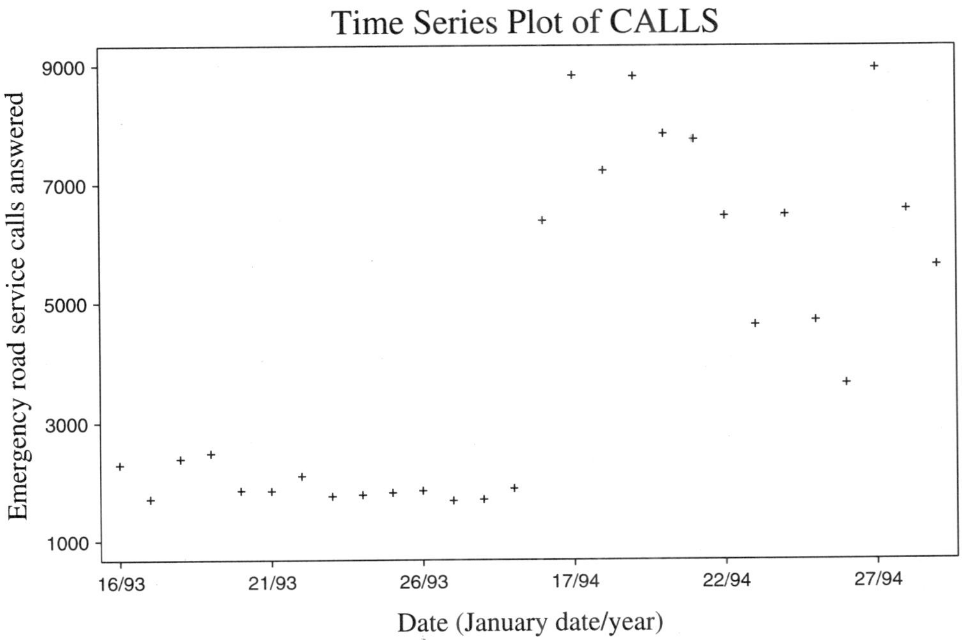

The first half of the plot (16/93 – 29/93) refers to values for 1993 and the latter half (16/94 – 29/94) refers to the corresponding period in 1994.

Several things are apparent from the plot. First, the number of calls in 1994 is much larger than the number of calls in 1993. The general level of calls is greater; the smallest number of calls in 1994 is far greater than the largest number of calls in 1993. Second, the variation in the number of calls from day to day in 1994 is far greater than the variation in the number of calls in 1993.

What could cause these differences? An obvious possibility is the low temperature for the day. In January, cold arctic air can sweep over the New York area bringing snow and freezing rain. Below freezing temperatures also reduce the effectiveness of car batteries.

Let's look at a scatter plot of the relationship between the number of calls and the low temperature:

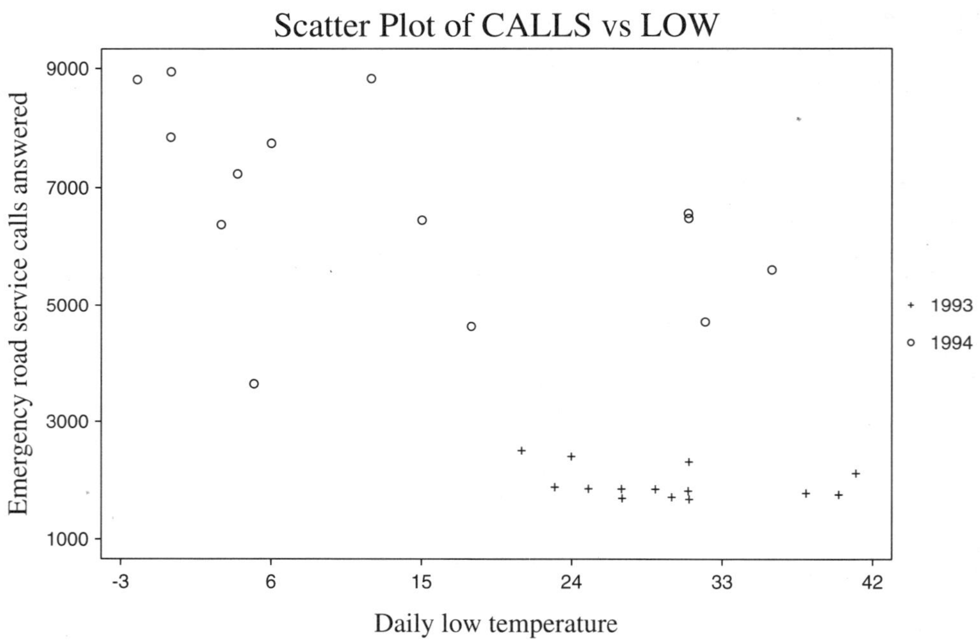

The values for 1993 have been represented by a + and those in 1994 have been represented by a ∘. It is clear that the temperature differences between the two years explain much of the difference in the number of calls. As the temperature decreases, the number of calls tends to increase. However, there is still a lot of variation.

Another factor is the cumulative effects of previous bad weather. Starting in mid–December 1993 a succession of heavy snow storms hit the New York area. The accumulated snow and ice did not melt away because the temperatures remained below normal. These cumulative effects, in addition to the daily low temperatures, could lead to a dramatic increase in the number of calls. In fact, this period in 1994 saw the largest number of calls in Club history: 167,492 in January 1994, compared with 64,132 in January 1993.

Our objective is to predict the levels of calls for the next day. It is not necessary to predict the total number (although that would be nice), but only to predict the level compared with other days in that month. The reason predicting the total number of calls is difficult is that it depends on the cumulative effects of the weather over time. This is difficult to measure and to take into account. An alternative is to model the percentage of the monthly ERS calls made on each day:

$$\texttt{PCALLS} = 100 \times \texttt{CALLS} \,/\, \texttt{TOTAL MONTHLY CALLS}.$$

For example, the percentages for 1/16/93 and 1/17/93 are 3.6% and 2.7%, respectively. This suggests that the resources necessary on the 17th would be about 75% = 2.7/3.6 of those necessary on the 16th. Thus 25% of those people working for the ERS on the 16th could be re–assigned or given a rest day. The advantage of considering PCALLS is that it adjusts for the total levels of calls for that month due to the cumulative effects of weather.

We want to be able to predict the usage for the next day. Hence, we should use the forecasted low temperature rather than the actual low temperature, because the latter number will not be available the previous day. The forecasted temperature and the actual temperature should be closely related. Let's look at a scatter plot of the relationship between the percentage of calls and the forecasted low temperature:

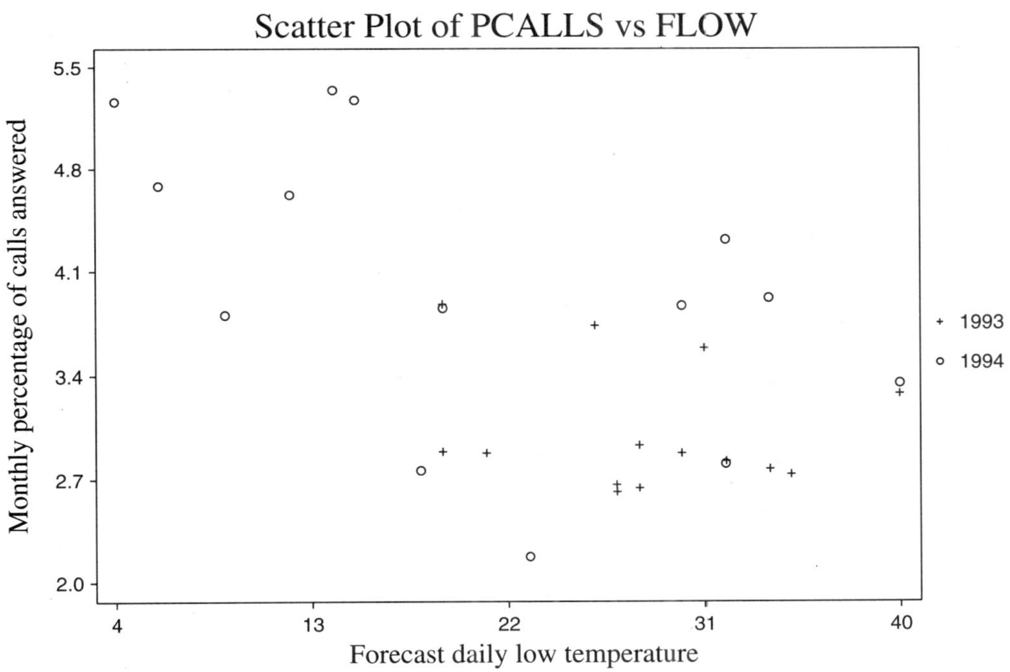

The values for 1993 have been represented by a + and those in 1994 have been represented by a ○. As the forecasted temperature decreases, the percentage of calls tends to increase. Although there is still a lot of variation, the relationship appears to be more linear than the relationship for the number of calls itself.

Let's do a **least squares regression** with PCALLS as the dependent (target) variable, and FLOW as the independent (predicting) variable:

```
LINEAR REGRESSION OF PCALLS  Monthly percentage of calls answered

PREDICTOR
VARIABLES     COEFFICIENT    STD ERROR     STUDENT'S T       P
---------     -----------    ---------     -----------     ------
CONSTANT         4.78953      0.38930         12.30        0.0000
FLOW            -0.05230      0.01481         -3.53        0.0016

R-SQUARED              0.3242    RESID. MEAN SQUARE (MSE)     0.56787
ADJUSTED R-SQUARED     0.2982    STANDARD DEVIATION           0.75357

SOURCE        DF        SS             MS          F          P
----------    ---    ----------    ----------    -----     ------
REGRESSION     1      7.08210       7.08210      12.47     0.0016
RESIDUAL      26      14.7645       0.56787
TOTAL         27      21.8466

CASES INCLUDED 28   MISSING CASES 0
```

The strength of the regression is quite modest, with an R^2 of .32, but the F–statistic is significant. There is certainly a relationship between the forecasted low temperature and the percentage of monthly calls. The entry given under "STANDARD DEVIATION" is the *standard error of the estimate*, and is an estimate of the standard deviation of the error term. Thus, we would estimate that roughly 95% of all daily percentages would be within $\pm 2 \times .75 = 1.50\%$ of the population regression line. Given that the original range of PCALLS was roughly $2.0 - 5.5\%$, this reinforces the modest strength of the observed regression relationship.

To see if the assumptions underlying the regression model are correct, let's look at some diagnostic plots. Here is a plot of standardized residuals versus fitted values:

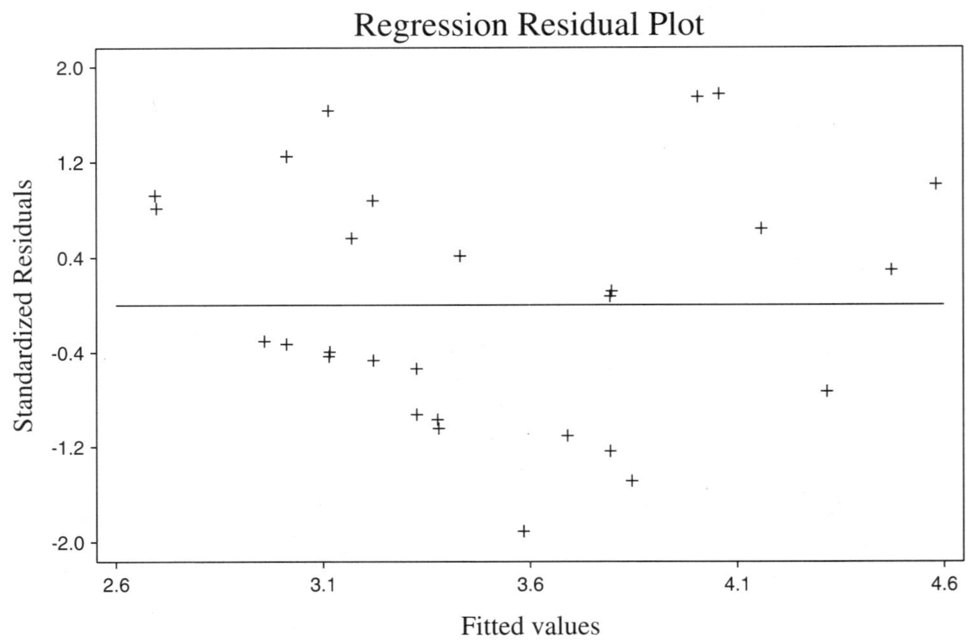

While there are no outlying values apparent in the plot, nor trends, it appears that the days with larger fitted values could have more variation than the days with smaller fitted values.

Here is a normal plot of the standardized residuals:

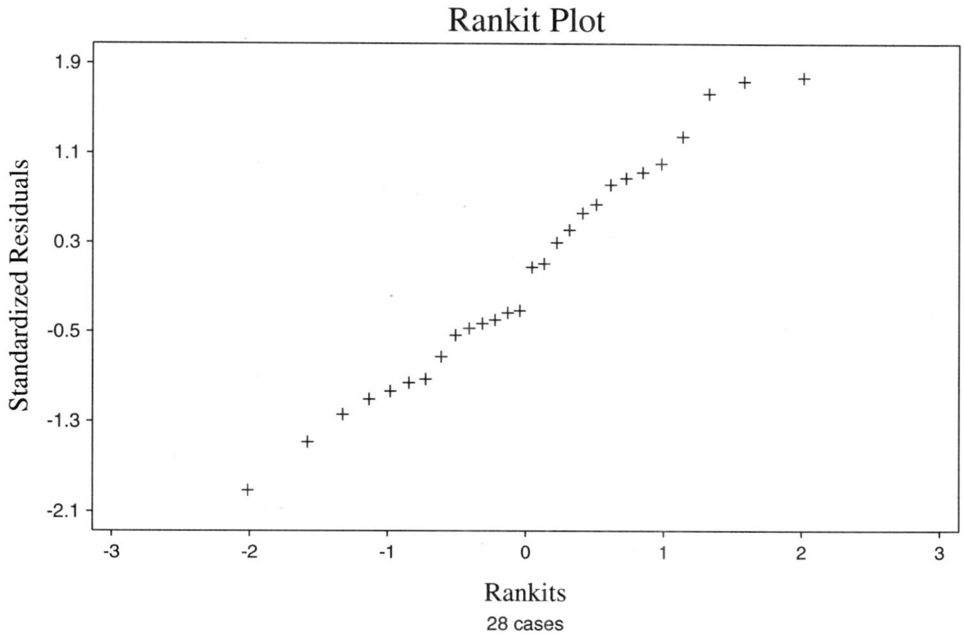

The standardized residuals appear to be roughly normally distributed. As a final diagnostic plot, let's look at a time series plot of the standardized residuals:

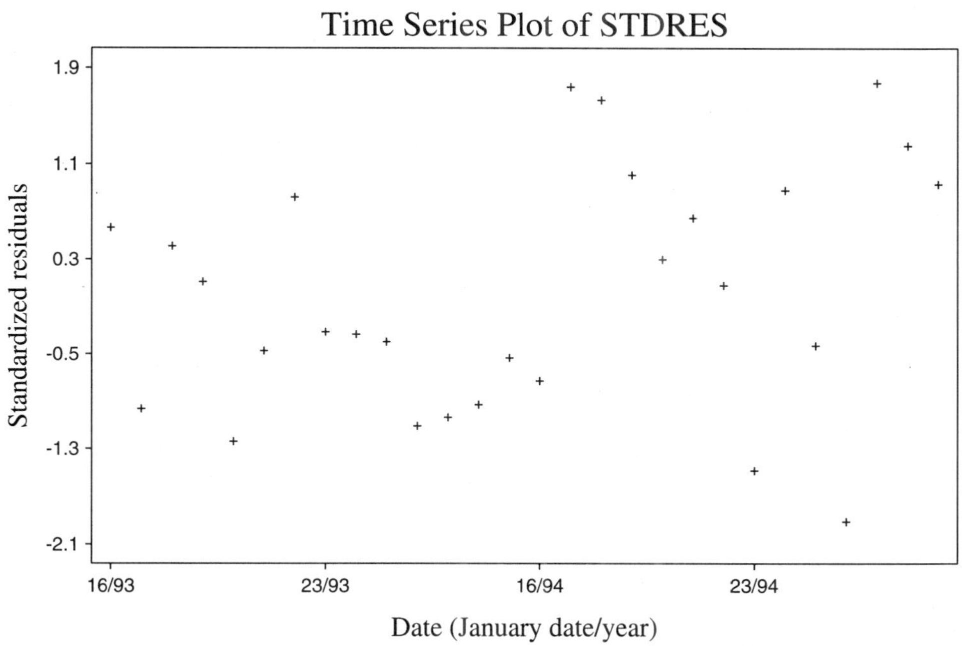

It appears that the variation in the residuals for the percentage usage in 1994 is greater than that for the values from 1993. However, the difference is not large, and we will not adjust for it here.

Thus, the regression assumptions, by and large, appear to hold. We can then interpret the coefficient of `FLOW` in the numerical output; each degree decrease in forecasted low temperature is associated with a 0.052% increase in the percentage of the monthly calls made on that day.

The busiest day of all in 1994 was January 27, when the daily low temperature was zero degrees and the ground was under six inches of snow. The Club answered 8947 calls, which was 5.3% of the monthly total (compared with 3.2% on the average day).

Could this have been predicted from the low temperature forecast only?

The forecast low for January 27 was 14 degrees. Let's predict the percentage of calls we would expect to be answered, given that the forecasted temperature was 14 degrees. To make the comparison fair, we refitted the above model with January 27 removed and then predicted the value of `PCALLS` based on the forecasted temperature:

```
PREDICTED VALUES OF PCALLS

LOWER PREDICTED BOUND      2.4034
PREDICTED VALUE            3.9486
UPPER PREDICTED BOUND      5.4939
SE (PREDICTED VALUE)       0.7503

PERCENT COVERAGE             95.0

PREDICTOR VALUES: FLOW = 14.000
```

The 95% prediction interval of `PCALLS` for a day with that forecasted low temperature is (2.40%, 5.49%). Thus, although the percentage of calls on January 27, 1994 was above the predicted level for such a day (`PCALLS` = 3.95%), the percent of calls is not inconsistent with the data as a whole. The Club can choose the confidence level to adjust how confident it wants to be to meet the demand and then staff accordingly. For example, if it wanted to be 95% confident of meeting the demand on that day, while also not assigning too many people to ERS, this interval suggests it should provide enough staff to meet between 2.4% and 5.49% of the monthly total. As the previous day's percentage was 2.17%, this suggests that the necessary resources would be about 253% = 5.49/2.17 of those necessary on the 26th.

Let's calculate a 93% prediction interval for January 27th:

```
PREDICTED VALUES OF PCALLS

LOWER PREDICTED BOUND      2.5284
PREDICTED VALUE            3.9486
UPPER PREDICTED BOUND      5.3688
SE (PREDICTED VALUE)       0.7503

PERCENT COVERAGE             93.0

PREDICTOR VALUES: FLOW = 14.000
```

The 93% prediction interval of `PCALLS` for a day with that forecasted low temperature is (2.53%, 5.37%). The observed value falls just inside the upper bound of this interval. Thus, we can be more precise; we would expect about 3.5% of days with a forecasted low temperature of 14 degrees to have a percentage monthly calls at or above the actual observed level.

Summary

The Automobile Club of New York would find it very useful to be able to forecast the demand for its Emergency Road Service (ERS) based on readily available information.

We can build a linear regression model for the daily percentage of the monthly number of calls to the ERS in terms of the daily low temperature forecasted the previous day.

We find that each degree decrease in forecasted low temperature is associated with a 0.052% increase in percentage of the monthly calls made on the next day.

Based on the model, we show that the busiest day of the year is under–predicted by the model. However, the model forecasts that about 3.5% of such days will have a percentage monthly calls at or above the observed level for that day. The model also predicts that the resources necessary to meet the demand on such a day would be at most 2.5 times the resources that were needed on the previous day, with 97.5% confidence.

Technical terms

F **–test**: a method used to assess the statistical significance of the linear relationship between the target variable and a set of predicting variables. The test is appropriate when the errors in the regression model can be considered to be a random sample from a (roughly) normal distribution.

Prediction interval: a confidence interval constructed for the predicted value of the target variable for a particular member of the population with given value(s) of the predicting variables. The interval has the usual t–based construction, using the standard error of the predicted value.

R^2 : an estimate of the proportion of the variability in the target variable accounted for by the regression. It equals the ratio of the regression sum of squares to the (corrected) total sum of squares.

Residual plot: a graphical test of the adequacy of regression assumptions. The residual for an observation is the difference between the observed target value and the value predicted by the fitted regression model (the **fitted value**). A **standardized residual** is a residual divided by its standard error,

$$\text{standardized residual} = \frac{\text{residual}}{\text{standard error of the residual}}.$$

The advantage of the standardized residuals is that, unlike the residuals, each one has the same variance. A successful fitting of a regression model should produce standardized residuals lying within $(-2.5, 2.5)$, roughly speaking, with no discernible pattern in their distribution. Thus, a scatter plot of standardized residuals versus fitted values should exhibit no apparent pattern. No regression analysis is complete without an examination of residuals.

Standard error of the estimate: an estimate of the standard deviation of the errors in a regression model. It equals the square root of the residual mean square, and can be used to assess the ability of the regression model to predict the target variable accurately.

Target variable: the goal of a regression model is to predict the value of a single **target** variable based on the value of one or more **predicting variables**. The ubiquity of the model has led to many synonym pairs, including dependent/independent, response/explanatory, and even simply Y/X.

Time series plot: a scatter plot of a variable versus time. The plot indicates (among other things) whether seasonal effects might be present in the variable.

Purchasing power parity: is it true?

Topics covered: Prediction. Residual plots. Simple regression. Tests of hypotheses.

Key words: F–statistic. Fitted value. Histogram. Normal plot. Outliers. p–value. Prediction interval. R^2. Regression. Residual. Scatter plot. t–statistic. Type I error.

Data File: `ppp.dat`

The principle of *purchasing power parity* (PPP) states that over long periods of time exchange rate changes will tend to offset the differences in inflation rate between the two countries whose currencies comprise the exchange rate. It might be expected that in an efficient international economy, exchange rates would give each currency the same purchasing power in its own economy. Even if it does not hold exactly, the PPP model provides a benchmark to suggest the levels that exchange rates should achieve. This can be examined using a **simple regression model**:

$$\begin{array}{c}\text{Average annual change in}\\ \text{the exchange rate}\end{array} = \beta_0 + \beta_1 \times \begin{array}{c}\text{Difference in average}\\ \text{annual inflation rates}\end{array} + \text{random error.}$$

The appropriateness of PPP can be tested using a cross–sectional sample of countries; PPP is consistent with $\beta_0 = 0$ and $\beta_1 = 1$, since then the regression model becomes

$$\begin{array}{c}\text{Average annual change in}\\ \text{the exchange rate}\end{array} = \begin{array}{c}\text{Difference in average}\\ \text{annual inflation rates}\end{array} + \text{random error.}$$

Data from a sample of 44 countries are the basis of the following analysis, which covers the years 1975 – 1990. The variable `CHNGEX75` is the target variable, and represents the estimated average annual rate of change of exchange rates (expressed as the country's currency per U.S. dollar), calculated as the difference of natural logarithms, divided by the number of years and multiplied by 100 to create percentage changes; that is,

$$\texttt{CHNGEX75} = \frac{\ln(1990 \text{ exchange rate}) - \ln(1975 \text{ exchange rate})}{15} \times 100\%.$$

(Based on the properties of the natural logarithm, this is approximately equal to the proportional change in exchange rate over all 15 years divided by 15, producing an annualized figure.) The variable `CHNGIN75` is the predicting variable, and represents the estimated average annual rate of change of the differences in wholesale price index values for the country versus the United States (with base year 1985 equaling 100). The data were derived and supplied by Professor Tom Pugel of New York University's Stern School of Business, based on information given in the *International Financial Statistics Yearbook*, which is published by the International Monetary Fund.

First, let's take a look at the data. Here are histograms of the two variables:

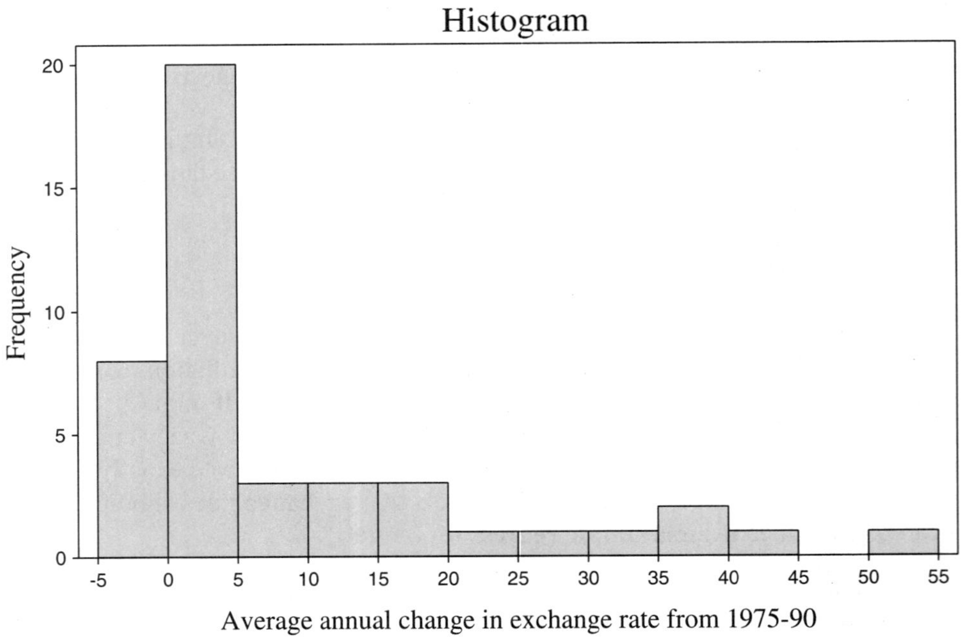

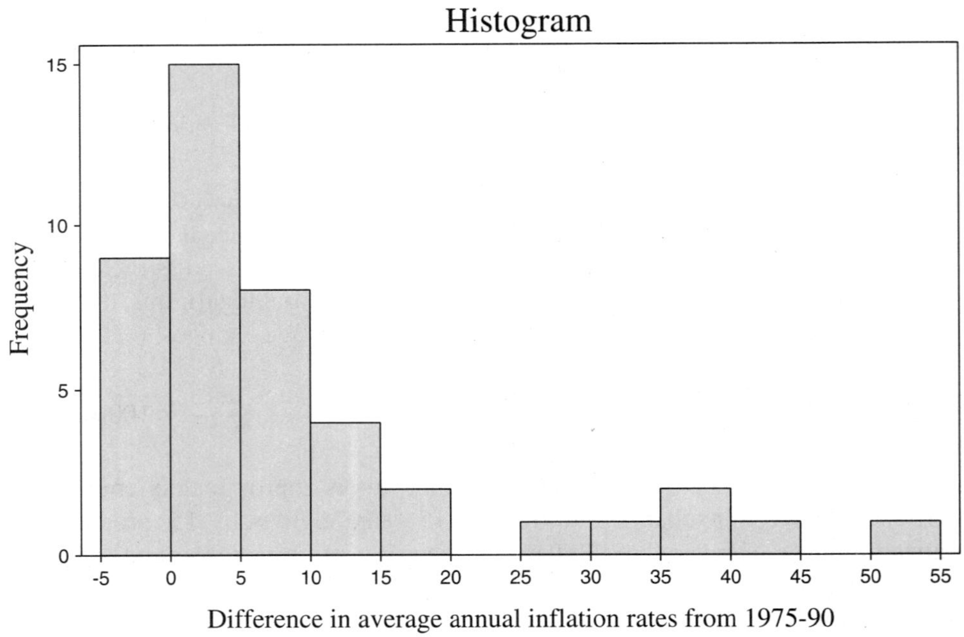

We notice that the distributions are right–skewed. In particular, six countries are somewhat distinct as having noticeably higher difference in inflation rates than the other countries. This need not cause problems, but sometimes suggests that the linear model being used is not appropriate. Let's look at a scatter plot of the two variables:

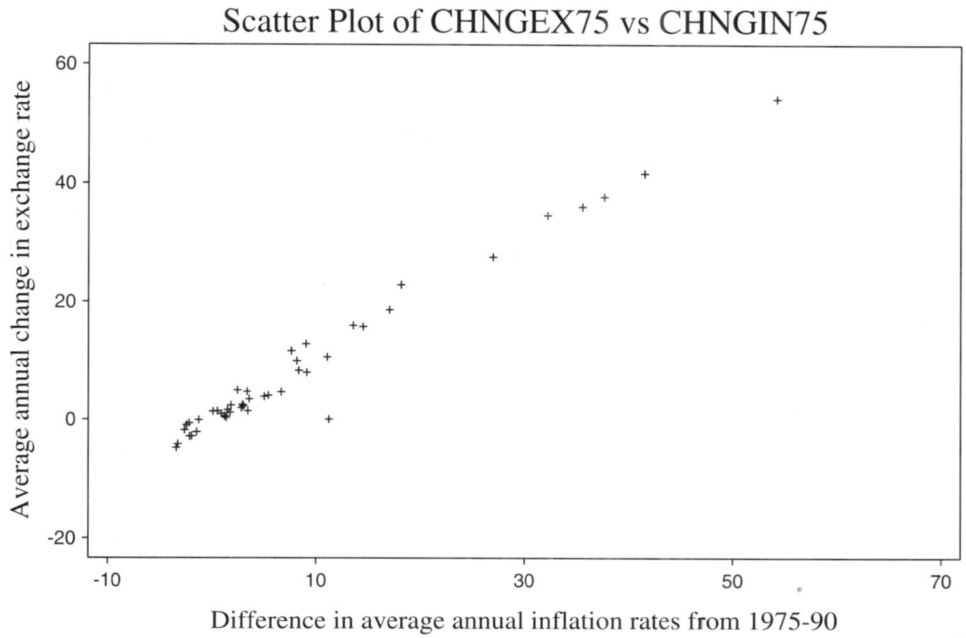

There appears to be a strong linear relationship here, except for one point that is clearly below the regression line. Let's ignore that for now and do a least squares regression with CHNGEX75 as the dependent (target) variable, and CHNGIN75 as the independent (predicting) variable:

```
LINEAR REGRESSION OF CHNGEX75 Average annual change in exchange rate

PREDICTOR
VARIABLES     COEFFICIENT    STD ERROR     STUDENT'S T        P
---------     -----------    ---------     -----------     ------
CONSTANT        -0.00928      0.42391         -0.02        0.9826
CHNGIN75         1.01759      0.02685         37.89        0.0000

R-SQUARED              0.9716      RESID. MEAN SQUARE (MSE)      5.49359
ADJUSTED R-SQUARED     0.9709      STANDARD DEVIATION            2.34384

SOURCE        DF       SS            MS          F         P
----------    ---   ----------    ----------   -----    ------
REGRESSION     1     7888.57       7888.57    1435.96   0.0000
RESIDUAL      42     230.731       5.49359
TOTAL         43     8119.30

CASES INCLUDED 44    MISSING CASES 0
```

The regression is quite strong, with an R^2 of .97, and the F–statistic is very significant. The t–statistic given in the output for CHNGIN75 tests the null hypothesis of $\beta_1 = 0$, which is strongly rejected. There is certainly a relationship between the change in exchange rates and the difference in inflation rates. Of course, it is not the hypothesis of $\beta_1 = 0$ that we're really interested in here. PPP says that $\beta_0 = 0$; the t–statistic for that hypothesis, listed in the line labeled CONSTANT, is given as –.02, which is of course not significant (that is, statistically significantly different from zero).

PPP also says that $\beta_1 = 1$; the t–statistic for this can be calculated manually:

$$t = \frac{1.01759 - 1}{.02685} = .655.$$

This is not significant at any reasonable Type I error level, so we do not reject this hypothesis either (the tail probability for this test is about .52). That is, these results **are** consistent with PPP (more correctly, since PPP represents the null hypothesis here, the results are not inconsistent with PPP).

The following simple regression scatter plot illustrates the use of **confidence intervals** and **prediction intervals**. Consider again the estimated regression line, CHNGEX75 = –.0093 + 1.0176 × CHNGIN75. Two possible uses of this line are: What is our best estimate for the true average CHNGEX75 for all countries in the population that have CHNGIN75 equal to some value? What is our best estimate for the value of CHNGEX75 for one particular country in the population that has CHNGIN75 equal to some value? Each of these questions is answered by substituting that value of CHNGIN75 into the regression equation, but the estimates have different levels of variability associated with them. The confidence interval (sometimes called the confidence interval for a fitted value) provides a representation of the accuracy of an estimate of the *average* target value for all members of the population with a given predicting value, while the prediction interval (sometimes called the confidence interval for a predicted value) provides a representation of the accuracy of a prediction of the target value for a *particular* observation with a given predicting value.

Simple Regression Plot

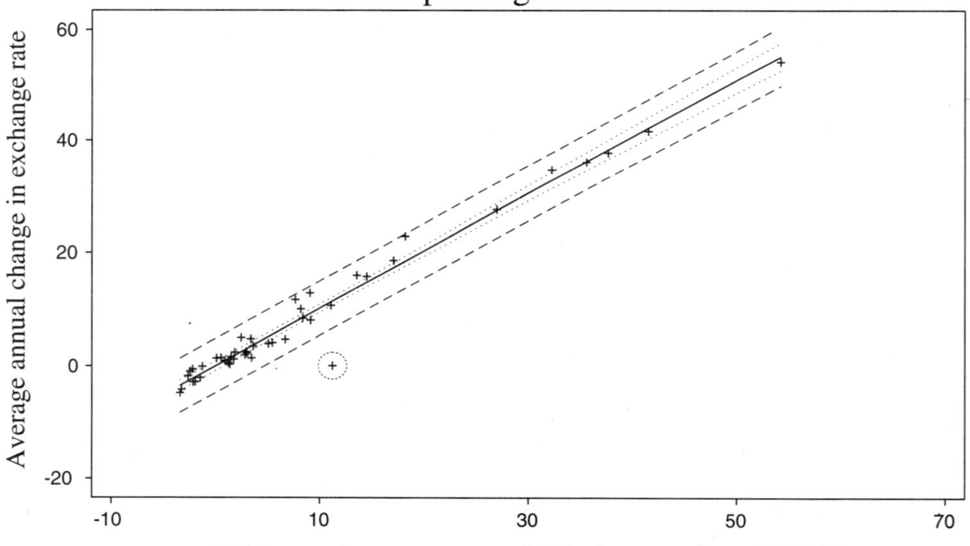

CHNGEX75 = -9.28E-03 + 1.0176 * CHNGIN75   95% conf and pred intervals

This plot illustrates a few interesting points. First, the pointwise confidence interval (represented by the inner pair of lines) is much narrower than the pointwise prediction interval (represented by the outer pair of lines), since while the former interval is based only on the variability of $\hat{\beta}_0 + \hat{\beta}_1 \times$ CHNGIN75 as an estimator of $\beta_0 + \beta_1 \times$ CHNGIN75, the latter interval also reflects the inherent variability of CHNGEX75 about the regression line in the population itself. Second, it should be noted that the interval is narrowest in the center of the plot, and gets wider at the extremes; this shows that predictions become progressively less accurate as the predicting value gets more extreme compared with the bulk of the points (this is more apparent in the confidence interval than it is in the prediction interval). Finally, it is clear that the point noted before is far outside the 95% prediction interval, indicating that it is very unusual.

Diagnostic plots also can help to determine whether this point is truly unusual, and to check whether the assumptions of least squares regression hold. Here is a plot of standardized residuals versus fitted values:

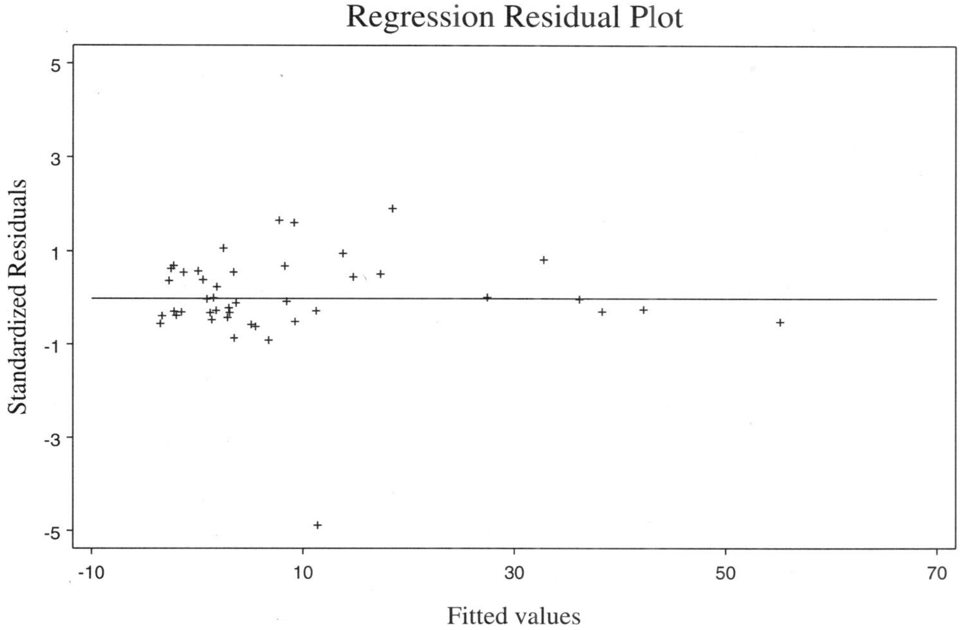

Here is a normal plot of the standardized residuals:

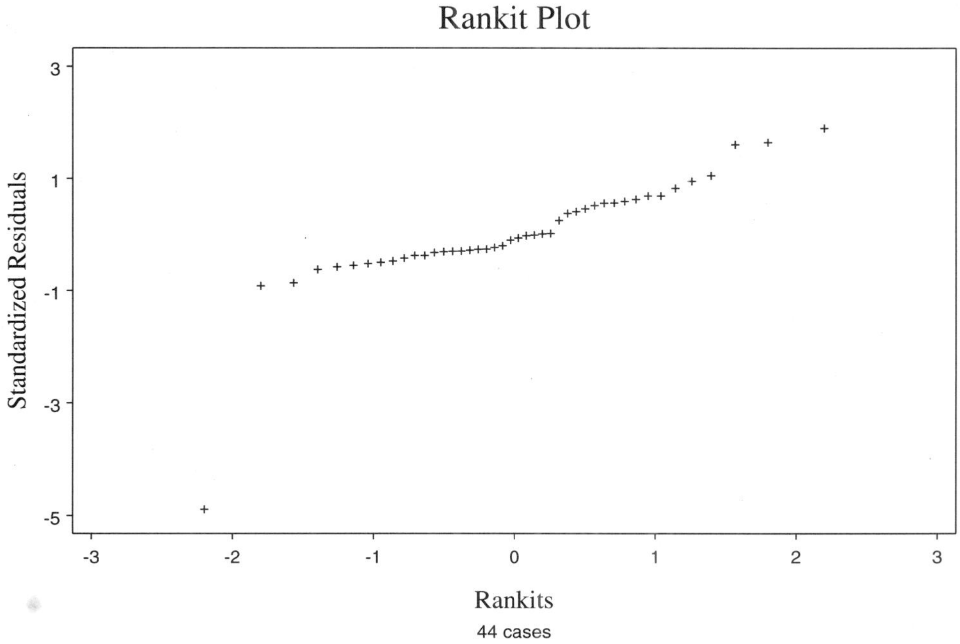

The one unusual case shows up in each plot. This is an outlier, and is a problem. An outlier can have a strong effect on the fitted regression, drawing the line away from the bulk of the points. It also can affect measures of fit like R^2, t–statistics, and the F–statistic. Finally, we can't very well claim to have a model that fits all of the data when this point isn't fit correctly!

Examination of the original data reveals that the outlier is Iran (case 21). The change in exchange rate for Iran is less than would have been expected, given the difference in inflation rates. Presumably this comes from two related issues: the overthrow of the government of the Shah of Iran in the late 1970's, which was replaced by a fundamentalist Islamic regime; and the U.S. – Iran hostage crisis in 1979–1980, which was followed by a period of tense relations between the two countries. It is not hard to imagine that the unusual situation in Iran over this period of time would lead to atypical behavior of exchange and/or inflation rate.

What should we do about this case? The unusual point has the potential to change the results of the regression, so we can't simply ignore it. We can remove it from the data set and analyze without it, *being sure to inform the reader about what we are doing*. That is, we can present results without Iran, while making clear that the implications of the model don't apply to Iran, or presumably to other countries with a similar political / economic situation.

Here are the results of the regression without Iran:

```
LINEAR REGRESSION OF CHNGEX75 Average annual change in exchange rate

PREDICTOR
VARIABLES     COEFFICIENT    STD ERROR      STUDENT'S T       P
---------     -----------    ---------      -----------     ------
CONSTANT         0.22161       0.28340          0.78        0.4387
CHNGIN75         1.02133       0.01785         57.22        0.0000

R-SQUARED                0.9876      RESID. MEAN SQUARE (MSE)      2.42514
ADJUSTED R-SQUARED       0.9873      STANDARD DEVIATION            1.55728

SOURCE          DF         SS            MS          F          P
----------      ---     ----------    ----------   -----      ------
REGRESSION       1      7940.26       7940.26    3274.15      0.0000
RESIDUAL        41      99.4306       2.42514
TOTAL           42      8039.70

CASES INCLUDED 43    MISSING CASES 0
```

We can see that the statistics haven't changed too much; the regression is a bit stronger. The two hypotheses we're interested in have similar results: $\beta_0 = 0$ is not rejected ($t = .78$), and $\beta_1 = 1$ is not rejected ($t = 1.19$). So, we still feel that PPP seems to be true over the 15 year period 1975–1990. The residual plots look better (actually, there is an indication of some structure in the standardized residual versus fitted values plot in relation to the highest fitted values; given the general strength of the regression fit, it is unlikely that eliminating this minor structure would have a very strong effect on the results for the entire data set, but we will examine those countries more closely in a later case), and the scatter plot with confidence and prediction intervals superimposed shows that the model fits the data well:

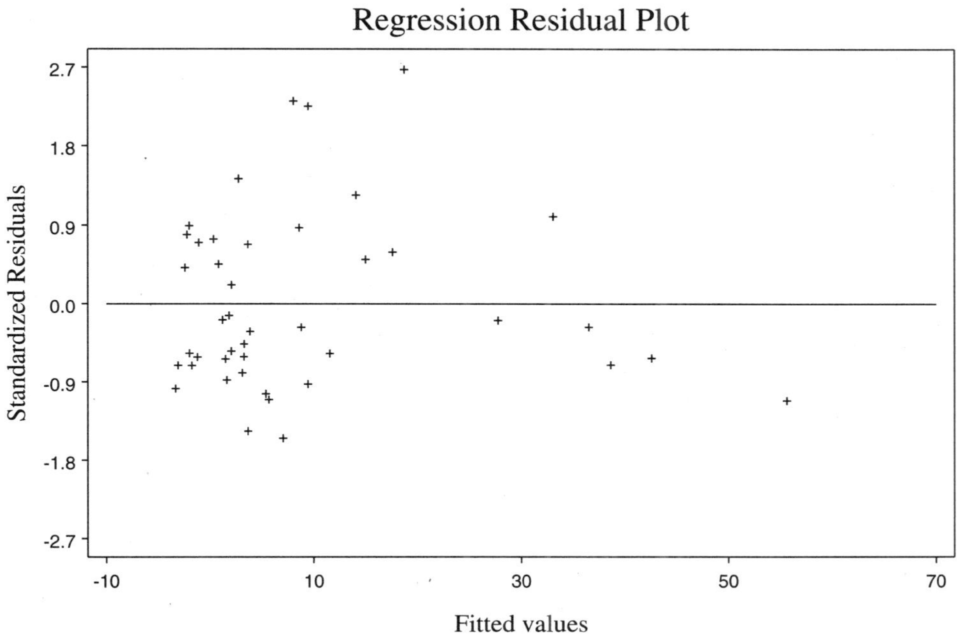
Regression Residual Plot
Standardized Residuals
2.7
1.8
0.9
0.0
-0.9
-1.8
-2.7
-10
10
30
50
70
Fitted values

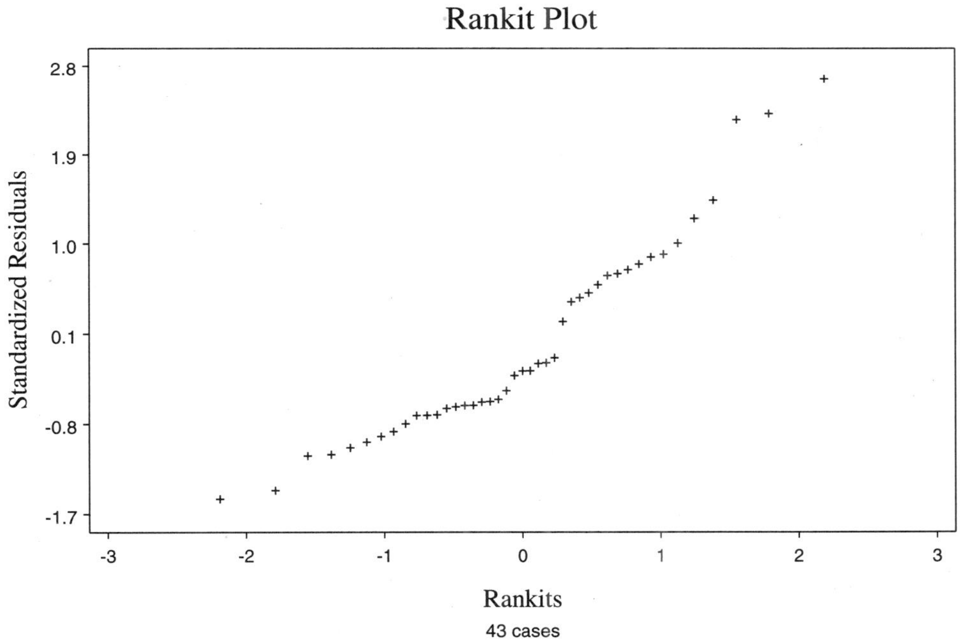
Rankit Plot
Standardized Residuals
2.8
1.9
1.0
0.1
-0.8
-1.7
-3
-2
-1
0
1
2
3
Rankits
43 cases

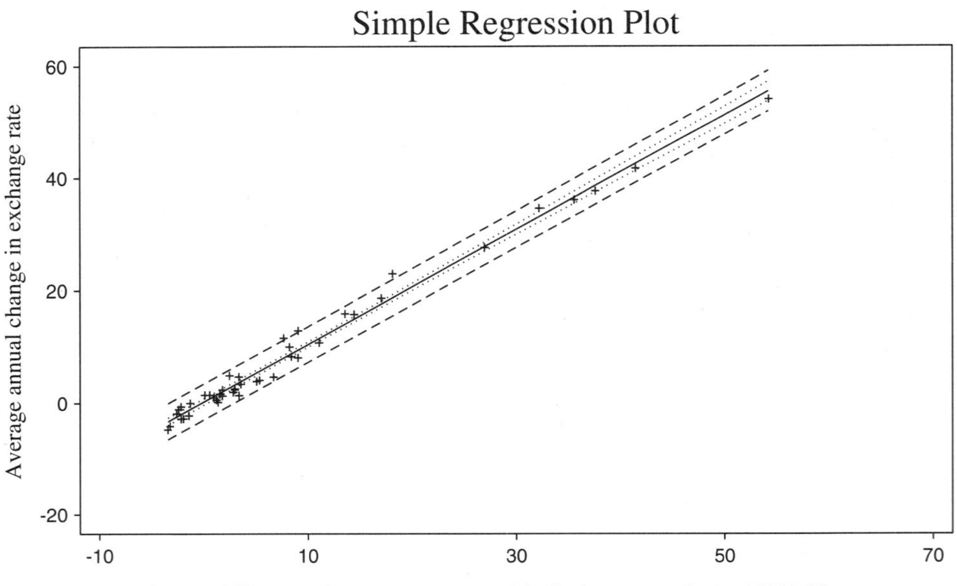

CHNGEX75 = 0.2216 + 1.0213 * CHNGIN75 95% conf and pred intervals

We can get an idea of the accuracy of the PPP model for predicting change in exchange rate using a prediction interval evaluated at a particular predicting value. What is the estimated change in exchange rate for a country whose annual difference in inflation rate is 15% (about what it is for Costa Rica, which has an average difference of 15.793%)? Here is output:

```
PREDICTED/FITTED VALUES OF CHNGEX75

LOWER PREDICTED BOUND     12.352    LOWER FITTED BOUND      15.010
PREDICTED VALUE           15.542    FITTED VALUE            15.542
UPPER PREDICTED BOUND     18.731    UPPER FITTED BOUND      16.073
SE (PREDICTED VALUE)      1.5793    SE (FITTED VALUE)       0.2630

PERCENT COVERAGE            95.0

PREDICTOR VALUES: CHNGIN75 = 15.000
```

"Predicted bound" refers to prediction of CHNGEX75 for one particular country with CHNGIN75 = 15, while "Fitted bound" refers to an estimate of the average CHNGEX75 value for all countries with CHNGIN75 = 15. So, our prediction interval for our Costa Rica–like country is (12.352%, 18.731%). In fact, the CHNGEX75 value for Costa Rica is 14.485%. Note that the interval given by the fitted bounds (on the right) of (15.010%, 16.073%) refers to an interval for the average CHNGIN75 value for all countries with CHNGIN75 = 15, so it is not what we want here.

Conventional wisdom says that while PPP should hold over the long run, it won't necessarily hold in the short–to–middle run. The reason for this is *overshooting*, where when there is a jump in the value of an asset, it is followed by a slow adjustment to equilibrium, in response to a change in expectations. This can lead to extreme volatility in exchange rates, and would correspond to a coefficient greater than one in the regression model. Here's a scatter plot for the variables referring to changes over the five–year period 1985–1990. We see that Iran is still an outlier, so we'll remove it from the start:

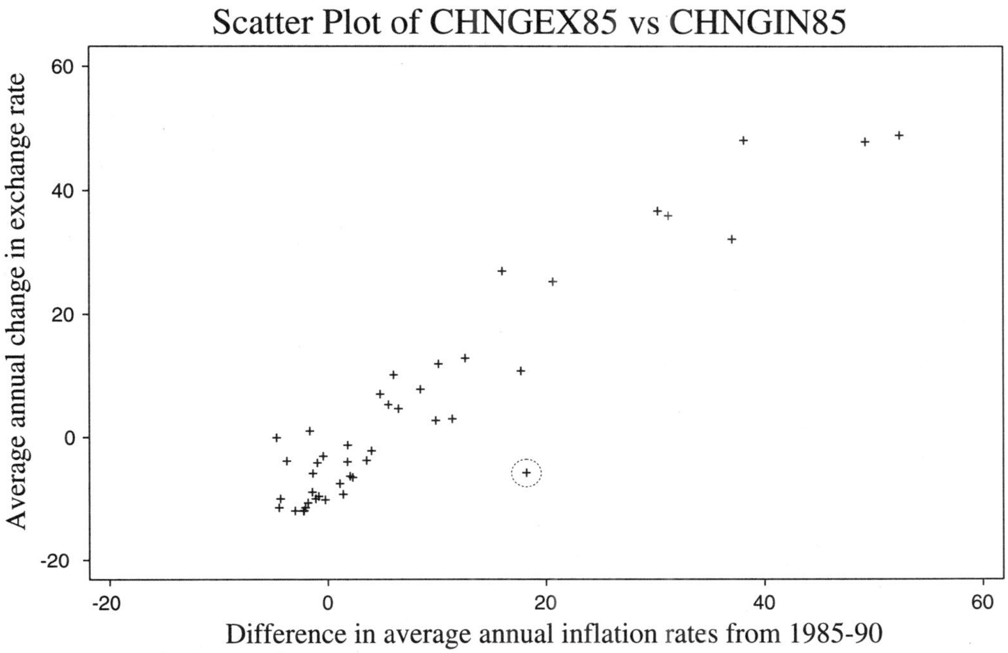

Here are the regression results:

```
LINEAR REGRESSION OF CHNGEX85 Average annual change in exchange rate

PREDICTOR
VARIABLES     COEFFICIENT    STD ERROR     STUDENT'S T       P
---------     -----------    ---------     -----------    ------
CONSTANT        -4.84027      0.92455         -5.24       0.0000
CHNGIN85         1.17444      0.05598         20.98       0.0000

R-SQUARED              0.9148      RESID. MEAN SQUARE (MSE)     27.8478
ADJUSTED R-SQUARED     0.9127      STANDARD DEVIATION           5.27711

SOURCE         DF        SS            MS          F         P
----------    ---    ----------    ----------    -----    ------
REGRESSION      1      12253.6       12253.6    440.02    0.0000
RESIDUAL       41      1141.76       27.8478
TOTAL          42      13395.4

CASES INCLUDED 43   MISSING CASES 0
```

We can see that the relationship between the change in exchange rate and the difference in inflation rate is still strong, although it is weaker than in the previous analysis. More interestingly, both of the hypotheses consistent with PPP are now rejected; the t–statistic for testing $\beta_0 = 0$ is –5.24 ($p < .00005$), and the t–statistic for testing $\beta_1 = 1$ is 3.12 ($p \approx .0033$). So, as we would have expected, PPP does not appear to hold over a five year period, although it does appear to hold over a fifteen year period.

The standardized residuals versus fitted values plot for this model looks adequate, although there is a suggestion of larger variability at larger fitted values (i.e., non–

constant variance). The normal plot suggests some non-normality, but since F– and t–statistics are insensitive to mild non–normality, this is unlikely to cause trouble:

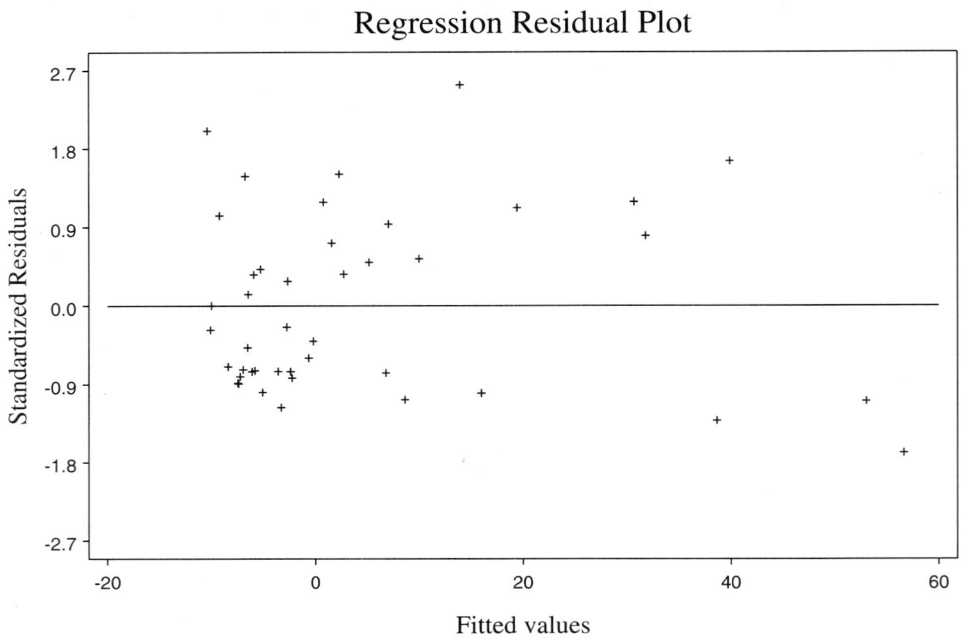

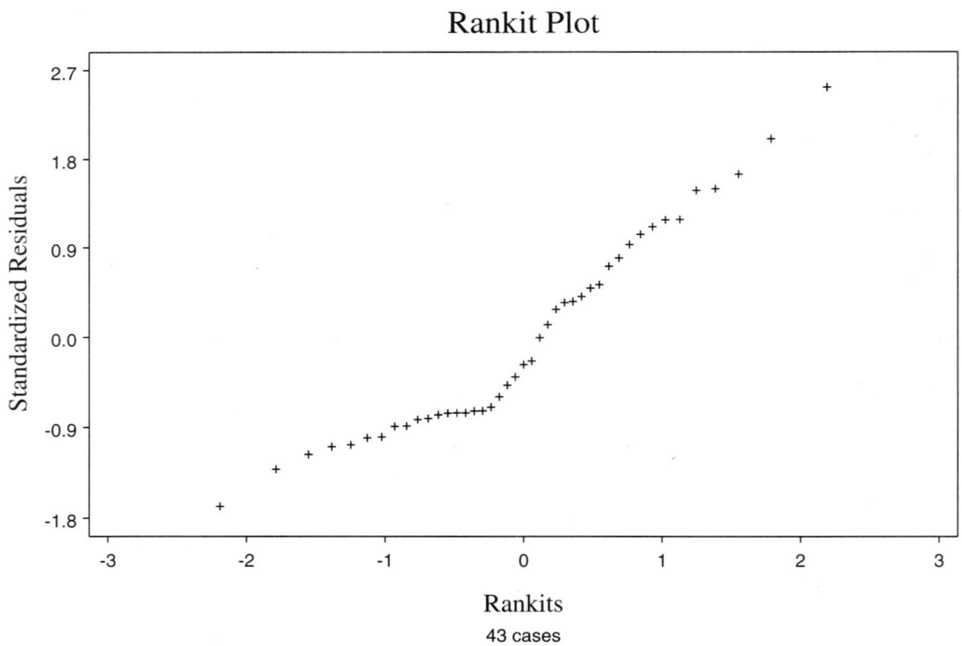

Summary

The principle of purchasing power parity (PPP) states that over long periods of time exchange rate changes will tend to offset the differences in inflation rate between the two countries whose currencies comprise the exchange rate. Using a sample of 44 countries, with the rates being defined over the fifteen year period 1975–1990, it was found that the data are consistent with PPP. An exception to this pattern is Iran, which had too low an average change in exchange rates for its average difference of inflation rates, probably reflecting its unusual political situation relative to the United States over this time period. PPP does not appear to hold, however, over the period 1985–1990, presumably because the five year time period is too short, and overshooting is occurring.

Technical terms

Confidence interval for a fitted value: a confidence interval constructed as an interval estimate of the *average* value of the target variable for *all* members of the population with given value(s) of the predicting variables.

Outlier: in the regression context, an observation whose observed relationship between the target and predicting variable(s) is different from that of the bulk of the observations; that is, a point far off the regression line. It is usually characterized by having a large absolute standardized residual.

PCB contamination of U.S. bays and estuaries

Topics covered: Prediction. Regression diagnostics. Residual plots. Simple regression. Tests of hypotheses. Transformation.

Key words: Cook's distance. Histogram. Leverage point. Masking effect. Outliers. Predicted value. Prediction interval. Regression. Residual. Scatter plot. Skewness. Transformation.

Data File: `pcb.dat`

Pollution of waterways is one of the most serious problems facing the world today. Billions of dollars have been spent on cleanup, tough antipollution laws have been passed, technological innovations have been sought to prevent pollution; still, the world is probably getting more, not less, polluted.

Pollutant levels in various bodies of water are important to study, to get a handle on the pollution problem. In particular, prediction of future levels of pollutants based on current characteristics is crucial for the determination of strategies to address pollution. The following analysis is based on 1984 and 1985 concentrations of PCBs (a family of hazardous industrial chemicals), measured in parts per billion, in samples from 37 U.S. bays and estuaries (data source: *Environmental Quality 1987 – 1988*, published by the Council on Environmental Quality).

Let's look at the data, starting first with histograms:

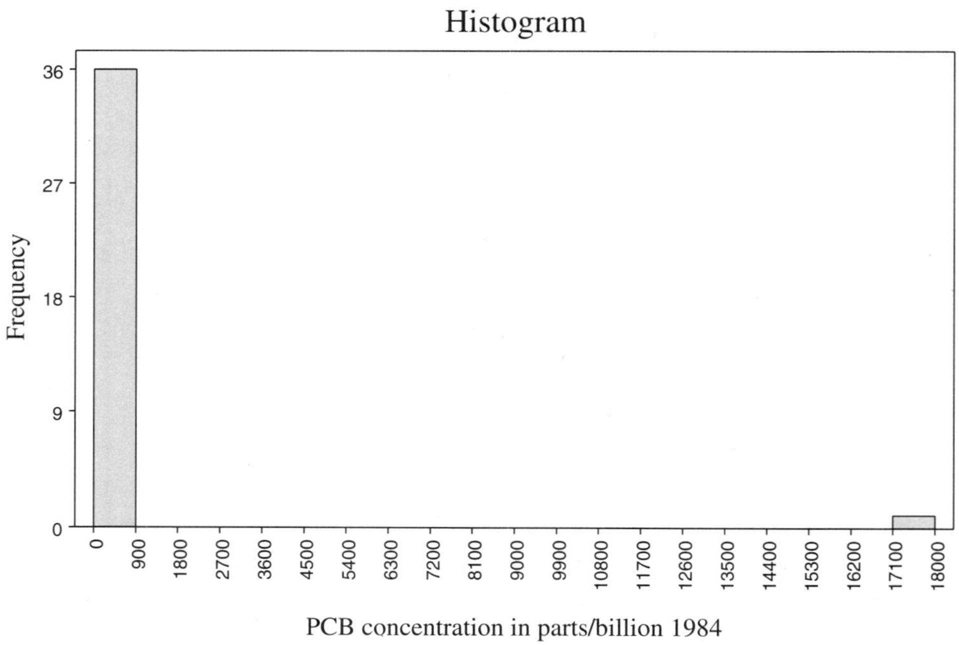

Histogram

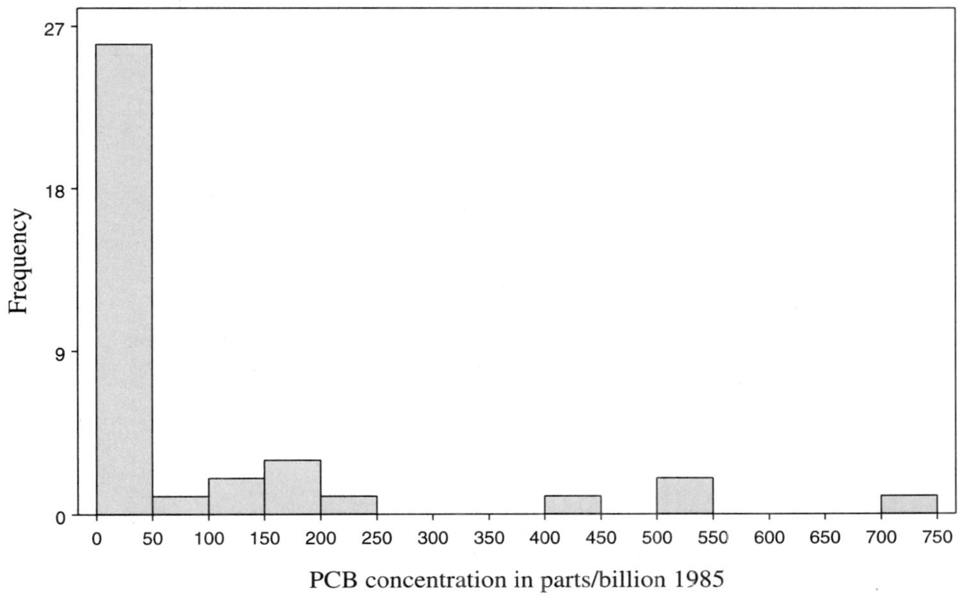

The distributions for both years have long right tails, and so some sort of transformation is probably going to be needed. We construct a scatter plot of the 1985 PCB level (PCB85) against the 1984 PCB level (PCB84) to see if there is a relationship between the levels from year to year:

Scatter Plot of PCB85 vs PCB84

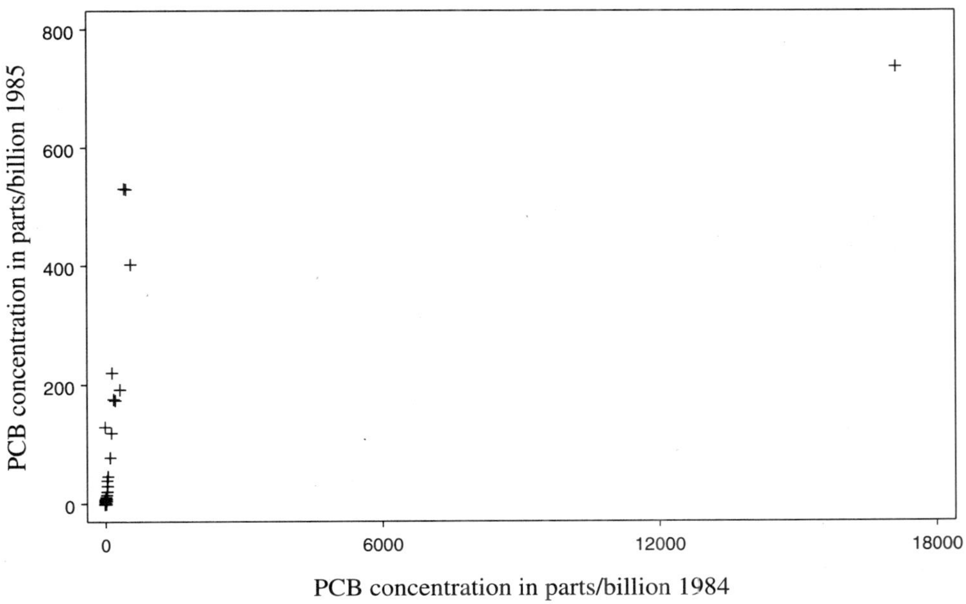

The plot is not very meaningful — we cannot see anything at all. Let's transform both variables by taking logarithms to the base 10. Since some of the observations are zero, we modify the data slightly by adding one; the new variables are LGPCB84 =

log(PCB84 + 1) and LGPCB85 = log(PCB85 + 1).

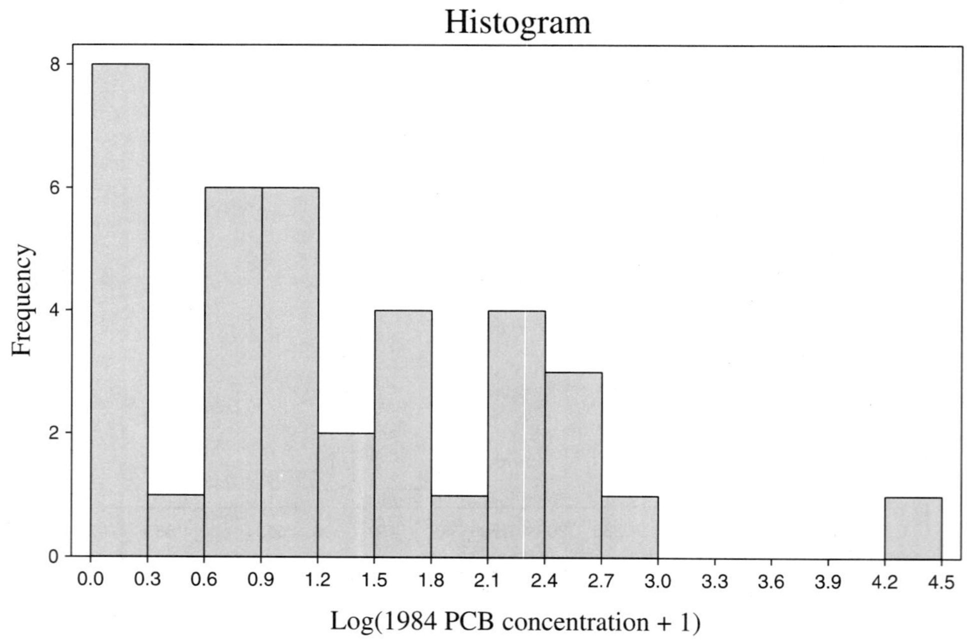

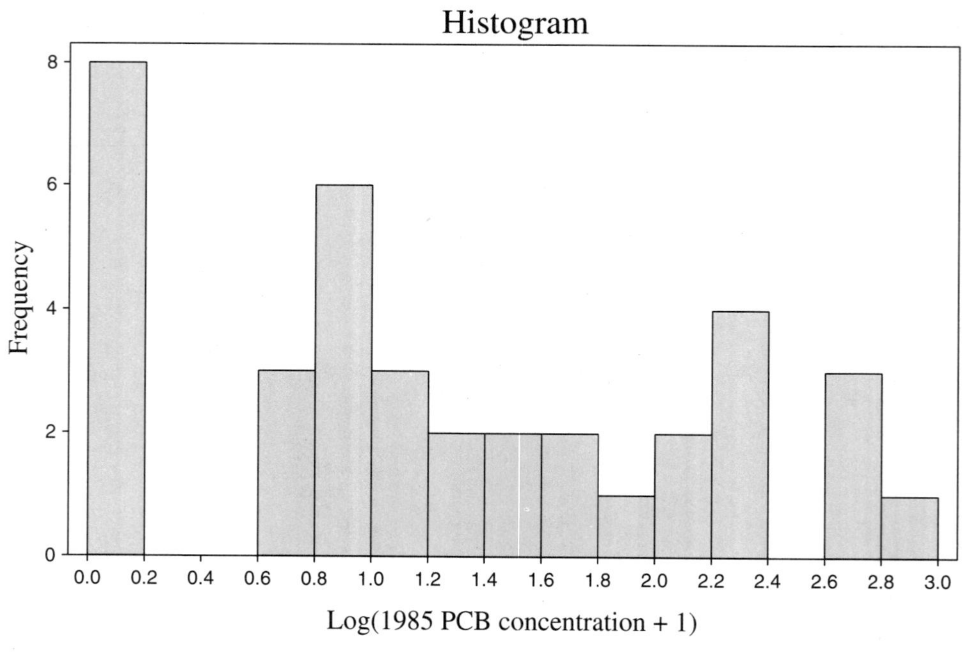

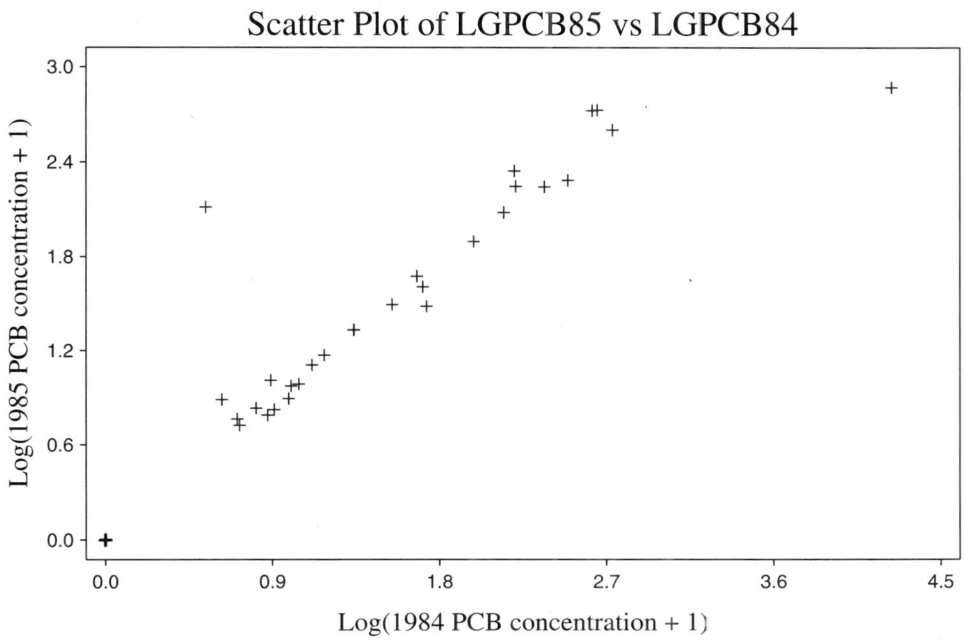

This looks better. However, there appear to be two clear outliers. We'll ignore them for now and fit a regression line to the entire data set:

```
LINEAR REGRESSION OF LGPCB85 Log(1985 PCB concentration + 1)

PREDICTOR
VARIABLES     COEFFICIENT    STD ERROR     STUDENT'S T       P
---------     -----------    ---------     -----------     ------
CONSTANT         0.18974      0.08660          2.19        0.0352
LGPCB84          0.84076      0.05418         15.52        0.0000

R-SQUARED              0.8731      RESID. MEAN SQUARE (MSE)     0.10676
ADJUSTED R-SQUARED     0.8695      STANDARD DEVIATION           0.32674

SOURCE        DF        SS            MS           F          P
----------    ---   ----------    ----------     -----      ------
REGRESSION     1     25.7113       25.7113      240.84      0.0000
RESIDUAL      35     3.73649       0.10676
TOTAL         36     29.4478

CASES INCLUDED 37   MISSING CASES 0
```

A strong relationship between the variables is apparent here, but are the assumptions of the regression valid?

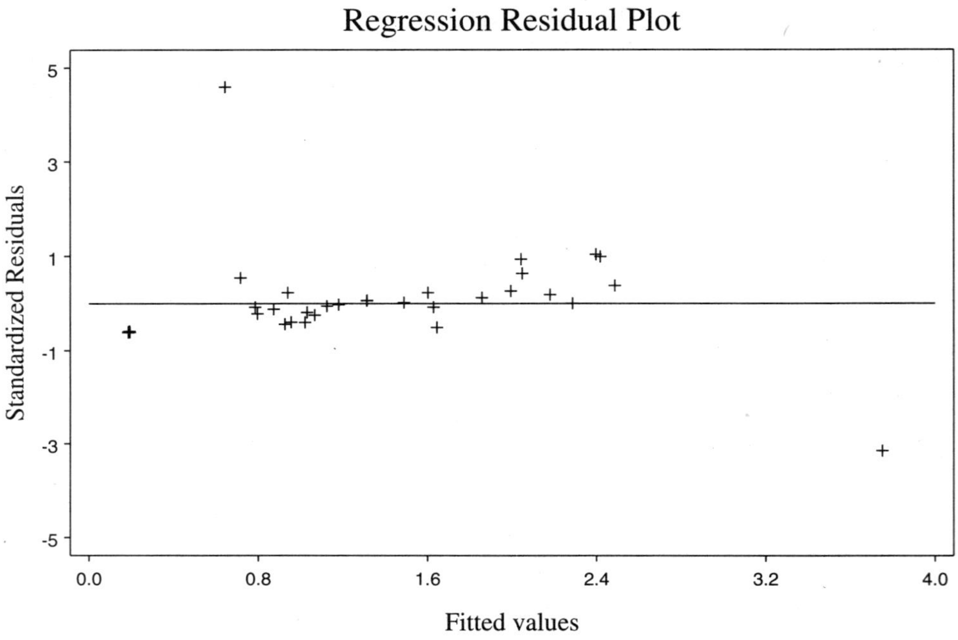

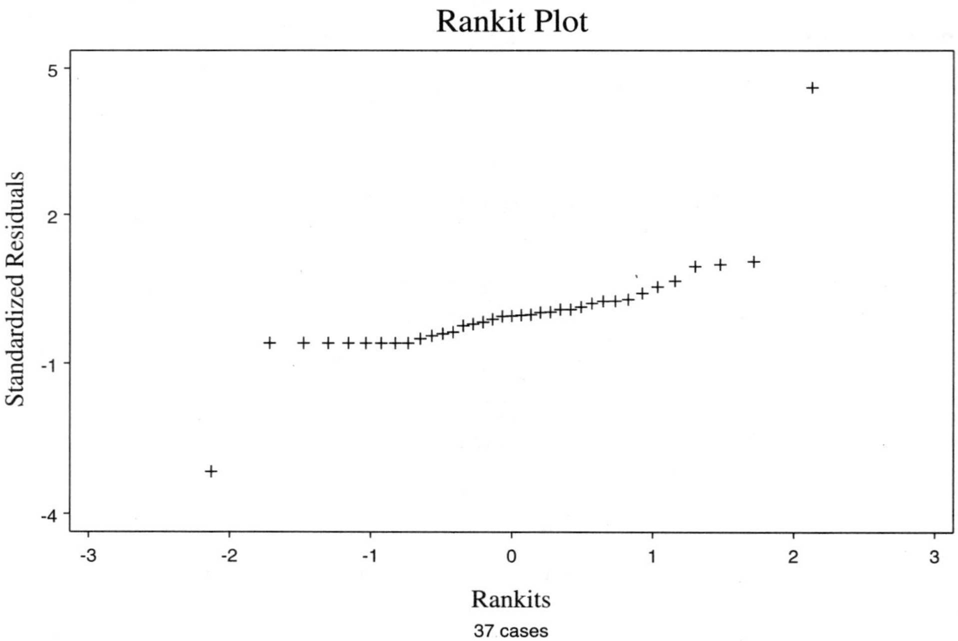

The two outliers show up clearly. They also show up when examining the diagnostics — leverage values (`LEVERAGE`), standardized residuals (`STDRES`) and (Cook's) distances (`COOKSD`).

CASE	BAY	LEVERAGE	STDRES	COOKSD
1	Casco Bay	0.0417	0.1179	0.0003
2	Merrimack River	0.0333	-0.5156	0.0046
3	Salem Harbor	0.0868	0.3944	0.0074
4	Boston Harbor	0.2710	-3.1593	1.8556
5	Buzzards' Bay	0.0691	0.0068	0.0000
6	Narragansett Bay	0.0520	0.9446	0.0245
7	E. Long Island Sound	0.0283	-0.2550	0.0009
8	W. Long Island Sound	0.0614	0.1892	0.0012
9	Raritan Bay	0.0805	0.9837	0.0423
10	Delaware Bay	0.0409	4.6011	0.4512
11	Lower Chesapeake Bay	0.0329	-0.0699	0.0001
12	Pamilico Sound	0.0703	-0.6023	0.0137
13	Charleston Harbor	0.0287	-0.1852	0.0005
14	Sapelo Sound	0.0703	-0.6023	0.0137
15	St. Johns River	0.0491	0.2706	0.0019
16	Tampa Bay	0.0703	-0.6023	0.0137
17	Apalachicola Bay	0.0276	-0.0456	0.0000
18	Mobile Bay	0.0703	-0.6023	0.0137
19	Round Island	0.0703	-0.6023	0.0137
20	Mississippi R. Delta	0.0293	0.0167	0.0000
21	Barataria Bay	0.0703	-0.6023	0.0137
22	San Antonio Bay	0.0703	-0.6023	0.0137
23	Corpus Christi Bay	0.0703	-0.6023	0.0137
24	San Diego Harbor	0.0788	1.0476	0.0469
25	San Diego Bay	0.0307	0.2359	0.0009
26	Dana Point	0.0303	-0.3826	0.0023
27	Seal Beach	0.0320	0.2345	0.0009
28	San Pedro Canyon	0.0519	0.6476	0.0115
29	Santa Monica Bay	0.0272	-0.0358	0.0000
30	Bodega Bay	0.0350	-0.0630	0.0001
31	Coos Bay	0.0380	0.5240	0.0054
32	Columbia River Mouth	0.0289	-0.4156	0.0026
33	Nisqually Beach	0.0349	-0.2218	0.0009
34	Commencement Bay	0.0272	0.0644	0.0001
35	Elliott Bay	0.0272	0.0644	0.0001
36	Lutak Inlet	0.0324	-0.1266	0.0003
37	Nahku Bay	0.0309	-0.4551	0.0033

The two outliers are Boston Harbor and Delaware Bay. Boston Harbor had a 1985 level of PCBs that was lower than expected, and corresponds to the isolated point in the lower right of the standardized residuals versus fitted values plot. Actually, it seems more like the level in 1984 was unbelievably high, while the level in 1985 was just very high. The city of Boston stopped dumping raw sewage into Boston Harbor in 1984, and opened a sewage treatment plant. Delaware Bay had a sharp increase in PCBs from 1984 to 1985 (it corresponds to the isolated point in the upper left of the standardized residuals versus fitted values plot). Why did the levels increase? It is difficult to draw conclusions using these data alone. However, it suggests questions to ask the oceanographers.

These two outliers have a potentially strong effect on the regression. They should be removed from the analysis. Note, however, that they suggest that our model won't be quite as effective as we might like, since there is apparently the potential for occasionally wild fluctuation in PCBs from year to year. Unless we can establish precisely when such fluctuations occur or what factors are associated with such fluctuations, we run the risk of being very far off in any predictions of future PCB level from present PCB level. Again, the information in these data is not sufficient to address these questions.

We remove those cases and redo the regression. Note the big changes in various statistics, showing that the two omitted points not only did not fit the model, but also had a strong effect on the fitted line.

```
LINEAR REGRESSION OF LGPCB85 Log(1985 PCB concentration + 1)

PREDICTOR
VARIABLES     COEFFICIENT    STD ERROR     STUDENT'S T       P
---------     -----------    ---------     -----------     ------
CONSTANT         0.01196      0.02675          0.45        0.6577
LGPCB84          0.97607      0.01811         53.90        0.0000

R-SQUARED              0.9888      RESID. MEAN SQUARE (MSE)     0.00880
ADJUSTED R-SQUARED     0.9884      STANDARD DEVIATION           0.09383

SOURCE        DF        SS             MS          F         P
----------    ---   ----------    ----------    -----     ------
REGRESSION     1     25.5768        25.5768    2904.89    0.0000
RESIDUAL      33     0.29056        0.00880
TOTAL         34     25.8673

CASES INCLUDED 35    MISSING CASES 0
```

The regression assumptions seem much more reasonable now:

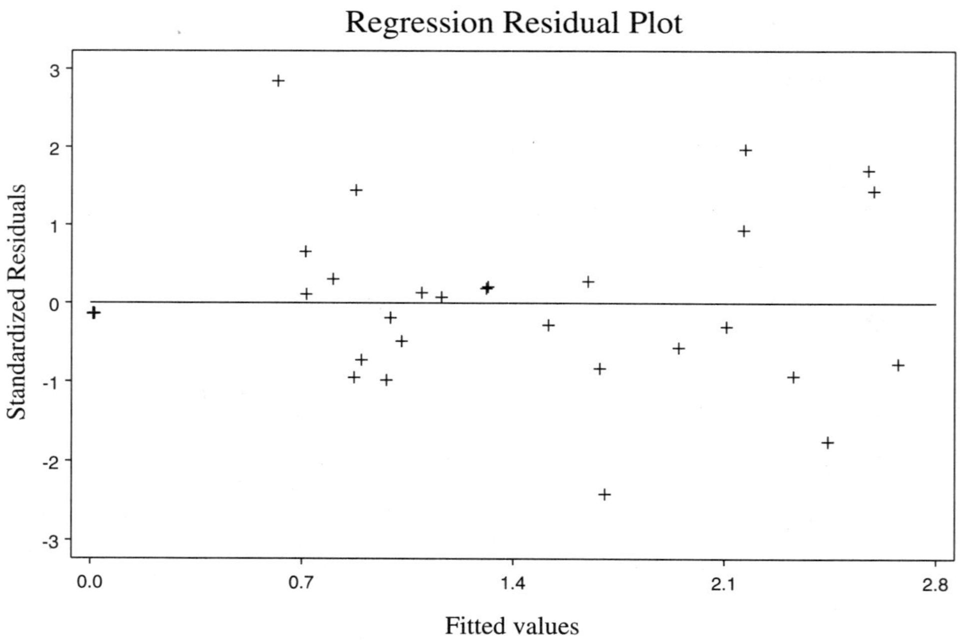

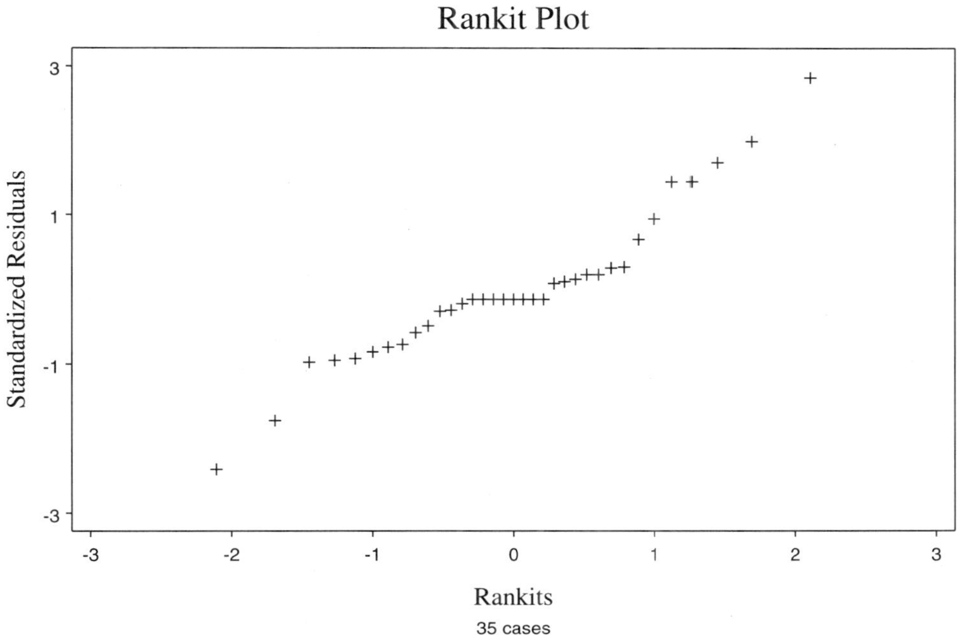

CASE	BAY	LEVERAGE	STDRES	COOKSD
1	Casco Bay	0.0521	-0.5788	0.0092
2	Merrimack River	0.0396	-2.4165	0.1202
3	Salem Harbor	0.1168	-0.7733	0.0395
5	Buzzards' Bay	0.0917	-1.7562	0.1556
6	Narragansett Bay	0.0671	1.9829	0.1415
7	E. Long Island Sound	0.0294	-0.4895	0.0036
8	W. Long Island Sound	0.0807	-0.9253	0.0376
9	Raritan Bay	0.1079	1.4406	0.1255
11	Lower Chesapeake Bay	0.0389	-0.8357	0.0141
12	Pamilico Sound	0.0812	-0.1330	0.0008
13	Charleston Harbor	0.0298	-0.1919	0.0006
14	Sapelo Sound	0.0812	-0.1330	0.0008
15	St. Johns River	0.0629	-0.2952	0.0029
16	Tampa Bay	0.0812	-0.1330	0.0008
17	Apalachicola Bay	0.0288	0.1336	0.0003
18	Mobile Bay	0.0812	-0.1330	0.0008
19	Round Island	0.0812	-0.1330	0.0008
20	Mississippi R. Delta	0.0333	-0.2791	0.0013
21	Barataria Bay	0.0812	-0.1330	0.0008
22	San Antonio Bay	0.0812	-0.1330	0.0008
23	Corpus Christi Bay	0.0812	-0.1330	0.0008
24	San Diego Harbor	0.1055	1.7006	0.1706
25	San Diego Bay	0.0319	1.4450	0.0344
26	Dana Point	0.0316	-0.7358	0.0088
27	Seal Beach	0.0375	0.2827	0.0016
28	San Pedro Canyon	0.0671	0.9419	0.0319
29	Santa Monica Bay	0.0286	0.0770	0.0001
30	Bodega Bay	0.0370	0.6616	0.0084
31	Coos Bay	0.0405	2.8455	0.1711
32	Columbia River Mouth	0.0300	-0.9737	0.0147
33	Nisqually Beach	0.0368	0.1017	0.0002
34	Commencement Bay	0.0294	0.1943	0.0006
35	Elliott Bay	0.0294	0.1943	0.0006
36	Lutak Inlet	0.0338	0.2937	0.0015
37	Nahku Bay	0.0321	-0.9510	0.0150

At this point, we could be satisfied with the fit, and proceed to look at the model and its implications. Before we do that, let's look again at the scatter plot, with Boston Harbor and Delaware Bay (the outliers) removed:

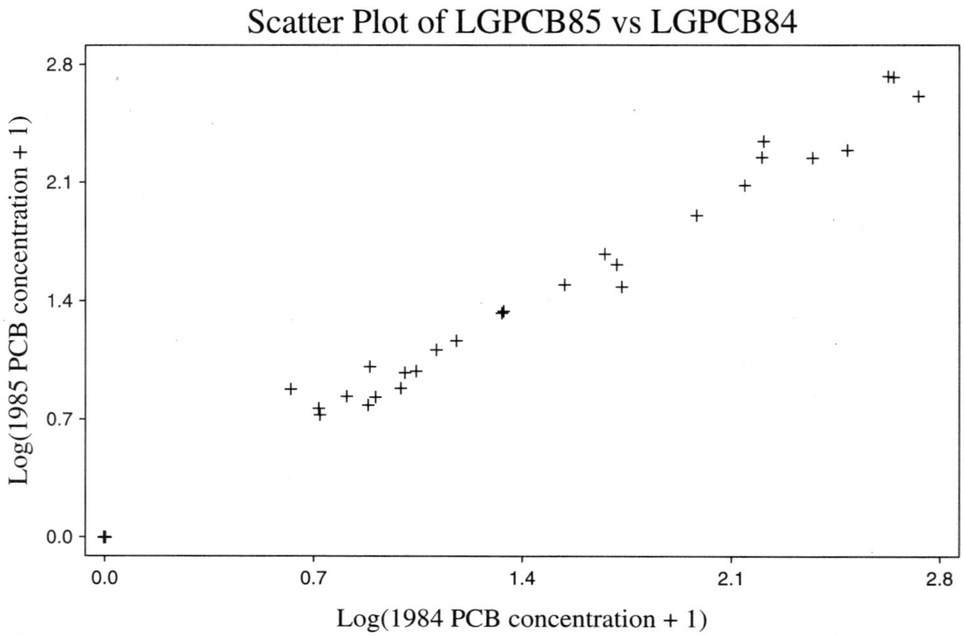

The values in the lower left corner stand out as unusual. These are the eight bays that had no PCBs in either 1984 or 1985. It seems reasonable to think that the mechanism governing changes in PCB concentration is different in bodies of water with no contamination at all. Let's see what happens if we restrict our study to only bays and estuaries where there was some PCB contamination in 1984 (we simply omit cases with PCB84 = 0). Since the zero values have been omitted, we can avoid the modification of adding 1, and just use logarithms of the variables themselves.

```
LINEAR REGRESSION OF LGPCB85 Log(1985 PCB concentration)

PREDICTOR
VARIABLES     COEFFICIENT    STD ERROR     STUDENT'S T       P
---------     -----------    ---------     -----------     ------
CONSTANT         0.04017      0.05301          0.76        0.4557
LGPCB84          0.95983      0.03185         30.14        0.0000

R-SQUARED              0.9732      RESID. MEAN SQUARE (MSE)      0.01344
ADJUSTED R-SQUARED     0.9721      STANDARD DEVIATION            0.11593

SOURCE         DF        SS            MS          F          P
----------    ---    ----------    ----------    -----     ------
REGRESSION      1     12.2055        12.2055     908.24    0.0000
RESIDUAL       25     0.33597        0.01344
TOTAL          26     12.5415

CASES INCLUDED 27    MISSING CASES 0
```

Note what happened. The coefficients have hardly changed, so in that sense it did not make any difference that we removed the bays and estuaries that had no PCB concentration in 1984 and 1985. The strength of the fit has decreased a bit. The observations we removed are what are sometimes called "good leverage points" — that is, cases that are far from the other cases, but are apparently on the same regression line as the other cases. Some statisticians suggest that such cases should be kept in the analysis, since they apparently improve the fit without changing the fitted regression line appreciably, but the dangers of doing this should be recognized. If the "good" leverage points are even slightly off the line, they immediately become "bad" leverage points. Also, it is probably not reasonable from a physical point of view to think that the mechanism that changes PCB levels from zero to nonzero is the same as the one that changes them from some nonzero level to another nonzero level.

One other question: if these cases were leverage points, why didn't they show up in the leverage values (diagnostics)? The reason is that the eight values all together "hid" each other, so that none of them looked unusual (this is called the *masking effect*, and can easily occur when performing a routine analysis).

How does the fitted model look? The residual plots and diagnostics are as follows:

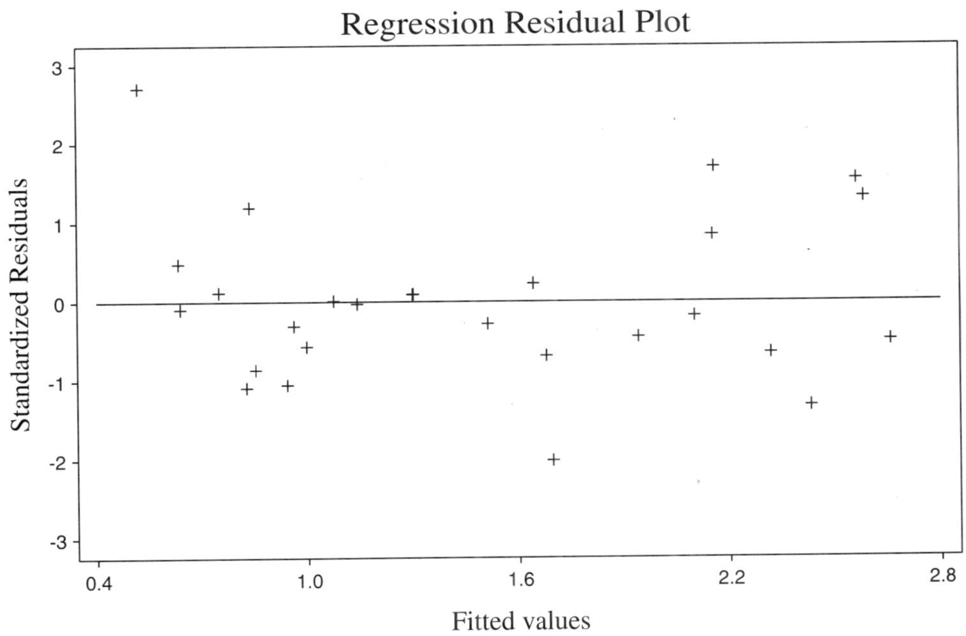

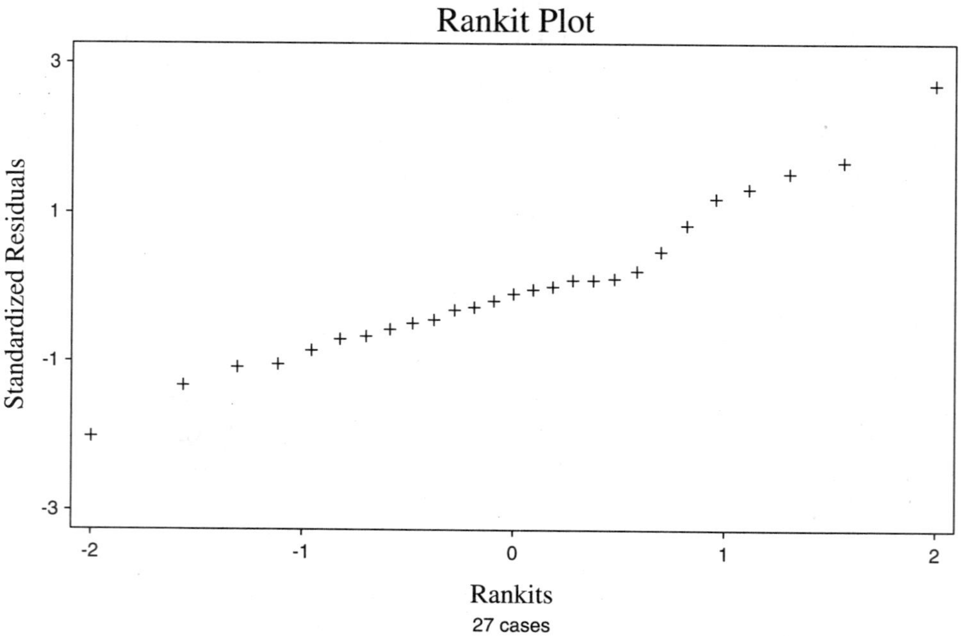

CASE	BAY	LEVERAGE	STDRES	COOKSD
1	Casco Bay	0.0536	-0.4441	0.0056
2	Merrimack River	0.0405	-2.0176	0.0859
3	Salem Harbor	0.1489	-0.4899	0.0210
5	Buzzards' Bay	0.1094	-1.3321	0.1090
6	Narragansett Bay	0.0734	1.6845	0.1124
7	E. Long Island Sound	0.0567	-0.5727	0.0099
8	W. Long Island Sound	0.0929	-0.6670	0.0228
9	Raritan Bay	0.1347	1.3223	0.1361
11	Lower Chesapeake Bay	0.0400	-0.7036	0.0103
13	Charleston Harbor	0.0599	-0.3101	0.0031
15	St. Johns River	0.0676	-0.1854	0.0012
17	Apalachicola Bay	0.0510	0.0057	0.0000
20	Mississippi R. Delta	0.0371	-0.2723	0.0014
24	San Diego Harbor	0.1310	1.5317	0.1768
25	San Diego Bay	0.0721	1.1905	0.0550
26	Dana Point	0.0700	-0.8584	0.0277
27	Seal Beach	0.0390	0.2170	0.0010
28	San Pedro Canyon	0.0733	0.8346	0.0275
29	Santa Monica Bay	0.0470	-0.0340	0.0000
30	Bodega Bay	0.0967	0.4803	0.0123
31	Coos Bay	0.1134	2.7102	0.4699
32	Columbia River Mouth	0.0613	-1.0441	0.0356
33	Nisqually Beach	0.0960	-0.0898	0.0004
34	Commencement Bay	0.0399	0.0924	0.0002
35	Elliott Bay	0.0399	0.0924	0.0002
36	Lutak Inlet	0.0817	0.1138	0.0006
37	Nahku Bay	0.0730	-1.0834	0.0462

The fit now appears satisfactory. What are the implications of this model? Our final model is

$$\texttt{LGPCB85} = .0402 + .9598 \times \texttt{LGPCB84}.$$

Remembering that logged models can be transformed back to the original scale (by exponentiating both sides) provides the relationship:

$$\begin{aligned}\text{PCB85} &= 10^{.0402} \times (\text{PCB84})^{.9598} \\ &= 1.097 \times (\text{PCB84})^{.9598}\end{aligned}$$

The slope coefficient is less than one (although not statistically significantly so; the t–statistic to test $H_0 : \beta_1 = 1$ is -1.26), which is important; it implies that the PCB levels of more polluted bays decreased more proportionally than the PCB levels of less polluted bays (that is, a waterway that had 2 times the PCB level in 1984 than another waterway is predicted to have 1985 PCB level that is only $2^{.9598} = 1.945$ times as high). We can see this by examining a couple of predictions. First, let's look at a prediction for a low PCB estuary, East Long Island Sound:

```
PREDICTED VALUES OF LGPCB85

LOWER PREDICTED BOUND      0.7546
PREDICTED VALUE            1.0000
UPPER PREDICTED BOUND      1.2454
SE (PREDICTED VALUE)       0.1192

PREDICTOR VALUES: LGPCB84 = 1.0000
```

The predicted value is 1.000, or PCB85 = 10 (note that PCB84 = 10). That is, the forecast is for no change in PCBs. In contrast, consider the case of Salem Harbor:

```
PREDICTED VALUES OF LGPCB85

LOWER PREDICTED BOUND      2.4019
PREDICTED VALUE            2.6578
UPPER PREDICTED BOUND      2.9137
SE (PREDICTED VALUE)       0.1243

PREDICTOR VALUES: LGPCB84 = 2.7272
```

The predicted value is 2.6578, or PCB85 = 454.779 (from PCB84 = 533.58). This is a forecasted decrease of about 15%. So, apparently the most heavily polluted waterways in 1984 improved the most in 1985, which is presumably a good thing.

This kind of data set is a situation where prediction intervals, rather than confidence intervals, are probably of most interest (we're more interested in what will happen in a particular place, rather than an average of all places). The prediction interval for East Long Island Sound is (.755, 1.245), or (5.69, 17.58) in the original scale (the true value was 8.62); the prediction interval for Salem Harbor is (2.402, 2.914), or (252.35, 820.35) in the original scale (the true value was 403.1). So, both prediction intervals contained the true value. We expect that about 95% of them will.

Summary

There is a strong relationship between PCB levels in 1984 and 1985 in the sample of bays and estuaries examined. However, two waterways (Boston Harbor and Delaware Bay) show up as clear outliers, which shows that large fluctuations in PCB level are possible as a result of unforeseen circumstances. This casts some doubt on the usefulness of this model for other waterways. Each of the eight waterways that had zero PCB contamination in 1984 also had zero contamination in 1985. The fitted regression model for the other waterways, based on logarithmic transformations of both concentration variables, implies that contamination dropped more (in a proportional sense) from 1984 to 1985 for waterways with high contamination than for those with low contamination.

Technical terms

Cook's distance: a diagnostic used to assess the influence of an observation on the fitted regression coefficients. A case whose deletion affects the fitted coefficients substantially is called an **influential point**, and is characterized by large values of Cook's distance. Observations with Cook's distances greater than one indicate major influence and should be examined. The effect of any point with large influence should be examined by refitting the model after omitting the influential point. A useful graphical representation of the Cook's distances is an index plot, where the distance is plotted against the case number.

Leverage value: a diagnostic that measures the potential for a point to exercise influence on the regression results. An observation with large leverage value lies far removed from most of the observations with respect to the predicting variables, and is called a **leverage point**. Observations with large leverage values often represent special situations. An operational rule is to examine all observations with a leverage value greater than $2.5(p+1)/n$, where p is the number of predicting variables in the model. An index plot is also useful to identify observations with excessive leverage.

Masking effect: a situation where two or more unusual observations, usually close to each other, "hide" each other, and cannot be identified using the usual regression diagnostics.

Postscript

In this case we followed up the primary source. The data for the years after 1985 are not reported in the subsequent annual versions of the secondary source (*Environmental Quality 1987 – 1988*). The 1984–85 data were abstracted from a report on the National Status and Trends Program for Marine Environmental Quality undertaken by the National Oceanic and Atmospheric Administration (NOAA). The program began in 1984. Data for 1986–87 are available from updates published by NOAA and are available in libraries that archive government publications.

The levels in 1986–87 for Boston Harbor were similar to the 1985 level. Because data were not available pre–1984, we cannot say if the 1984 level was an anomaly or if there was a shift in level between 1984 and 1985. The levels in Delaware Bay equaled or exceeded the 1985 level. This suggests that the 1984 level was low, rather than the 1985 level being high.

The report also makes clear that the concentrations reported do not represent each waterway as a whole. In particular, concentrations at other sites within Boston Harbor and Delaware Bay showed great variation, and tended to be lower than those reported here. It was also clear that sediment characteristics such as grain size were factors. Sites where the sediment was muddy rather than sandy tended to have higher levels of contamination. This suggests that shifts in sedimentation over time could have a big effect on the recorded levels.

Electricity usage, temperature and occupancy

Topics covered: Transformation. Using a model to identify unusual behavior.

Key words: Histogram. Leverage. Mean. Median. Normal plot. Outliers. Regression. Residual. Residual plots. Scatter plot. Time series plot.

Data File: `elusage.dat`

In Westchester County, north of New York City, Consolidated Edison bills residential customers for electricity consumption on a monthly basis. The utility would like to be able to predict residential usage, in order to plan purchases of fuel and budget revenue flow; residents would like to be able to predict usage in order to control costs and identify atypical usage. What can we learn about the average electricity usage (in kilowatt–hours per day)? We would expect that it is related to temperature. Is this so? The data set includes information on usage and average monthly temperature for 55 consecutive months (August 1989 — February 1994) for an all–electric home. The usage is measured in kilowatt–hours (i.e., the energy needed to run a 1000 watt appliance for an hour). The average monthly temperature is the average of the daily mean temperatures in degrees Fahrenheit. The daily mean temperature is the average of the daily maximum and the daily minimum temperatures.

Here are some descriptive statistics (KWH is the usage variable, while TEMPERATURE is the temperature variable):

DESCRIPTIVE STATISTICS

	KWH	TEMPERATURE
N	55	55
MEAN	43.275	53.818
SD	24.009	15.543
MINIMUM	10.414	24.000
1ST QUARTILE	24.207	39.000
MEDIAN	38.621	54.000
3RD QUARTILE	59.375	69.000
MAXIMUM	101.66	79.000

Here are histograms of the variables:

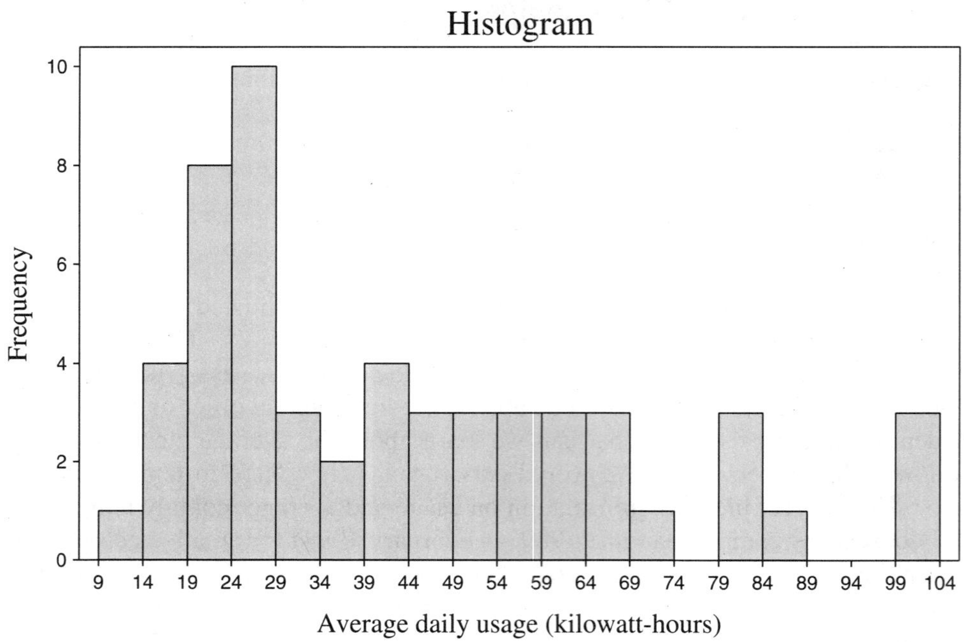

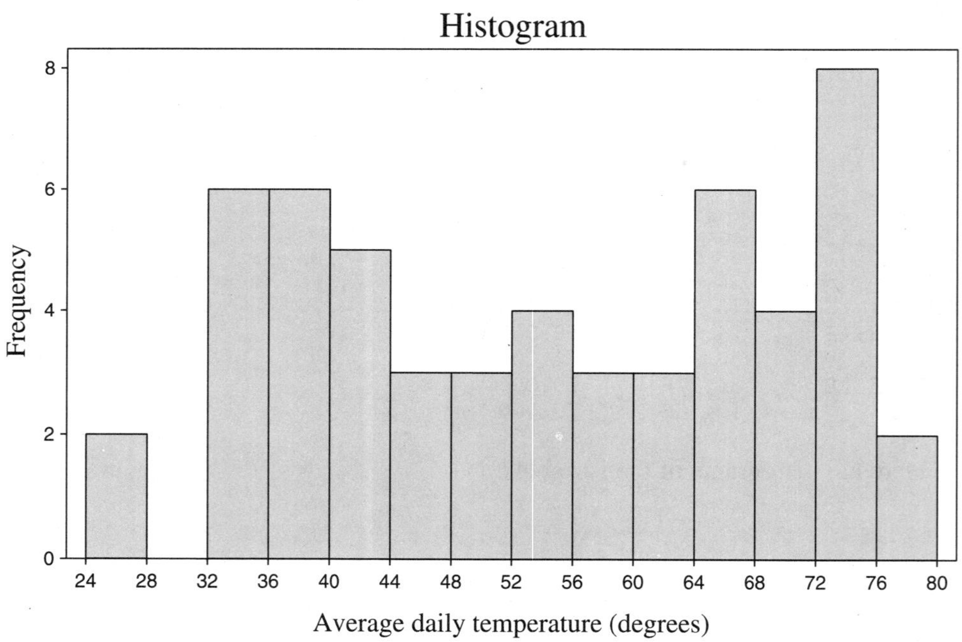

The usage is right–skewed, as we might expect, reflecting a few months with unusually high usage. The mean usage is 43.3 Kwh, while the median usage is 38.6 Kwh. The average daily temperature has a relatively flat distribution, with a mean of about 54 degrees.

Let's look at a plot of the usage over time (i.e., a time series plot), to see how it varies:

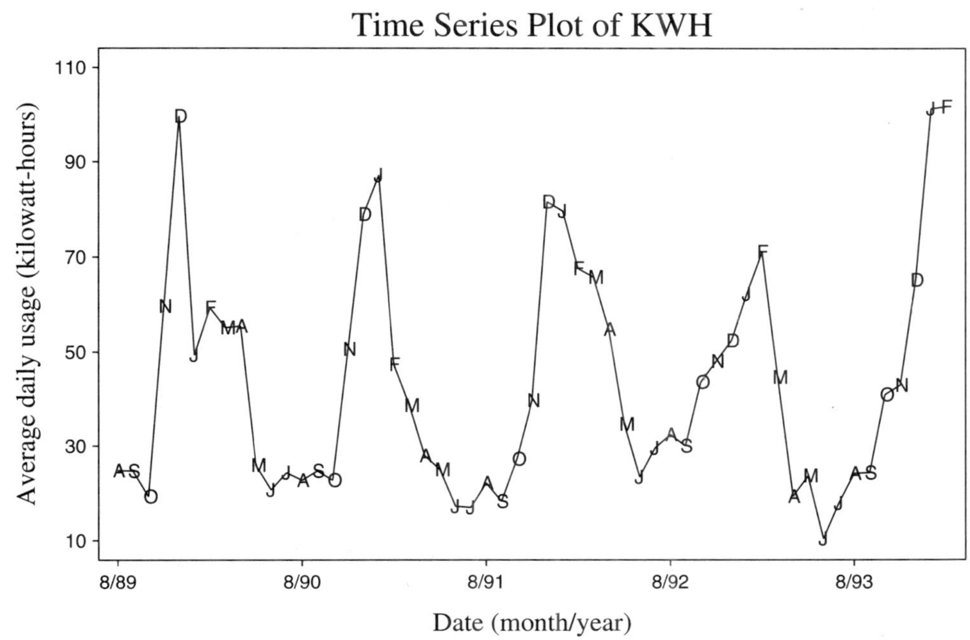

The plot symbols are one–character designations of the name of the month. There is a strong cyclical pattern here and it is a good guess that much of the variation in the usage is understandable as variations with the season. Let's overlay on top of this plot the average temperature for the month to make sure that this is actually true:

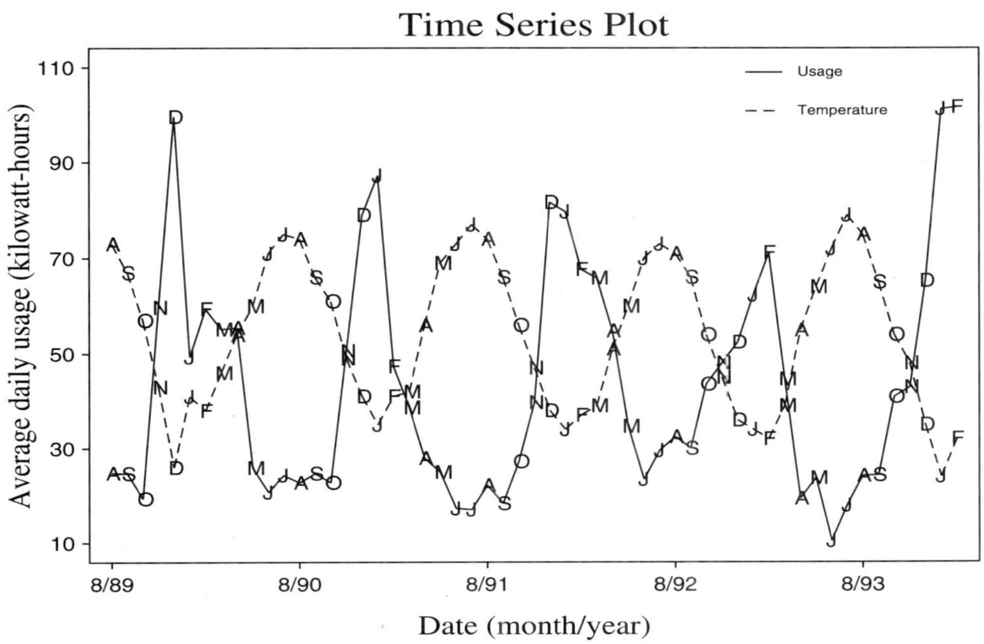

The temperature has a much more regular pattern than the usage. The two variables appear to be highly (negatively) correlated. Let's look at a scatter plot of usage against temperature:

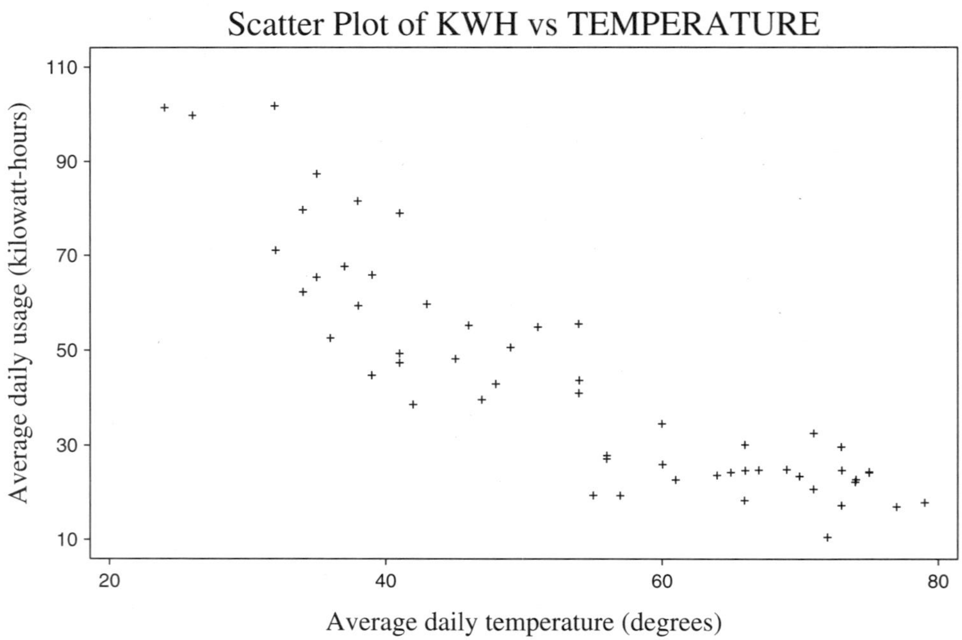

The strong relationship is clear here, although it does not appear to be linear over the entire range of temperature. Below about 60 degrees the usage appears to increase in an roughly linear fashion with decreasing temperature. Above 60 degrees it appears to be constant or perhaps decreasing slightly with increasing temperature. A little thought makes it clear why this should be: electricity heats the home, and the amount of electricity required to maintain the house at a constant temperature increases as the temperature outside drops. Above about 60 degrees the electric heating is not needed and so the usage levels off. Interestingly the usage does not increase as the temperature climbs above 80 degrees. This suggests that the house is probably not cooled by electric air–conditioning in the summer.

How can we build a model to explain the variation in usage in terms of the average temperature? Clearly a linear regression model in terms of temperature alone is not appropriate, as the relationship between the two is not linear over the entire range of temperatures. Given the right–skewness of `KWH`, let's try taking logarithms of both `KWH` and `TEMPERATURE`. Here is the corresponding scatter plot:

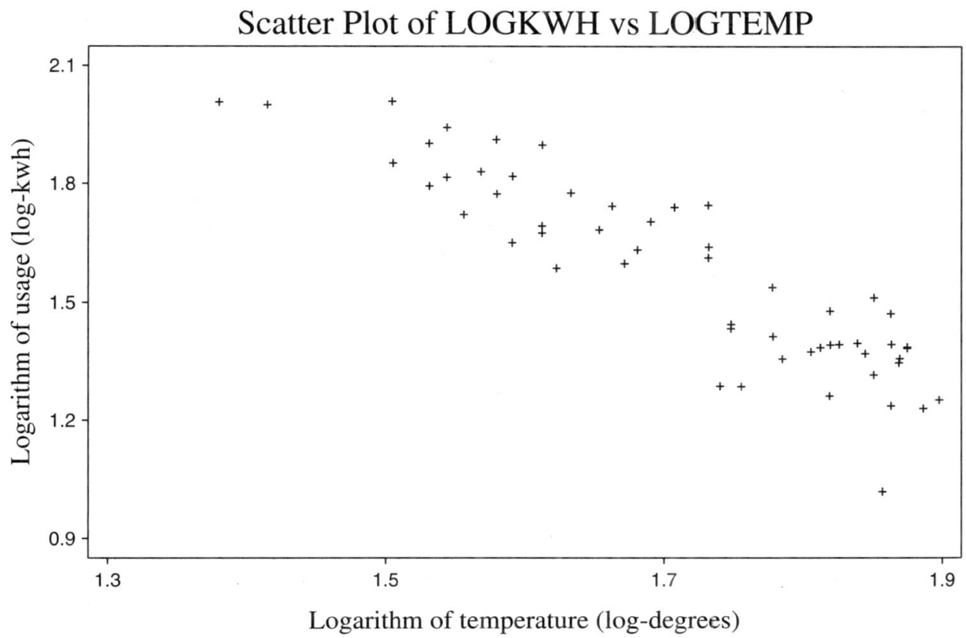

The relationship appears to be much more linear on this scale.

Let's look at the regression of log–usage on log–temperature. The numerical regression summary is as follows:

```
LINEAR REGRESSION OF LOGKWH  Logarithm of usage (log-kwh)

PREDICTOR
VARIABLES     COEFFICIENT    STD ERROR     STUDENT'S T       P
---------     -----------    ---------     -----------     ------
CONSTANT          4.30834      0.18186        23.69        0.0000
LOGTEMP          -1.59886      0.10594       -15.09        0.0000

R-SQUARED            0.8112   RESID. MEAN SQUARE (MSE)     0.01105
ADJUSTED R-SQUARED   0.8077   STANDARD DEVIATION           0.10512

SOURCE        DF        SS            MS          F         P
----------    ---    ----------    ----------   -----    ------
REGRESSION     1      2.51670       2.51670    227.77    0.0000
RESIDUAL      53      0.58561       0.01105
TOTAL         54      3.10230

CASES INCLUDED 55   MISSING CASES 0
```

Here is the standardized residuals versus fitted values plot:

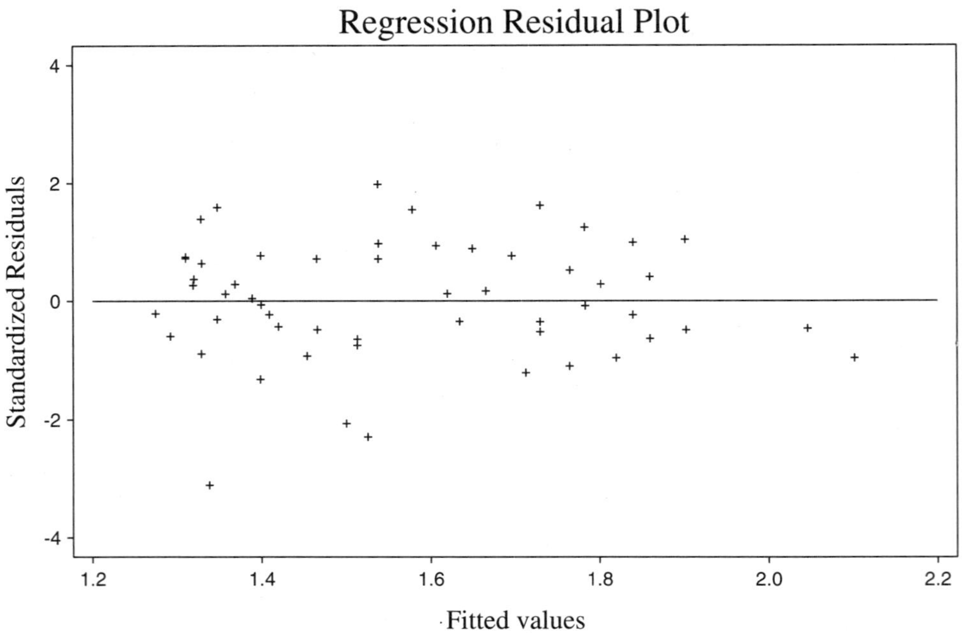

There is a large outlier (June 1993) with a standardized residual of –3.11. The leverage and influence values (not given here) are modest. During this month the house was empty, as the occupants were overseas. The refrigerator was not opened and no hot water was used. Thus we might expect substantially reduced usage during this period.

Let's look at a plot of the standardized residuals as a time series plot:

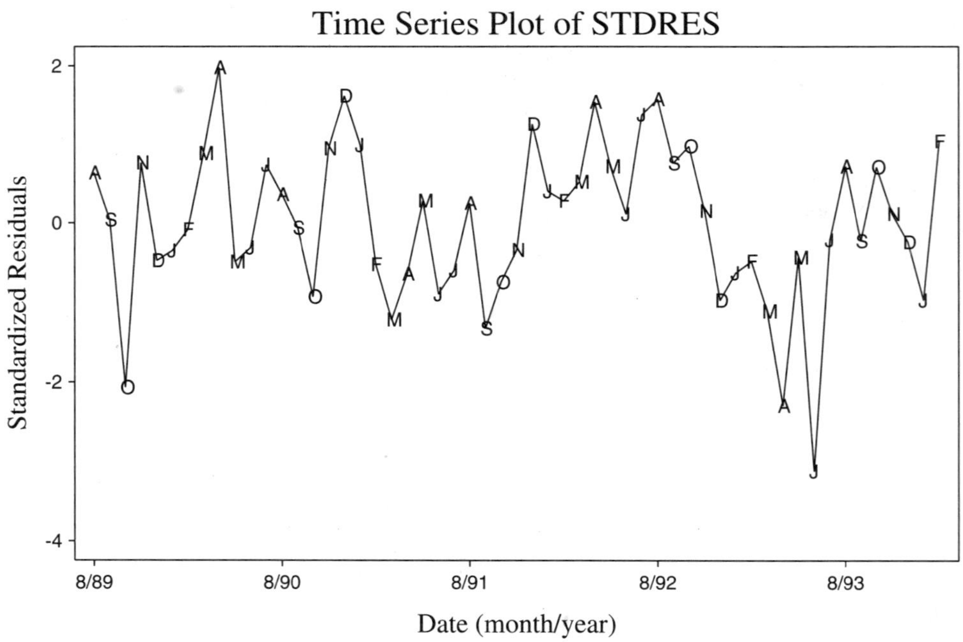

The patterns are a little clearer here. The plot emphasizes how the observed usages deviate from the pattern defined by the model. During December 1992–May

1993, only one person occupied the house, which could have reduced the electricity consumption. From this plot we see that these months have lower usage than expected from the model. The other notably negative residual corresponds to October 1989, which has no clear explanation; it is just low. Let's omit the period that the house was partially or completely empty (i.e., December 1992–June 1993) and refit the regression model. The numerical regression summary is then the following:

```
LINEAR REGRESSION OF LOGKWH  Logarithm of usage (log-kwh)

PREDICTOR
VARIABLES     COEFFICIENT    STD ERROR     STUDENT'S T       P
---------     -----------    ---------     -----------     ------
CONSTANT         4.32969      0.17027         25.43        0.0000
LOGTEMP         -1.59995      0.09873        -16.21        0.0000

R-SQUARED              0.8509    RESID. MEAN SQUARE (MSE)     0.00816
ADJUSTED R-SQUARED     0.8477    STANDARD DEVIATION           0.09031

SOURCE        DF        SS             MS           F         P
----------    ---   ----------     ----------     -----    ------
REGRESSION     1     2.14188        2.14188      262.62    0.0000
RESIDUAL      46     0.37517        0.00816
TOTAL         47     2.51705

CASES INCLUDED 48    MISSING CASES 0
```

Here is a plot of the standardized residuals versus the fitted values:

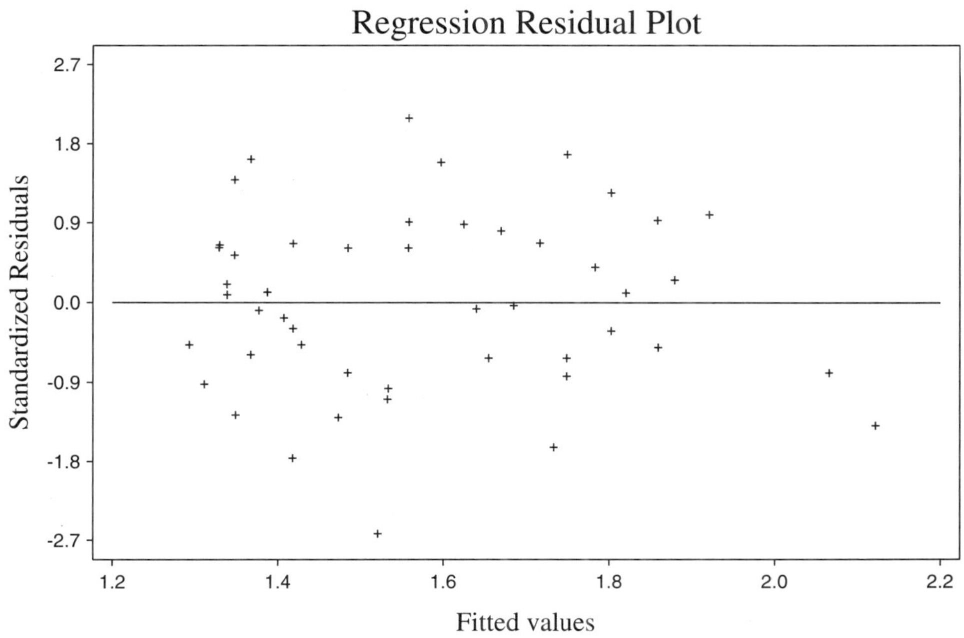

The usage in October 1989 is still a low outlier. In addition, the two points to the right of the plot show up as potential leverage points. They correspond to December 1989 and January 1994, the only two months where the average temperature was below 32°. The individual influence of each point is modest, based on Cook's distance values (not given here). Note, by the way, that since both points have negative residuals, there is an indication that the log–linear relationship of usage with temperature might not be appropriate for very cold months. Apparently, there is an upper limit to the electricity usage, even for temperatures as much as 10 degrees below the typical lower value, which would account for the observed usage being less than the predicted usage.

The normal plot of the standardized residuals is as follows:

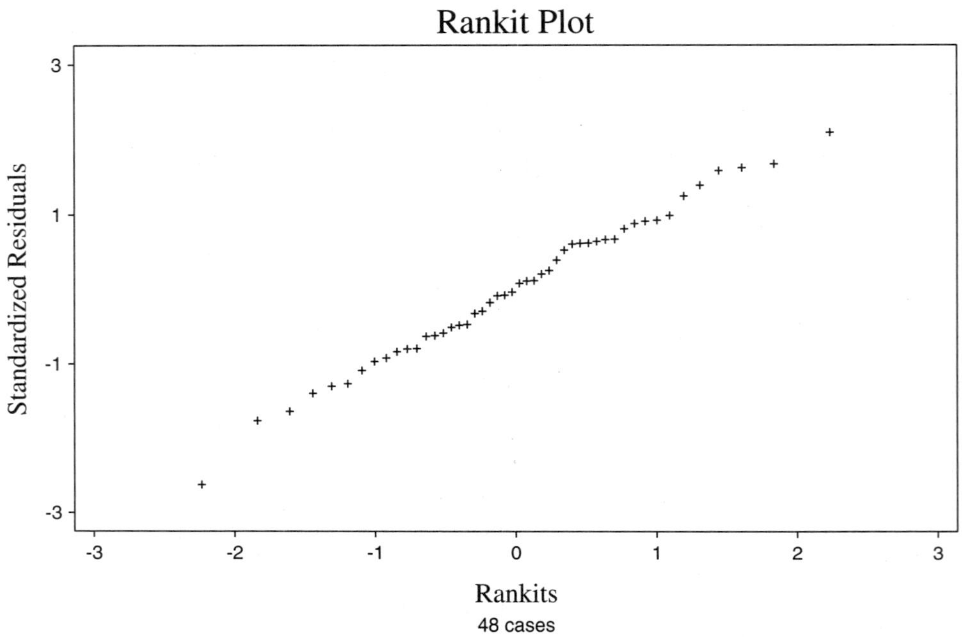

Except for the usage in October 1989, the standardized residuals appear to be distributed reasonably close to that expected from a Gaussian distribution.

Summary

The monthly electricity usage of a house is closely related to the average monthly temperature, using a log–log model. Usage was consistently lower during a period of time when the house was partially or completely unoccupied. This becomes apparent from the statistical analysis, which shows a pattern of deviation of the usages during this period quite different from the periods of full occupancy.

Long and short term performance of stock mutual funds

Topics covered: Influential subgroups in regression.

Key words: Masking effect. Outliers. Regression diagnostics. Scatter plot.

Data Files: `funds.dat`

Refer again to the stock fund data previously discussed in the case "The performance of stock mutual funds." In this case we will attempt to quantify the relationship between short–term performance (as measured by the 1992 return) and long–term performance (as measured by the five–year return).

Construct a scatter plot of the 1992 return (vertical axis) versus the five–year return (horizontal axis). Does it appear that a linear regression model would be appropriate for these data?

Now, calculate the least squares regression line for predicting the 1992 return from the five–year return. Is there a significant relationship between long– and short–term performance? Construct a scatter plot with the fitted regression line superimposed. Does it appear to be a reasonable representation of the observed relationship? Do you notice any apparent systematic violations of the regression assumptions?

Construct and examine regression diagnostics for this regression model. You will notice an observation with large positive standardized residual. It is Skyline Special Equity Fund, which had a remarkable 41% 1992 return. Remove this point and rerun the regression. Is the newly obtained regression line satisfactory? Do you notice any apparent systematic violations of the regression assumptions?

Let's start over again, using what we learned earlier about these data. From the side–by–side boxplots, we know that International–and–Global funds had noticeably poor performance on both the long– and short–term measures, and seemed quite different from the other types of funds. Omit the International–and–Global funds from the data, and reanalyze the regression relationship. You can see that the regression relationship is now much weaker. Apparently long– and short–term performance are almost completely unrelated, except that the International–and–Global funds were poor on both measures.

It is noteworthy that even though the 29 International–and–Global funds were influential on the regression jointly, their regression diagnostics were not noteworthy. This is because of the masking effect, where nearby unusual observations hide each other, making regression diagnostics ineffective.

Comments to the Instructor

A scatter plot of RETURN92 versus FIVEYR seems to indicate a weak, but noticeable, direct relationship. A linear regression seems to support this, as the resultant model is highly significant ($F = 69.04$), with an R^2 of about 27%. However, a scatter plot with the fitted regression line superimposed indicates a problem, in that the line consistently overestimates RETURN92 for the funds with low values of FIVEYR.

Case 121 (Skyline Special Equity Fund) is an outlier, with a standardized residual of 3.58. However, removing this point does not materially affect the fitted line, or the fact that it systematically overestimates RETURN92 for the funds with low values of FIVEYR.

If the 29 International–and–Global funds are omitted from the analysis, the results change dramatically. The resultant fitted regression model only has an R^2 of 3.7%, with a weakly significant regression line ($F = 5.80$). Thus, virtually all evidence for a (linear) relationship comes from the poor performance of International–and–Global funds with respect to both short– and long–term performance.

To maintain consistency with the unannotated edition,
the next case begins on the following page.

Estimating rates of return of investments

Topics covered: Estimation of effective return rate. Semilog model.

Key words: Logarithmic transformation. Regression. Residual plots. Scatter plot.

Data File: `return.dat`

The rate of return is a fundamental property of any investment. Consider, for example, a fixed–rate investment, such as a certificate of deposit issued by a bank. The amount of money, P_t, in the investment at the end of time period t then has the exact form

$$P_t = P_0(1 + i)^t, \tag{1}$$

where P_0 is the initial investment and i is the interest rate per time period (this is simply compound interest at work). Most investments, of course, do not give a fixed rate of return, but it is possible to use this structure in order to estimate an *effective interest rate* using the observed behavior of the investment over time.

The simplest way to do this is just to use the first and last period's values. Say, for example, that an investment started with \$10,000, and after 10 time periods had a value of \$19,671.51. The estimated effective interest rate is just the solution i to the formula

$$19671.51 = (10000)(1 + i)^{10},$$

which turns out to be $i = .07$; thus, the estimated effective interest rate is 7%. This isn't a very good estimate, however, because it is only based on one data value (and the initial principal); we could do better by using information about the cumulative performance of the investment for all 10 time periods, rather than just the last one.

Regression gives us a way to do this. Suppose we had data giving us the value of the investment at the end of each of ten time periods. If we take logarithms of both sides of equation (1), we get

$$\log(P_t) = \log(P_0) + \left[\log(1 + i)\right] \times t. \tag{2}$$

If we do a regression of $\log(P_t)$ on t, the slope term is an estimate of $\log(1 + i)$, so we immediately get an estimated effective interest rate based on all of the data. To see this compare the representation

$$\log(P_t) = \beta_0 + \beta_1 \times t$$

to (1). The functional form (1), with an error term, is sometimes called a semilog model, since it is consistent with a linear relationship between $\log(P_t)$ and t.

Consider the performance of the stock market as a whole, as measured by the Standard and Poor's 500 Index value. Using the year–end values of an initial \$10,000 investment, an effective annual return of a completely diversified portfolio can be estimated. Here is output for the regression based on model (2), using year–end values for the years 1976 – 1993:

```
LINEAR REGRESSION OF LGSANDP Log (S & P Index)

PREDICTOR
VARIABLES     COEFFICIENT    STD ERROR     STUDENT'S T        P
---------     -----------    ---------     -----------     ------
CONSTANT          3.91128      0.01694          230.95     0.0000
PERIOD            0.06190      0.00170           36.40     0.0000

R-SQUARED                0.9881     RESID. MEAN SQUARE (MSE)     0.00140
ADJUSTED R-SQUARED       0.9873     STANDARD DEVIATION           0.03743

SOURCE          DF       SS              MS          F          P
----------     ---   ----------     ----------    -----     ------
REGRESSION       1     1.85625         1.85625  1324.72     0.0000
RESIDUAL        16     0.02242         0.00140
TOTAL           17     1.87867

CASES INCLUDED 18    MISSING CASES 0
```

Here PERIOD is the number of years since 1976 and LGSANDP is the logarithm of the year–end value. Examination of residual plots and diagnostics (not given here) suggests no problems with the fit of the model (although 1976 is a marginal outlier).

Thus, the estimated effective interest rate satisfies $\log(1+\hat{\imath}) = .0619$, or $1+\hat{\imath} = 10^{.0619}$, or $\hat{\imath} = .1532$. That is, a fully diversified portfolio has been increasing at a rate of about 15.32% per year. The naive estimate, which solves the equation $8.7832 = (1+\hat{\imath})^{17}$, gives $\hat{\imath} = .1363$, which understates the return rate somewhat.

A scatter plot can illustrate why this is so. We can see that the initial returns for the S & P were low (in fact, the return was negative in 1977), implying that the initial principal value in 1976 is a bit above the regression line, leading to underestimation of the return rate by the naive estimate. Note also the near–perfect linearity in the plot, indicating that the semilog model is appropriate here.

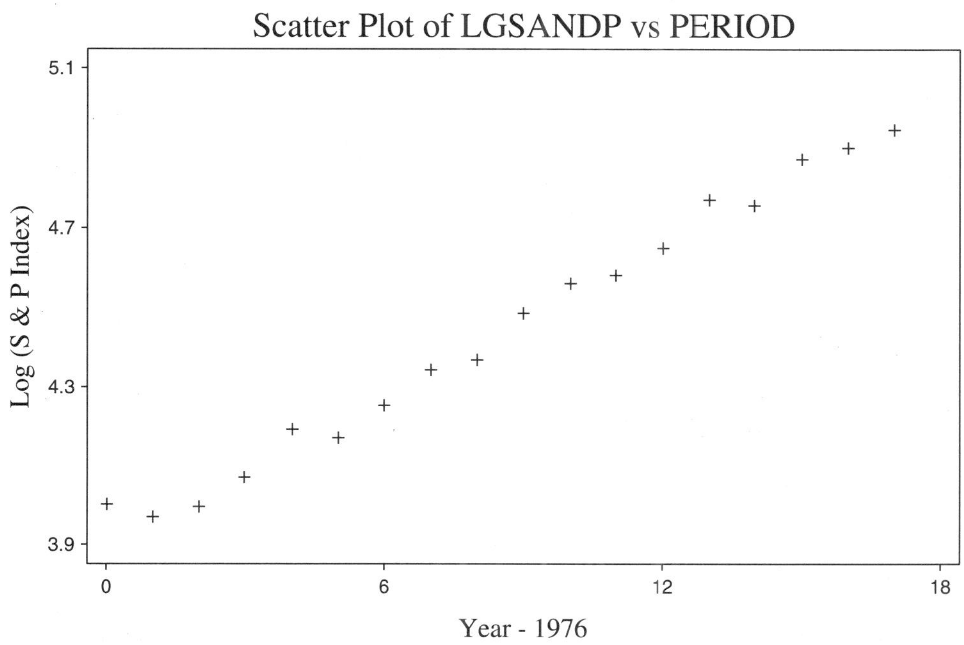

Still another way to estimate the effective annual return rate would be to calculate the sample mean of the 17 observed annual return rates. This turns out to be 14.35%, which is similar to the other estimates. This is clearly not a very sensible estimate, however. Say an investment started with $10,000, and achieved a 10% gain in its first year, followed by a 10% loss in the second year. At the end of the two years the cumulative value of the investment would be $9900 (a net loss), even though the average–based estimate of the cumulative performance would be zero.

Many stock funds (called index funds) attempt to provide investors with conservative performance by tracking the market as a whole. Do they succeed? The following output refers to cumulative performance data for the Vanguard Index Trust–500 Portfolio, an example of such a fund:

```
LINEAR REGRESSION OF Log(Vanguard Index Trust-500)

PREDICTOR
VARIABLES     COEFFICIENT    STD ERROR     STUDENT'S T       P
---------     -----------    ---------     -----------     ------
CONSTANT        3.90853       0.01709        228.73        0.0000
PERIOD          0.06033       0.00172         35.16        0.0000

R-SQUARED              0.9872      RESID. MEAN SQUARE (MSE)     0.00143
ADJUSTED R-SQUARED     0.9864      STANDARD DEVIATION           0.03777

SOURCE         DF        SS             MS          F         P
----------     ---   ----------    ----------    -----     ------
REGRESSION      1     1.76344       1.76344    1236.09    0.0000
RESIDUAL       16     0.02283       0.00143
TOTAL          17     1.78626

CASES INCLUDED 18    MISSING CASES 0
```

The estimated effective rate of return for this portfolio is $\hat{\imath} = 10^{.06033} - 1 = .149$, which can be contrasted with the 15.32% rate of the S & P 500 Index. The rates are close, with the small difference probably reflecting fees charged to investors by the fund. Examination of residual plots and diagnostics (not given here) suggests no problems with the fit of this model, either (although again, 1976 is a marginal outlier).

Summary

The effective rate of return of a variable–rate investment can be estimated using a semilog model. For the time period 1976–1993, the Standard and Poor's 500 Index exhibited an estimated annual rate of return of 15.32%. The Vanguard Index Trust–500 Portfolio, a stock fund designed to mimic the performance of the market as a whole, exhibited a slightly lower estimated return rate of 14.9%, presumably reflecting fees charged to investors by the fund.

Predicting international adoption visas from earlier years

Topics covered: Effects of outliers on regression.

Data File: `adopt.dat`

Recall from our earlier investigation of adoption visas ("International adoption rates") that while there was some stability year–to–year, there also were instances of large fluctuations in visas issued for a particular country.

Build a regression model to try to forecast 1992 visas from the 1988 or 1991 visa values. Pay careful attention to violations of assumptions, and any transformations that might be sensible here.

Is it possible to forecast future visa values at all accurately?

Comments to the Instructor

A regression of `VISA92` on `VISA88` looks good ($R^2 = 92.6\%$, $F = 439.65$). This is illusory, however, as examination of plots and diagnostics indicates that South Korea is an extreme leverage point. Similarly, a regression of `VISA92` on `VISA91` appears useful ($R^2 = 37.5\%$, $F = 20.97$), but two large outliers/leverage points (Romania and South Korea) completely invalidate the results.

As was suggested in the case "International adoption rates," it is sensible to take logs of both variables (arbitrarily setting the two zero values of `VISA88` to have zero `LOGV88` values). Now, a regression of `LOGV92` on `LOGV88` is significant ($R^2 = 59.7\%$, $F = 51.84$), but the four clear outliers noted in the earlier case (Cambodia, Romania, South Korea and Vietnam) invalidate these results. If these four countries are omitted from the analysis, the model assumptions appear to be reasonable, and the model fits fairly well ($R^2 = 81.2\%$, $F = 133.86$).

A regression of `LOGV92` on `LOGV91` appears strong ($R^2 = 81.76\%$, $F = 156.84$), but two countries are again flagged as outliers/leverage points (Romania and South Korea). Omitting these points yields a reasonable model ($R^2 = 86.9\%$, $F = 218.95$).

A multiple regression using both `LOGV88` and `LOGV91` might be considered, but it is not worth the trouble for these data. All four outlying countries with respect to the model using only `LOGV88` are still outliers and/or leverage points, and the resultant two variable model only accounts for a small amount of variability over the model using only `LOGV91` (with the advantage that Cambodia and Vietnam do not need to be considered outliers for the simpler model).

Thus, statistical analysis supports the observation that for the bulk of countries, the number of adoption visas granted is stable, and future values can be predicted from past ones. However, for certain countries (particularly Romania and South Korea for these data), the process can be quite unstable, making accurate prediction difficult or impossible.

The pattern in employment rates over time

Topics covered: Estimation of rate of change. Semilog model.

Data File: `empl.dat`

The following table gives the employment rates in the United States of 25– to 34–year old males with 9–11 years of schooling, by year, as given in *The Condition of Education, 1991*, which is published by the U.S. Department of Education:

Year	Employment rate
1971	87.9
1972	88.5
1973	88.8
1974	90.2
1975	78.1
1976	79.6
1977	81.5
1978	82.4
1979	80.5
1980	77.7
1981	76.7
1982	73.2
1983	69.3
1984	72.2
1985	76.0
1986	73.3
1987	75.0
1988	75.5
1989	77.6
1990	75.9

How has the employment rate changed over the years? That is, what is the rate of change of employment rate for this group? Compare a naive estimate of this number with a more effective one. What accounts for the difference in the estimates?

Comments to the Instructor

The rate of change of the employment rate can be estimated using a semilog model. The fitted regression model is

$$\text{Log}(\widehat{\text{Employment}}\text{ rate}) = 10.2922 - .00424 \times \text{Year}.$$

Thus, the estimated rate of change of the employment rate equals $10^{-.00424} - 1 = -.00972$, or a decrease of about 1% per year.

A naive estimate of the rate of change satisfies

$$\left(\frac{\text{Employment rate}_{1990}}{\text{Employment rate}_{1971}}\right)^{\frac{1}{19}} - 1 = \left(\frac{75.9}{87.9}\right)^{\frac{1}{19}} - 1 = -.00770.$$

The naive estimate is less negative than the regression–based estimate because the employment rate actually increased from 1971 to 1974; since the initial value for 1971 is below the regression line, the naive estimate corresponds to a "flattening out" of the estimated slope.

Note, by the way, that there is autocorrelation apparent in the residuals from the regression model (the Durbin–Watson statistic equals 0.86).

To maintain consistency with the unannotated edition,
the next case begins on the following page.

Estimating a demand function

Topics covered: Added variable plot. Multicollinearity. Multiple regression.

Key words: Confidence interval. Correlation. F–statistic. Model selection. Normal plot. Partial correlation. Partial regression. Regression coefficients. Regression diagnostics. Residual plots. Scatter plot. t–statistic. Time series plot. Variance inflation factor.

Data File: `demand.dat`

A fundamental concept in economics is the idea of supply and demand curves. Simply put, the supply offered for sale of a commodity is directly related to its price, while the demand for purchase of a commodity is inversely related to its price. Can this pattern be quantified and empirically verified? Let's consider gasoline consumption in the United States. These data are based on information given in the *Economic Report of the President, 1987*, and are based on annual statistics for the years from 1960 through 1986. The variables included in the data set are:

```
G       Total gasoline consumption (in tens of millions of
          1967 gasoline-dollars)
PG      Price index for gasoline (in 1967 dollars)
I       Per capita real disposable income (in 1967 dollars)
PNC     Price index for new cars  (in 1967 dollars)
PUC     Price index for used cars  (in 1967 dollars)
PPT     Price index for public transportation (in 1967 dollars)
PD      Aggregate price index for durable goods (in 1982 dollars)
PN      Aggregate price index for nondurable goods (in 1982 dollars)
PS      Aggregate price index for consumer services (in 1982 dollars)
YR      Year (years since 1900)
YRSQ    Year squared (i.e., YR*YR)
```

Gasoline consumption is computed as the current dollar expenditure divided by the price index of gasoline. Since the price index is normalized to be 100 in 1967, it represents gasoline expenditures in units of 1967 "gasoline dollars." The price indices for new cars, used cars and public transportation are also normalized to be 100 in 1967, while those for durable goods, nondurable goods and services are normalized to be 100 in 1982. The variable `YR` represents the number of years since 1900.

The gasoline consumption (`G`) should be a reasonable measure of demand while the price index for gasoline (`PG`) should be a good measure of the price (in real terms). According to economic theory we would expect these two variables to be inversely related. To check this, let's look at a scatter plot of demand versus price:

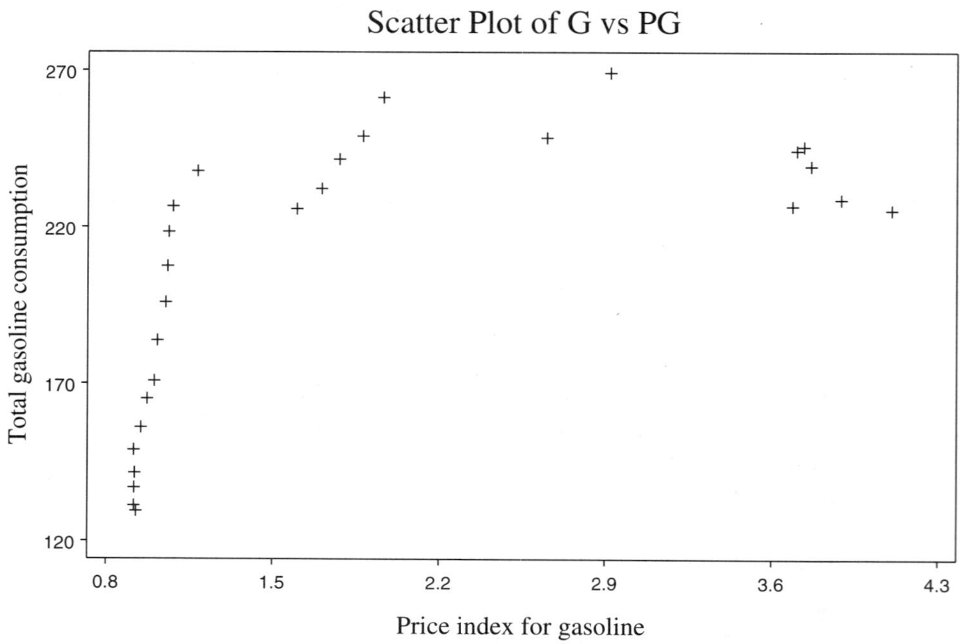

Wait a second! There is apparently a *direct* relationship here, not an inverse one! Could the economic theory be all wrong in practice?

Well, no. We haven't taken into account that potential determinants of gasoline consumption other than price have also changed over the years. We can do this using multiple regression to model G in terms of the other variables. Let's take a look at the time series plot of consumption versus year for a start:

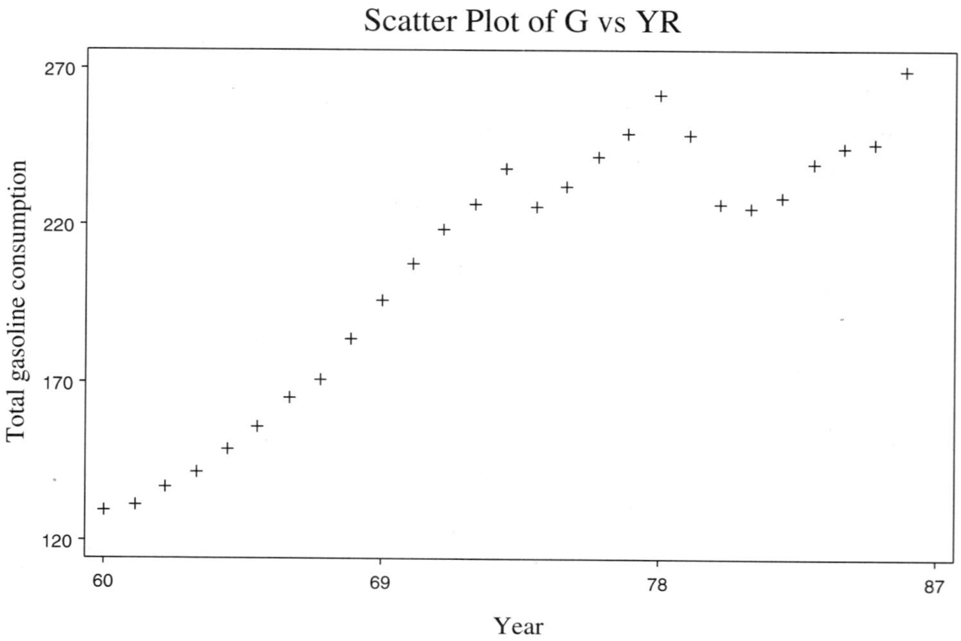

We see that gasoline consumption increased steadily until 1973, after which there is apparently a leveling off effect. One way to capture this behavior is to hypothesize that gasoline consumption has a quadratic relationship with time. We can do this by including both YR and YR^2 (YRSQ) as predicting variables in the multiple regression model.

Examination of the data shows that these predicting variables are highly correlated with each other. A regression using all of the variables will exhibit high **multicollinearity**, and will include several redundant variables. Here are the pairwise correlation coefficients:

CORRELATIONS (PEARSON)

	G	PG	I	PNC	PUC	PPT	PD	PN	PS	YR
PG	0.6361									
I	0.9588	0.8122								
PNC	0.6963	0.9525	0.8519							
PUC	0.6722	0.9108	0.8318	0.9790						
PPT	0.7049	0.9229	0.8569	0.9814	0.9880					
PD	0.7596	0.9633	0.8987	0.9911	0.9696	0.9765				
PN	0.7693	0.9648	0.9070	0.9881	0.9634	0.9725	0.9984			
PS	0.7496	0.9379	0.8915	0.9932	0.9855	0.9913	0.9930	0.9913		
YR	0.9144	0.8801	0.9881	0.9141	0.8947	0.9166	0.9516	0.9565	0.9438	
YRSQ	0.8988	0.8923	0.9817	0.9315	0.9134	0.9330	0.9634	0.9676	0.9581	0.9989

CASES INCLUDED 27 MISSING CASES 0

Here is the regression output. Just as we suspected, we have many redundant variables (i.e., ones with coefficients that are not statistically significantly different from zero, as measured by their t–statistics). Note the absurdly high variance inflation factor (VIF) values, by the way, indicating tremendously high multicollinearity.

LINEAR REGRESSION OF G Total gasoline consumption

PREDICTOR VARIABLES	COEFFICIENT	STD ERROR	STUDENT'S T	P	VIF
CONSTANT	2084.17	697.202	2.99	0.0087	
PG	-23.6541	7.28924	-3.25	0.0051	236.8
I	0.03091	0.00688	4.49	0.0004	327.5
PNC	12.3119	37.8893	0.32	0.7494	859.3
PUC	-3.12157	6.88753	-0.45	0.6565	147.3
PPT	34.7454	8.53921	4.07	0.0009	292.7
PD	450.342	90.9472	4.95	0.0001	1346.2
PN	-28.5600	101.857	-0.28	0.7828	2529.8
PS	-732.502	144.456	-5.07	0.0001	6244.8
YR	-69.2705	21.3945	-3.24	0.0052	94234.5
YRSQ	0.56381	0.17742	3.18	0.0058	138453.2

R-SQUARED	0.9974	RESID. MEAN SQUARE (MSE)	7.95624
ADJUSTED R-SQUARED	0.9959	STANDARD DEVIATION	2.82068

SOURCE	DF	SS	MS	F	P
REGRESSION	10	49749.7	4974.97	625.29	0.0000
RESIDUAL	16	127.300	7.95624		
TOTAL	26	49877.0			

CASES INCLUDED 27 MISSING CASES 0

Here is the regression using only the variables with significant t–statistics (i.e., p–values < 0.05):

```
LINEAR REGRESSION OF G  Total gasoline consumption

PREDICTOR
VARIABLES     COEFFICIENT    STD ERROR     STUDENT'S T      P        VIF
---------     -----------    ---------     -----------    ------    -----
CONSTANT        2141.86       383.769          5.58       0.0000
PG             -24.5627       2.79510         -8.79       0.0000     38.9
I               0.03143       0.00592          5.31       0.0000    270.7
PPT             34.6922       6.86060          5.06       0.0001    210.9
PD              454.356       57.6670          7.88       0.0000    604.2
PS             -748.238       93.6662         -7.99       0.0000   2930.6
YR             -70.7234       11.7576         -6.02       0.0000  31768.5
YRSQ            0.57248       0.09512          6.02       0.0000  44424.8

R-SQUARED              0.9973      RESID. MEAN SQUARE (MSE)    7.12779
ADJUSTED R-SQUARED     0.9963      STANDARD DEVIATION          2.66979

SOURCE         DF        SS            MS           F          P
----------     ---    ----------    ----------    -----     ------
REGRESSION      7      49741.6       7105.94     996.93     0.0000
RESIDUAL       19      135.428       7.12779
TOTAL          26      49877.0

CASES INCLUDED 27    MISSING CASES 0
```

The key thing to recognize here is that the coefficient for the price of gasoline (PG) in the equation is –24.56. Recall what this means: **given that income, transportation price index, durable goods price index, consumer services price index and year are held fixed, a one unit increase in the gasoline price index is associated with a 24.56 unit decrease in gasoline consumption**. This, of course, is the inverse relationship we expect to see in a demand curve. The variance inflation factors are still very high, which will make the standard error of the coefficient of PG higher. A 95% confidence interval for the decrease in gasoline consumption associated with a one unit increase in the price index is

$$\begin{aligned} &\widehat{\beta}_{\text{PG}} \pm t^{n-8}_{.025} \times \widehat{\text{SE}}(\widehat{\beta}_{\text{PG}}) \\ &= -24.563 \pm t^{19}_{.025} \times 2.795 \\ &= (-30.41, \ -18.71), \end{aligned}$$

where $n = 27$ is the number of cases, $\widehat{\text{SE}}(\widehat{\beta}_{\text{PG}})$ is the estimated standard error of $\widehat{\beta}_{\text{PG}}$, and $t^{19}_{.025} = 2.093$ is the t–table value for a two-sided 95% confidence interval on 19 degrees of freedom. The values for $\widehat{\beta}_{\text{PG}}, n,$ and $\widehat{\text{SE}}(\widehat{\beta}_{\text{PG}})$ can be read from the numerical summary above.

Here are the leverage values, standardized residuals and influence diagnostics from the fit:

YEAR	LEVERAGE	STDRES	COOKSD
1960	0.5879	-0.2969	0.0157
1961	0.2667	0.2318	0.0024
1962	0.1743	0.8134	0.0175
1963	0.2670	1.5889	0.1150
1964	0.1399	-0.8925	0.0162
1965	0.1459	-1.0720	0.0245
1966	0.1849	-0.3448	0.0034
1967	0.1670	-1.5603	0.0610
1968	0.1260	-1.1723	0.0248
1969	0.1088	0.8759	0.0117
1970	0.1441	0.7557	0.0120
1971	0.1799	1.1223	0.0345
1972	0.1940	0.9188	0.0254
1973	0.4337	0.1179	0.0013
1974	0.3565	0.5971	0.0247
1975	0.2576	-1.2745	0.0704
1976	0.2594	-2.2679	0.2252
1977	0.1846	0.6254	0.0111
1978	0.4479	1.4707	0.2193
1979	0.4177	0.4246	0.0162
1980	0.4039	-0.3410	0.0098
1981	0.4209	0.0254	0.0001
1982	0.3358	0.3678	0.0085
1983	0.4075	0.1133	0.0011
1984	0.2942	-0.7588	0.0300
1985	0.4555	1.1609	0.1409
1986	0.6387	-1.1417	0.2880

They do not suggest any problems.

Here is the plot of the standardized residuals versus the fitted values and also the normal plot for the standardized residuals:

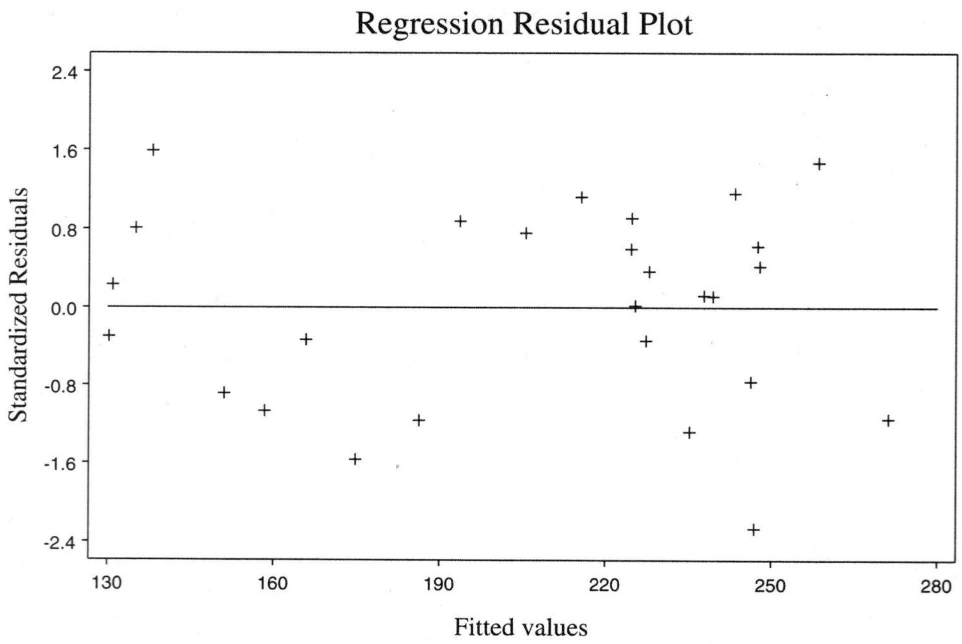

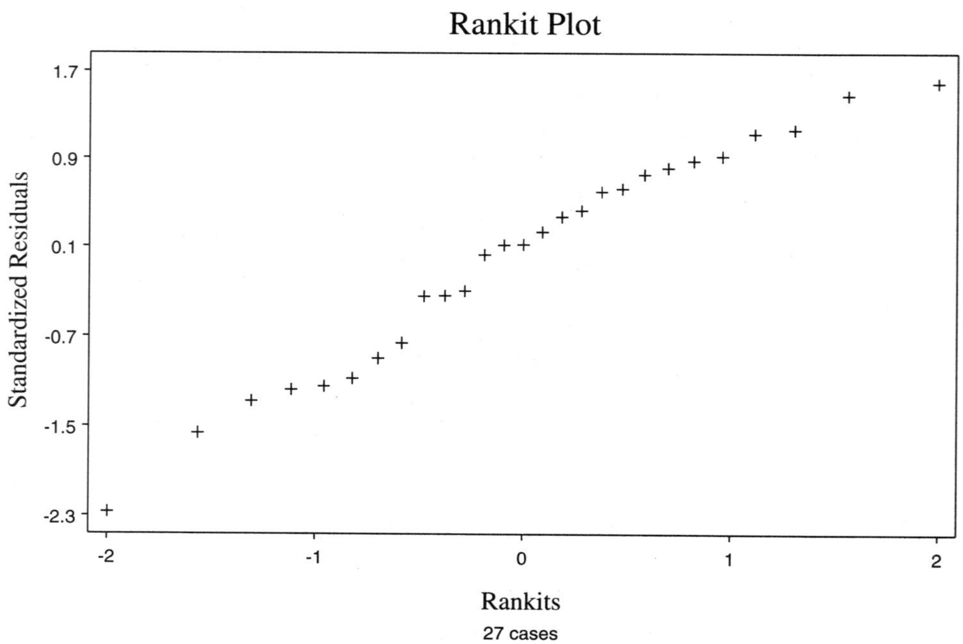

These are not "textbook," but are adequate. Let's look at the time series plot of the standardized residuals:

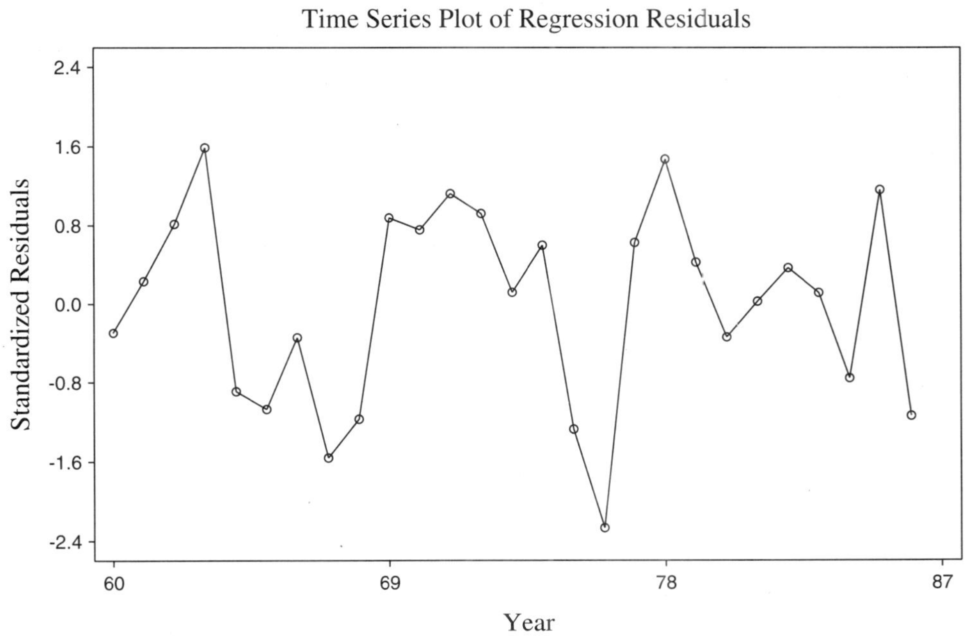

There is a pattern in the standardized residuals related to the time ordering, although not a very strong one (we will say more about this in a later case).

We can separate out the demand relationship using the concept of *partial regression*. The partial correlation coefficient represents the correlation between two variables, after removing the effect of other predictors on them. Here is the partial correlation of G with PG taking I, PPT, PD, PS, YR and YRSQ into account:

```
PARTIAL CORRELATIONS WITH G  Total gasoline consumption
CONTROLLED FOR I PPT PD PS YR YRSQ

PG            -0.8958

CASES INCLUDED 27   MISSING CASES 0
```

Note how large (and negative) the value is. Another way to see what is going on is to look at a plot of the "adjusted demand" versus "adjusted gasoline price index." That is, the partial correlation of –.896 tells us that there's a strong inverse relationship between demand for gasoline and price after all of the other variables have been taken into account; can we see it graphically? The answer is yes, using an *added variable plot*. Some computer packages can do it automatically. Here's how to do it manually:

(1) Fit a regression of the target variable (here G) on all of the predictors except the one you're interested in (here PG). Save the residuals (call it RESI1).

(2) Fit a regression of the predicting variable you're interested in (here PG) on the other predicting variables. Save the residuals (call it RESI2).

(3) If you regressed RESI1 on RESI2 you'd notice something remarkable — the coefficient for RESI2 is precisely the same as the coefficient for the predictor of interest in the full multiple regression model (that is, the coefficient for RESI2 here is exactly –24.5627).

(4) If you'd like a graphical representation of the adjusted relationship between

the predictor of interest and the target variable, taking the other predictors into account, just do a scatter plot of RESI1 on RESI2. The correlation between RESI1 and RESI2 is the partial correlation given above (check it!).

Completing these four steps, we have the (adjusted) demand curve for gasoline, which clearly shows the inverse relationship:

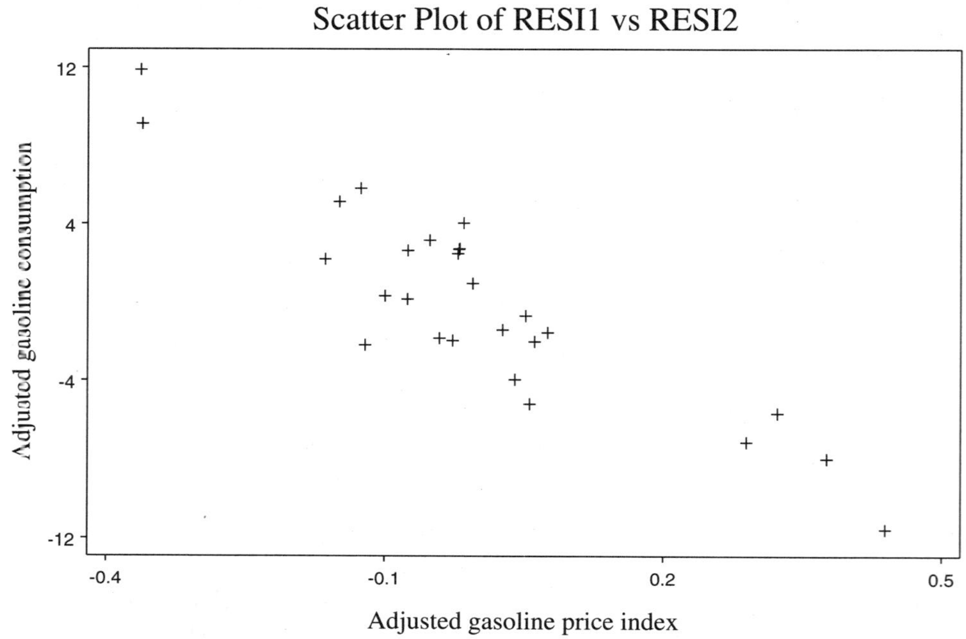

There's one more point worth making about these data. Multicollinearity is a serious problem here, as the high VIFs show. This means that the coefficients in the regression can be somewhat unstable. In other words, our estimate of the change in consumption associated with a change in price holding all other variables constant might not be accurate, because we have very little data consistent with that occurrence. This is because price is highly correlated with the other variables, so that there are little data on situations where price alone changes while all of the other variables remain unchanged. The adjusted demand curve is therefore not estimated very precisely.

One way around this instability is to try to build a reasonable model that doesn't exhibit multicollinearity. One possibility is a three variable model for the gasoline consumption based on the price index of gasoline (PG), per capita real disposable income (I) and the price index of used cars (PUC). The model fits very well, with little collinearity.

```
LINEAR REGRESSION OF G  Total gasoline consumption

PREDICTOR
VARIABLES     COEFFICIENT    STD ERROR     STUDENT'S T       P        VIF
---------     -----------    ---------     -----------    ------     -----
CONSTANT        -103.506      9.68365        -10.69       0.0000
PG              -10.9454      2.47158         -4.43       0.0002      6.2
I                0.04058      0.00147         27.52       0.0000      3.4
PUC             -8.25978      3.11278         -2.65       0.0142      6.9

R-SQUARED              0.9839      RESID. MEAN SQUARE (MSE)     34.8246
ADJUSTED R-SQUARED     0.9818      STANDARD DEVIATION           5.90124

SOURCE        DF       SS            MS          F         P
----------    ---   ----------   ----------    -----    ------
REGRESSION     3     49076.0      16358.7     469.75    0.0000
RESIDUAL      23     800.966      34.8246
TOTAL         26     49877.0

CASES INCLUDED 27    MISSING CASES 0
```

Note that the coefficients have meaningful signs. The coefficient for per capita disposable income is positive, indicating that gasoline consumption goes up with an increase in per capita disposable income when the other variables are unchanged. The negative value of the coefficient for the price index of gasoline (PG) indicates the inverse relationship predicted by theory (holding all else fixed). The negative coefficient for PUC (price index of used cars) is also in accord with theory; as the price of used cars goes up (with other predictors remaining constant), fewer are bought and hence less gasoline is consumed.

The coefficient for PG is still negative, although it is half the magnitude of what it was before. As before we can construct a 95% confidence interval for the decrease in gasoline consumption with a one unit increase in the price index:

$$\begin{aligned}&\widehat{\beta}_{\mathrm{PG}} \pm t^{n-4}_{.025} \times \widehat{\mathrm{SE}}(\widehat{\beta}_{\mathrm{PG}})\\ &= -10.945 \pm t^{23}_{.025} \times 2.472,\\ &= (-16.06, \ -5.83)\end{aligned}$$

where $t^{23}_{.025} = 2.069$. This interval has smaller width than the interval based on the larger model; this estimate is more trustworthy, making us pretty confident that demand for gasoline actually is inversely related to its price. Note that the two coefficients $\widehat{\beta}_{\mathrm{PG}}$ discussed here are measuring slightly different things, as the predictors that are held fixed differ between the two models.

Summary

The demand for gasoline in the United States can be modeled using the price index for gasoline, in addition to other price indices and a time trend represented by a quadratic relationship with Year. Given the other predictors, the estimated change in gasoline consumption associated with a one unit increase in the gasoline price index is roughly an 18 – 30 unit decrease. If the time trend is not explicitly taken into account and we adjust only for the per capita real disposable income and the price index of used cars, the estimated decrease in gasoline consumption associated with a one unit increase in the gasoline price index is roughly 6 – 16 units.

Technical terms

Added variable plot: a graphical representation of the relationship between the target variable and a given predictor variable, adjusting for the effects of the other predictor variables. Explicitly, it is a plot of the target variable with the linear effects of the other variables removed against the given predictor variable with the linear effects of the other predictor variables removed. In each case the variable with the linear effects removed is the residual variable from the multiple linear regression of the variable on the other predictor variables. This plot is sometimes called a **partial regression plot.**

Multicollinearity: the existence of strong linear relationships between, and among, the predicting variables in a multiple regression. The condition is often simply termed **collinearity**. Collinearity can lead to instability in the regression estimates, with small changes in the data leading to large changes in regression coefficients and t–statistics. The significance of individual variables is often understated in the presence of collinearity, due to inflation of the estimated standard errors of their associated estimated coefficients.

Variance inflation factor: a diagnostic used to assess the level of multicollinearity in a set of predicting variables. The VIF is calculated as follows: if R_j^2 is the R^2 value for the regression of variable X_j on the other predictors, then the VIF for variable X_j is $1/(1 - R_j^2)$. The VIF is an estimate of the proportional increase (or inflation) in the variance of the estimated regression coefficient for X_j over the value that would have occurred if the predictors had been perfectly uncorrelated. Values of VIF over 10 are often considered indicative of potential collinearity problems.

The effectiveness of National Basketball Association guards

Topics covered: Finding structure and unusual values. Multiple regression modeling. Prediction.

Key words: Histogram. Median. Normal plot. Outliers. Predicted value. Prediction interval. Quartiles. Regression. Residual. Residual plots. Scatter plot. Skewness. Stem–and–leaf.

Data File: `nba.dat`

The National Basketball Association (NBA) is a professional league of 27 teams. It has grown in prominence in recent years with revenue and player salaries rising dramatically. What factors affect the (statistical) performance of players that play the "guard" position on each team? A typical team plays two guards at the same time, but keeps more than that on its 12 man roster. *The Pro Basketball Bible* publishes the statistical totals for the year for each player, including the number of games in which the player appeared, minutes played, assists made, points scored and rebounds made. It also records the percentage of free throws made and the percentage of field goals made.

Here are some descriptive statistics for all 105 guards for the 1992– 1993 season. Variables include minutes per game (`MPG`), points per minute (`PPM`), field goal percentage (`FGP`), assists per minute (`APM`) and free throw percentage (`FTP`).

DESCRIPTIVE STATISTICS

	AGE	HEIGHT	GAMES	MINUTES	MPG
N	105	105	105	105	105
MEAN	27.533	190.17	64.152	1656.5	24.303
SD	3.3686	7.0744	19.059	885.61	9.8574
MINIMUM	22.000	160.00	8.0000	55.000	4.3500
1ST QUARTILE	25.000	185.00	50.500	874.00	16.679
MEDIAN	27.000	191.00	73.000	1622.0	24.277
3RD QUARTILE	30.000	196.00	79.000	2433.5	33.691
MAXIMUM	37.000	210.00	82.000	3117.0	40.711

	PPM	FGP	APM	FTP
N	105	105	105	105
MEAN	0.4236	45.132	0.1599	77.855
SD	0.1159	4.3922	0.0604	9.1764
MINIMUM	0.1593	34.800	0.0494	37.500
1ST QUARTILE	0.3320	42.650	0.1070	74.050
MEDIAN	0.4187	45.300	0.1577	79.300
3RD QUARTILE	0.4884	47.850	0.2106	84.000
MAXIMUM	0.8291	59.500	0.3437	94.800

How are these factors related to each other? Are there patterns that might help us predict the various measures of performance of the players?

The first issue is the choice of measures of performance. Performance for a guard is not unidimensional. For many players, the primary responsibility is to distribute the ball to the other players on the team who score (an "assist"). These players are often called "point" guards. Often for these players scoring points and rebounding are secondary issues. Others players look to score points first and distribute the ball second.

These players are often called "shooting" guards. Clearly we need to be conscious of this.

Another issue is whether we should be focusing on the overall performance (i.e., *production*) as measured by the statistical totals (e.g., points scored), or the performance rate (i.e., *effectiveness*). In determining the performance rate of a player one can consider the production per opportunity to produce (e.g., points scored per minute). We will consider some alternative measures later. Here we focus on scoring rate, that is, the points scored per minute, because the number of minutes per game varies by player.

Let's take a look at the opportunities the players have to perform. First let's look at the distribution of the number of games played:

```
STEM AND LEAF PLOT OF GAMES   Games played

   LEAF DIGIT UNIT = 1
   1  2  REPRESENTS 12 GAMES

          STEM  LEAVES
       2   +0. 89
       4    1* 03
       4    1.
       4    2*
       5    2. 8
       9    3* 0133
      14    3. 55799
      17    4* 001
      25    4. 55678999
      30    5* 01334
      35    5. 55799
      39    6* 0134
      48    6. 688889999
     (11)   7* 00223444444
      46    7. 55566677777888899999999
      23    8* 00000011111111222222222

105 CASES INCLUDED     0 MISSING CASES
```

The distribution is left–skewed with four players playing in only 13 games or less. The median number of games played is 73 out of a total of 82 games in the season. The "wall" at 82 games reflects that there were 82 games in the season. From the last row of the plot, we see that 23 out of the 105 players played in 80 or more of the games.

The distribution of minutes played is also interesting:

```
STEM AND LEAF PLOT OF MINUTES   Minutes played

   LEAF DIGIT UNIT = 100
   1  2  REPRESENTS 1200 MINUTES

          STEM  LEAVES
      6    +0*  000111
     10    +0T  2233
     14    +0F  4555
     19    +0S  66677
     32    +0.  8888888889999
     36     1*  0111
     40     1T  2223
     50     1F  4555555555
   (11)     1S  66666666777
     44     1.  89
     42     2*  0000011
     35     2T  2222333
     28     2F  444444455
     19     2S  6677
     15     2.  8888889999
      5     3*  00011
105 CASES INCLUDED     0 MISSING CASES
```

This distribution is not skewed to the right or left, but rather exhibits several peaks and valleys. Interestingly, a large number of players played just less than 1000 minutes. This works out to about 12 minutes per scheduled game. There is also a scarcity of players who played just less than 2000 minutes. There is some variation in the number of games played and the total number of minutes played. Let's look at minutes played per game (MPG) as a measure of the contribution per game played:

```
STEM AND LEAF PLOT OF MPG  Minutes played per game

   LEAF DIGIT UNIT = 1
   1  2  REPRESENTS 12.

          STEM  LEAVES
      2    +0* 44
      9    +0. 6666799
     23     1* 01122223333344
     36     1. 6667777788899
   (20)     2* 00000111122223344444
     49     2. 55677778899
     38     3* 001111122333334444
     20     3. 5555556666666888899
      1     4* 0

105 CASES INCLUDED     0 MISSING CASES
```

Each stem now represents five values rather than the two in the previous plot, and thus the plot is more compressed. Players with MPG in the range between 20 and 24 are more common than those in the groups on either side. Based on observing this pattern we could, as a working rule, group the players into two categories of MPG: part–timers (MPG $<$ 25) and regulars (MPG $\geq$ 25).

Let's see how this factor is related to the number of games played by plotting the minutes played per game versus the number of games played:

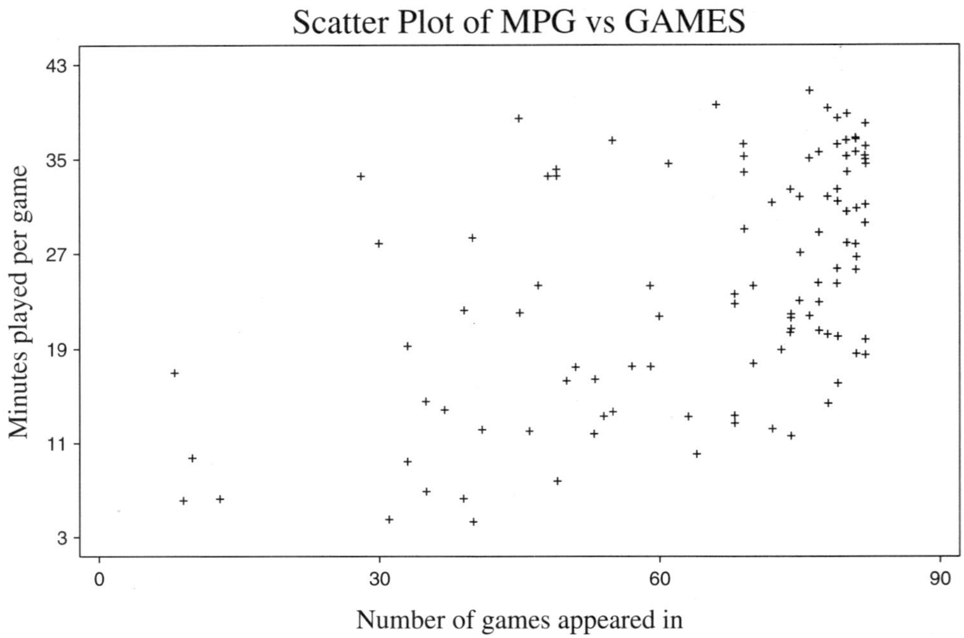

As we might expect, there is a positive relationship between the variables. The variability in MPG for players in the middle is reasonable constant. The variability decreases slightly for those who played in almost all games (do not be fooled by the figure; the density of players increases at the right edge of the plot).

How are the player's ages distributed? Let's look at a stem–and–leaf display:

```
STEM AND LEAF PLOT OF AGE  Player's age

   LEAF DIGIT UNIT = 0.1
   1  2  REPRESENTS 1.2

          STEM  LEAVES
      1    22   0
     12    23   00000000000
     24    24   000000000000
     35    25   00000000000
     44    26   000000000
   (13)    27   0000000000000
     48    28   000000000000
     36    29   0000
     32    30   0000000000
     22    31   0000000
     15    32   000000
      9    33   000
      6    34   0000
      2    35   0
      1    36
      1    37   0

105 CASES INCLUDED     0 MISSING CASES
```

Here the leaves are the same (0), as the ages are recorded to the nearest year, and the stem–and–leaf display looks a lot like the corresponding frequency histogram. We note that the number of players tapers off above 32 years of age. Interestingly, there is a deficit of players aged 29.

Now let's look at performance:

```
STEM AND LEAF PLOT OF PPM  Points scored per minute

   LEAF DIGIT UNIT = 0.01
   1  5  REPRESENTS 0.15

          STEM  LEAVES
      1     1   5
      6     2   13344
     12     2   567889
     30     3   000000111222333444
     45     3   556667788889999
    (23)    4   00000111223333333344444
     37     4   55667777888999
     23     5   0112244
     16     5   5556777788
      6     6   023
      3     6   6
      2     7
      2     7   7
      1     8   2

 105 CASES INCLUDED     0 MISSING CASES
```

The distribution is interesting, as it suggests two modes and has a couple of high values.

Let's see how AGE is related to PPM. We could use a scatter plot, although the discreteness of age will make it difficult to see the general picture. Instead we will use a set of side–by–side boxplots, one for each age:

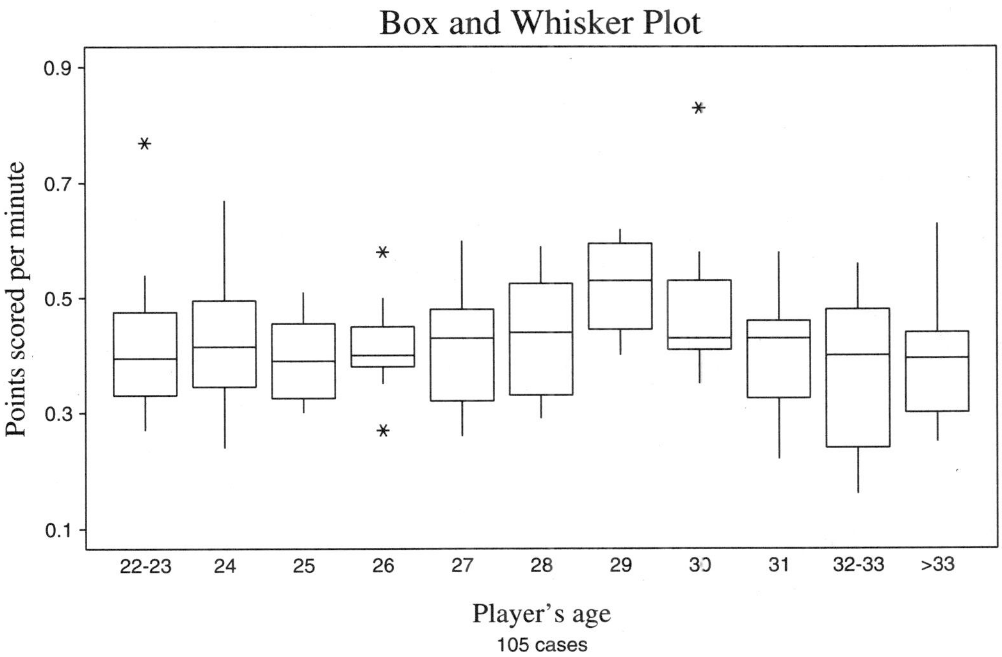

Note that we have pooled a few ages to ensure the individual boxplots are based on sufficient observations. We see that the level of PPM is reasonably stable up to age 26 years and then increases up to age 29 years. After that there is a gradual decline. There does not seem to be a pattern in the variability with age. We note four outliers. The two of them at age 26 seem reasonable. The upper outlier in the youngest group is Sean Green, who played a total of 81 minutes in the entire season. Green did not follow up on this performance very effectively, and has since been cut from his team. The upper outlier at age 30 is Michael Jordan. He seems to be in a different league, and is very different from the other 30 year olds. However, the general pattern does suggest that he could be at his peak and that decline could now begin. Perhaps it is time to try baseball!

We know that some of the variation apparent in these plots can be understood in terms of other factors, such as the number of games played or number of minutes played per game. Let's consider the latter by looking at a scatter plot:

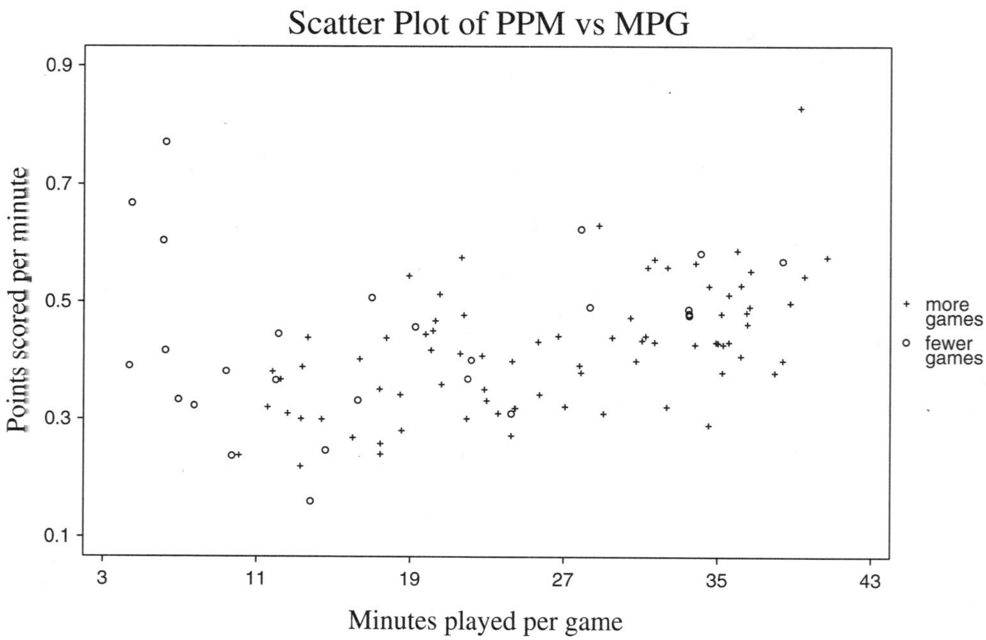

In this plot we have used the plot symbol to categorize the players based on the number of games in which they appeared. If a player appeared in fewer than the lower quartile number of games (i.e., less than 51 games) the plot symbol is a ○. Otherwise, the plot symbol is the usual +. In this way we can represent three variables in the plot. We see a generally positive relationship between the number of minutes played per game and the points per minute. Usually the team's game plan is centered on the regular players; they play more minutes per game and have scoring plays designed around them. However, the plot provides much more. Three part–time players with low MPG scored a relatively large number of points in those minutes. From the data–table these are found to be Sean Green (81 minutes), Bobby Phills (139 minutes), and Bo Kimble (55 minutes). These players usually played at the end of games when the outcome had been decided.

Interestingly, there appears to be a hint of a downward shift in PPM at about

MPG = 25. The pattern suggests that those playing less than 25 minutes per game tend to be more effective scoring than those playing more than 25 minutes per game, given the generally downward trend in the scoring rate as the minutes decrease for the regular players. The decreasing trend in PPM with increasing MPG for the players with MPG $\geq$ 25 might be a result of fatigue; another possibility is that players who play a lot of minutes pace themselves so as not to be too tired at the end of the game. In addition, their effectiveness appears to be less variable than those playing less than 25 minutes per game.

Let's look at the plot symbols and identify the players from the original data. We see seven players who appeared in fewer games who were regular players in the games they did play. Jimmy Jackson was a rookie player involved in an extended contract dispute that restricted his play to the last 28 games in the season. The other players were key veterans who were injured and played in limited games: Sarunas Marciulionis (30 games), Kevin Edwards (40 games), Mitch Richmond (45 games), Steve Smith (48 games), Kevin Johnson (49 games), and Clyde Drexler (49 games). Sarunas Marciulionis is particularly interesting because he was injured 30 games into the season and was scoring at a relatively high rate. Indeed, as a group, these players appear to have a higher scoring rate than we would expect given the number of minutes played.

Finally, there is an outlier at the upper right corner of the plot that is more apparent here when the relationship between PPM and MPG is taken into account. Michael Jordan played the third highest minutes per game and had by far the highest value of PPM. The next closest regular was Ricky Pierce with 0.63 points per minute against Jordan's 0.83. Pierce is a 34 year old player who played 29 minutes per game against Jordan's 39. Pierce also had higher PPM than would be expected from the number of minutes played per game, but not nearly as dramatically as did Jordan.

How well can we predict PPM based on the other factors? Which factors should we use? Variables such as AGE and MPG are naturally thought of as *explanatory* variables. That is, they are entirely determined by unrelated factors (e.g., age) or can be largely controlled directly by the coach or players. Variables such as assists per minute, field goal percentage and PPM are the outcomes of interest or *target* variables. When we predict PPM we will not use other target variables, because they cannot be manipulated easily by the coach or players. The division between the two groups is hazy (e.g., free throw made percentage [FTP]). Although the decision to use only variables that are under direct control as predictors is not the only possibility here, it does permit us to investigate scoring effectiveness apart from other types of performance.

We have seen that players playing very few minutes are quite different from those who play a sizable part of the season. From now on we will be focusing on those players playing 10 or more minutes per game and appearing in 10 or more games. This excludes 10 players. We see that Michael Jordan is an outlier in terms of PPM, so we will omit him (for the moment) from the data.

Let's start by looking at a simple regression model predicting PPM from MPG. The numerical regression summary is as follows:

```
LINEAR REGRESSION OF PPM  Points scored per minute

PREDICTOR
VARIABLES     COEFFICIENT     STD ERROR      STUDENT'S T        P
---------     -----------     ---------      -----------      ------
CONSTANT          0.24168       0.02765             8.74      0.0000
MPG               0.00669       0.00101             6.60      0.0000

R-SQUARED               0.3212      RESIDUAL MEAN SQUARE (MSE)     0.00700
ADJUSTED R-SQUARED      0.3138      STANDARD ERROR OF ESTIMATE     0.08367

SOURCE          DF         SS              MS            F          P
----------     ---     ----------     ----------     -----     ------
REGRESSION       1       0.30473        0.30473      43.53     0.0000
RESIDUAL        92       0.64405        0.00700
TOTAL           93       0.94878

CASES INCLUDED 94     MISSING CASES 0
```

The standardized residuals versus fitted values plot can then be constructed:

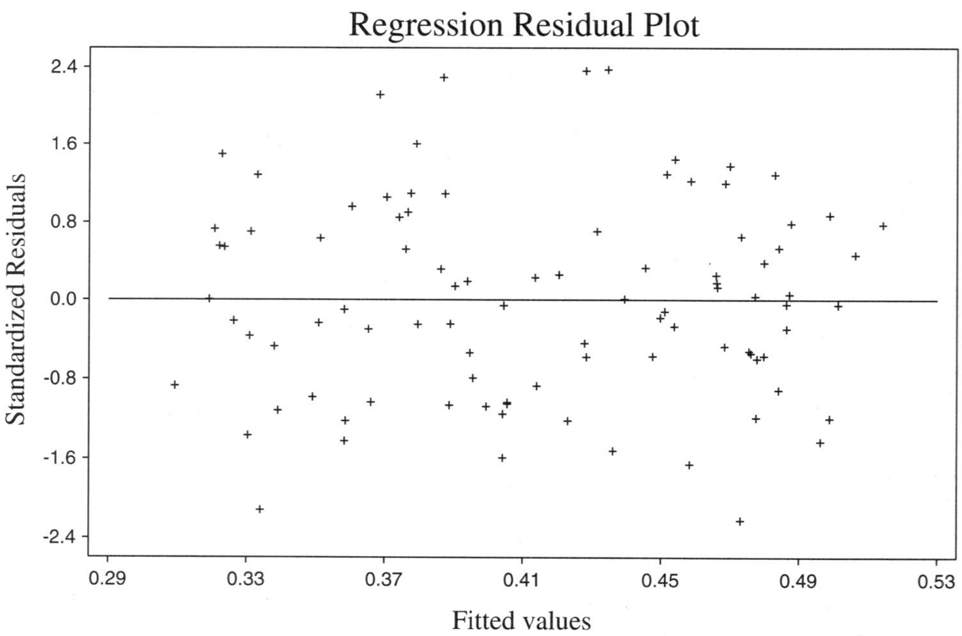

An inspection of the standardized residuals indicates none outside $(-2.5, 2.5)$, although the shift we noted in the scatter plot is visible. This feature might be explainable in terms of other variables. Thus the relationship between PPM and MPG could be linear for fixed levels of the other variables, even though it may not be so unconditionally. The Cook's distances and leverages of the players are modest (not reported here).

Let's add now the player's height. Here is the numerical regression summary:

```
LINEAR REGRESSION OF PPM  Points scored per minute

PREDICTOR
VARIABLES     COEFFICIENT    STD ERROR     STUDENT'S T       P        VIF
---------     -----------    ---------     -----------    ------     -----
CONSTANT        -0.47710       0.23768        -2.01       0.0477
MPG              0.00663     9.720E-04         6.82       0.0000       1.0
HEIGHT           0.00379       0.00125         3.04       0.0031       1.0

R-SQUARED              0.3839      RESIDUAL MEAN SQUARE (MSE)     0.00642
ADJUSTED R-SQUARED     0.3703      STANDARD ERROR OF ESTIMATE     0.08015

SOURCE        DF        SS            MS          F         P
----------   ---    ----------    ----------    -----    ------
REGRESSION     2     0.36422       0.18211      28.35    0.0000
RESIDUAL      91     0.58456       0.00642
TOTAL         93     0.94878

CASES INCLUDED 94   MISSING CASES 0
```

The standardized residual versus fitted values plot now has the following form:

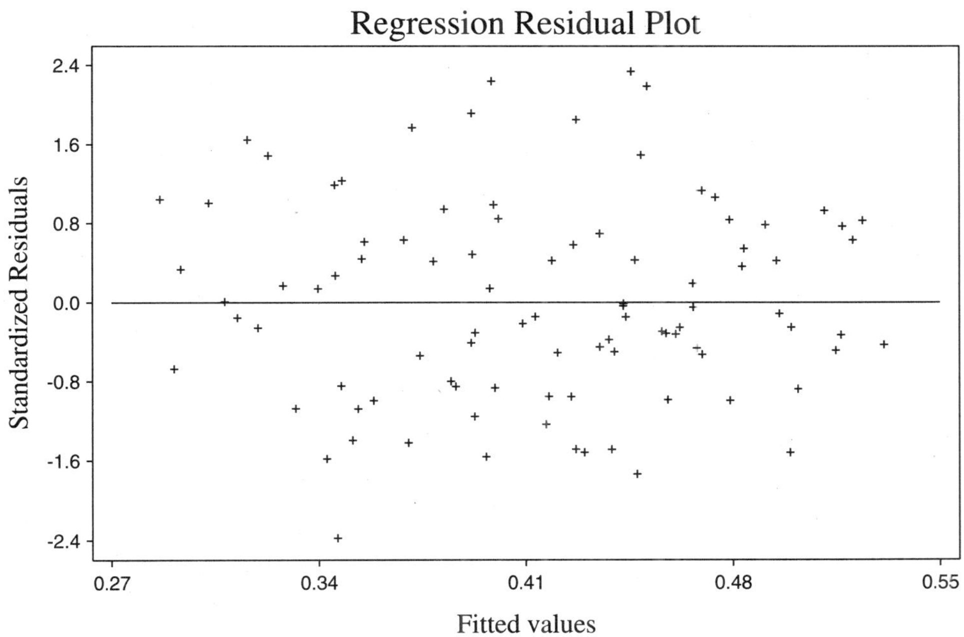

Darryl Walker has a standardized residual of -2.37. He is a veteran player who has a defensive role, as indicated by his low scoring rate. Ricky Pierce has a standardized residual of 2.34, commensurate with his high–scoring rate. Some other residuals are large, although again none outside of $(-2.5,\ 2.5)$. Spud Webb and Muggsy Bogues have modest standardized residuals (0.101 and 0.012, respectively) and, as we might expect, elevated leverages (0.118 and 0.241, respectively). At 5'7" and 5'3", respectively, they are special players. However, their influence on the model is modest, suggesting that they conform to the general pattern (Cook's distances of 0.01 and 0.10, respectively). The relationship looks reasonably linear in these two variables.

Both variables have coefficients that are statistically significantly different from zero and are, as expected, positive. The model explains about 38.4% of the variation in PPM. Let's see if adding FTP will improve the model. We would expect PPM and FTP to be positively related, given the other variables.

The numerical regression summary of the new model is the following:

```
LINEAR REGRESSION OF PPM  points scored per minute

PREDICTOR
VARIABLES    COEFFICIENT    STD ERROR    STUDENT'S T       P      VIF
---------    -----------    ---------    -----------    ------   -----
CONSTANT       -0.78138       0.24458        -3.19      0.0019
MPG             0.00565     9.724E-04         5.81      0.0000     1.1
HEIGHT          0.00414       0.00119         3.48      0.0008     1.0
FTP             0.00337       0.00103         3.26      0.0016     1.1

R-SQUARED            0.4489     RESIDUAL MEAN SQUARE (MSE)     0.00581
ADJUSTED R-SQUARED   0.4305     STANDARD ERROR OF ESTIMATE     0.07622

SOURCE       DF        SS          MS          F        P
----------   ---   ----------  ----------   -----   ------
REGRESSION    3     0.42588     0.14196     24.43   0.0000
RESIDUAL     90     0.52290     0.00581
TOTAL        93     0.94878

CASES INCLUDED 94    MISSING CASES 0
```

Here is the standardized residuals versus fitted values plot:

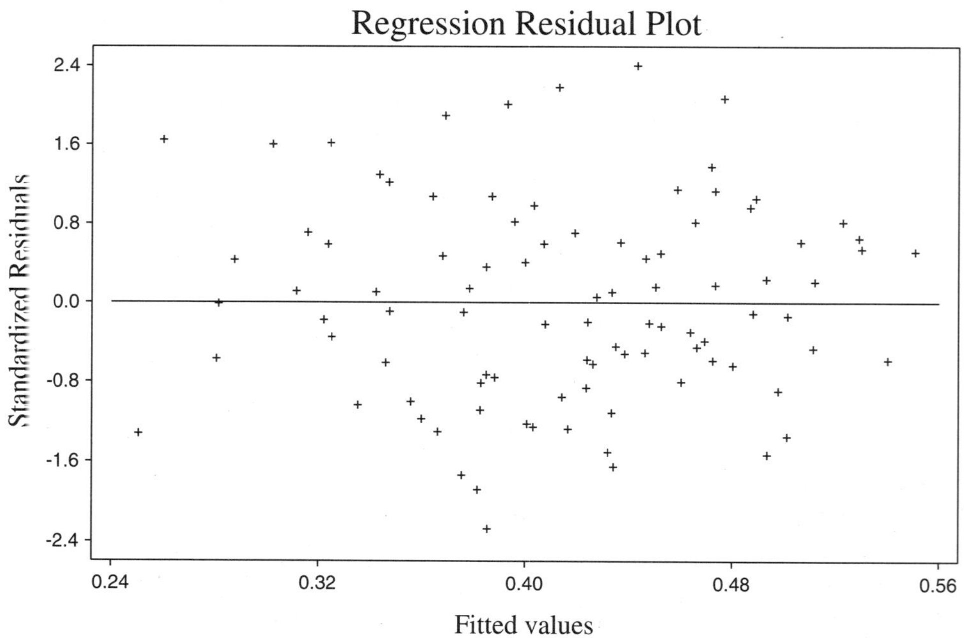

The larger standardized residuals are due to Sarunas Marciulionis (2.40), who scored more than expected, and Darryl Young (−2.27), who scored less. Neither of these residuals are troublesome, given the number of cases considered. The leverages and Cook's distances are generally well behaved, with Muggsy Bogues and Spud Webb still notably different.

The normal plot for these standardized residuals is as follows:

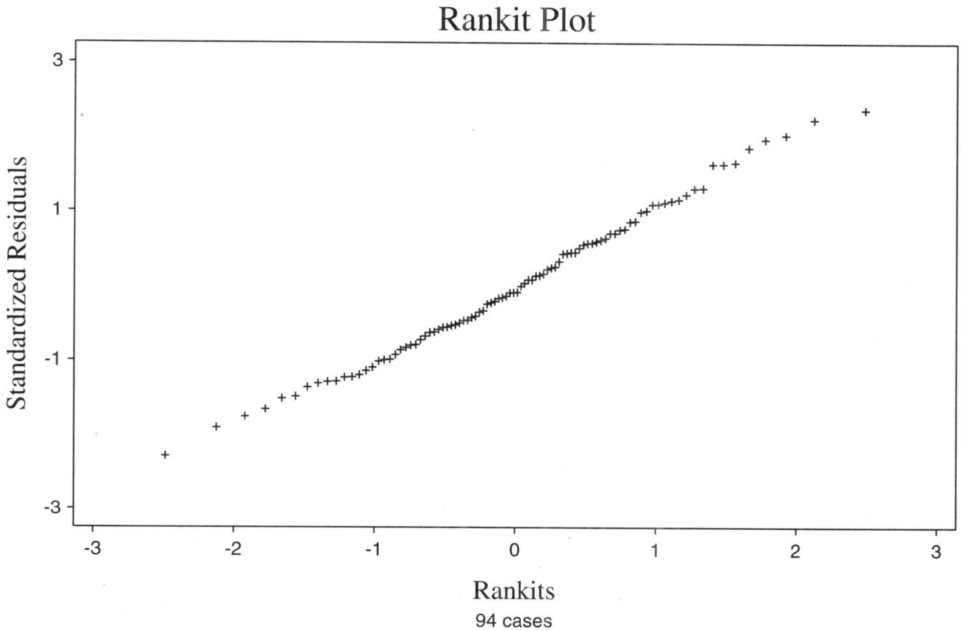

The standardized residuals seem reasonably well–behaved and their distribution appears to be satisfactorily normal (i.e., Gaussian).

As mentioned before, Marciulionis played only 30 games at the beginning of the season; clearly he was scoring relatively heavily during these games. It would be interesting to have known how effective he would have been over the entire season. Let's predict how Sarunas Marciulionis and Michael Jordan would be expected to do, given their characteristics. To make the comparison fair (i.e., not to use the data to fit the model and then use it to predict), we refitted the previously determined model with Marciulionis removed, and then predicted his value of PPM based on his other characteristics.

```
PREDICTED VALUES OF PPM

LOWER PREDICTED BOUND      0.2625
PREDICTED VALUE            0.4400
UPPER PREDICTED BOUND      0.6176
SE (PREDICTED VALUE)       0.0749

PERCENT COVERAGE             98.0
CORRESPONDING T              2.37

PREDICTOR VALUES: MPG = 27.867, HEIGHT = 196.00, FTP = 76.100
```

The 98% prediction interval of PPM for a player with Marciulionis' height, free–throw percentage and minutes per game is (0.26, 0.62). His observed effectiveness was PPM = 0.62. Thus, although Marciulionis performed way above the predicted level for such a player (PPM = 0.440), his performance is not inconsistent with these characteristics. To be more precise, we would expect about 1% of players with his characteristics to score at or above his level over a season.

Consider now the situation for Michael Jordan:

```
PREDICTED VALUES OF PPM

LOWER PREDICTED BOUND      0.2788
PREDICTED VALUE            0.5378
UPPER PREDICTED BOUND      0.7969
SE (PREDICTED VALUE)       0.0761

UNUSUALNESS (LEVERAGE)     0.0534
PERCENT COVERAGE             99.9
CORRESPONDING T              3.40

PREDICTOR VALUES: MPG = 39.321, HEIGHT = 198.00, FTP = 83.700
```

The 99.9% prediction interval for PPM for a player with Jordan's height, free–throw percentage and minutes per game is (0.28, 0.80). His observed effectiveness was PPM = 0.83 . Thus, not only is Jordan performing way above the predicted level for such a player (PPM = 0.538), his performance is above a level we would expect to be attained by only 0.05% of such players (in fact, his value is at the limit of a 99.96% prediction interval, so only 0.02%, or 1 in 5000, players would attain or exceed his level). Indeed, had Jordan's effectiveness been only half of what is was (i.e., PPM = 0.41), he would have still been consistent with a player with these characteristics.

Summary

There are many ways to measure the (statistical) performance of guards in the NBA. Some of these measures represent overall performance (i.e., *production*) by the statistical totals (e.g., assists made) while others represent the relative performance of a player (i.e., *effectiveness*) by considering the production per opportunity to produce (e.g., assists made per minute). Both kinds can be legitimate measures and the choice of measure should be determined by the objective of the comparison. In this case we focused on the scoring rate as measured by the points scored per minute (PPM). Our motivation is to predict PPM based on other factors at least partially under the control of the coach or players. The analysis described is carried out for the 1992–1993 season.

The points scored per minute are related to the minutes played per game, player height and their free–throw percentage. Each centimeter increase in height is associated with an extra 0.00414 points per minute played. Each percent increase in free–throw percent is associated with an extra 0.00337 points per minute played and each minute increase in minutes played per game is associated with an extra 0.00565 points per minute played. Note that each of these interpretations is assuming that the other factors are held fixed. Over a season in which the player played the median number of minutes (1522), the total increases were 6.7, 5.5 and 9.2 points, respectively, so they are not of great practical importance.

The multiple regression model using these three factors explains 45% of the variation in PPM. Predictions based on this model show that Michael Jordan is extraordinarily more productive than other guards with the same characteristics.

The possibility of voting fraud in an election

Topics covered: Prediction interval. Regression diagnostics. Residual plots. Simple regression. Tests of hypotheses. Transformation.

Key words: Outliers. Predicted value. Residual. Scatter plot.

Data File: `vote.dat`

It is a basic principle of a democratic society that the election of a public official should not be allowed to be determined by voter fraud. However, even if fraud is strongly suspected (or even proven), how can it be determined that fraud was the decisive factor in the outcome of the election? What degree of certainty is necessary to overturn the results of an election?

Questions of this type arose after a special election in Philadelphia in 1993. In the special election to fill a State Senate vacancy in the Second District, the Republican candidate, Bruce Marks, received 19,691 votes on voting machines, compared to 19,127 votes on voting machines for the Democratic candidate, William Stinson. However, in absentee ballots, Mr. Stinson out–polled Mr. Marks by 1,396 to 371, giving him a victory by 461 votes.

Allegations of voting fraud were made immediately after the election. Among the charges leveled by either the Republicans, or local newspapers, were that Democrats were given absentee ballots and told that they need not go to the polls to vote; that in other cases, they were specifically guided to vote Democratic; that homeless people were hired to circulate absentee ballots; and that these people were compensated at the rate of $1 per completed absentee ballot. All these actions would be illegal. In February 1994, Federal District Judge Clarence Newcomer ruled that many of the absentee ballots had been improperly obtained and processed by the Democratic–controlled Philadelphia County Board of Elections, and declared Mr. Marks to be the winner of the election. On March 12, 1994, a Federal appellate court ruled on the case, letting the decision to void the results of the election stand, but ordering Judge Newcomer to reconsider his decision to seat Mr. Marks, rather than call a new election. Shortly thereafter, Mr. Stinson and two Democratic campaign workers were charged with election fraud by the Pennsylvania Attorney General (a Republican).

How might it be determined if the observed pattern of absentee ballots was consistent with voter fraud? Historical data can help here.

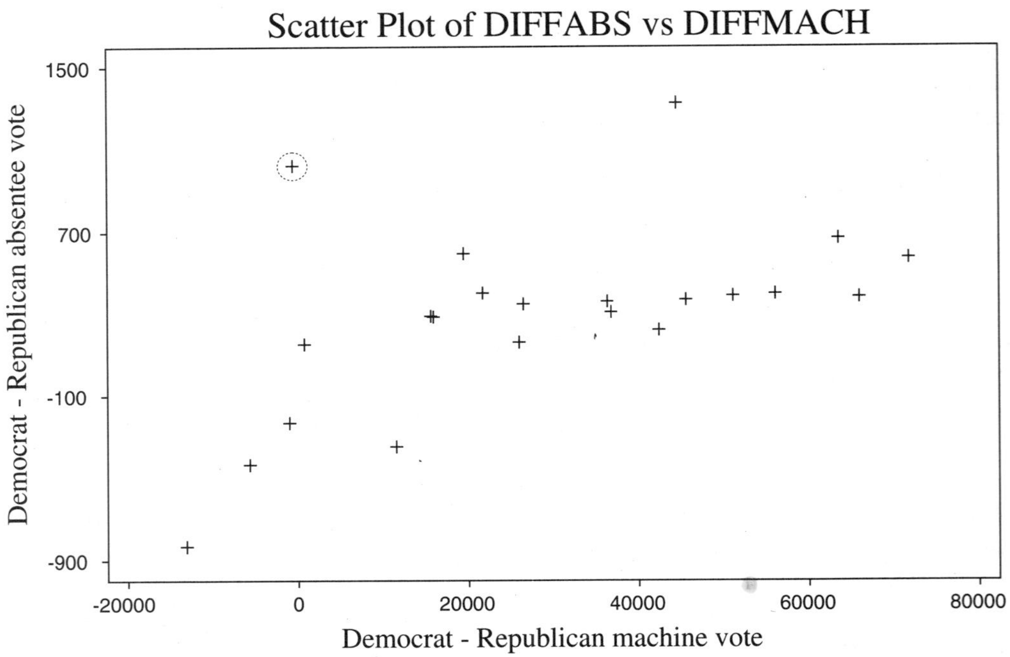

The above scatter plot relates the difference between the Democratic and Republican votes on voting machines (DIFFMACH) to the difference between the Democratic and Republican votes on absentee ballots (DIFFABS), for 22 elections in Philadelphia's senatorial districts from 1982 through 1993. The data were gathered by Orley Ashenfelter of Princeton University, who was hired by Judge Newcomer to advise on the statistical issues involved here.

The circled point corresponds to the challenged 1993 election. It is apparent that the observed difference in absentee ballot votes is somewhat out of line with the general relationship between machine and absentee votes for that election. The question remains, however, whether that difference could just be a result of random chance. Note also that another election, in which no fraud was alleged, is also somewhat out of line with the others (this is the 1992 election for the First Senatorial District).

We can investigate this question using regression. Here are the results of a regression fitting the difference in absentee ballot votes from the difference in machine votes, for the 21 unchallenged elections (since the 1993 election is known *a priori* to be in question, it is reasonable to omit it from the data in order to model the usual relationship between the variables):

```
LINEAR REGRESSION OF DIFFABS Democrat-Republican absentee vote

PREDICTOR
VARIABLES     COEFFICIENT    STD ERROR     STUDENT'S T        P
---------     -----------    ---------     -----------     ------
CONSTANT        -125.904      114.255         -1.10        0.2842
DIFFMACH         0.01270      0.00298          4.26        0.0004

R-SQUARED               0.4888     RESID. MEAN SQUARE (MSE)   1.055E+05
ADJUSTED R-SQUARED      0.4619     STANDARD DEVIATION           324.838

SOURCE        DF       SS            MS           F        P
----------   ---   ----------    ----------    -----    ------
REGRESSION     1   1.917E+06     1.917E+06     18.17    0.0004
RESIDUAL      19   2.005E+06     1.055E+05
TOTAL         20   3.922E+06

CASES INCLUDED 21    MISSING CASES 0
```

The difference in machine votes and difference in absentee ballot votes are directly related, as we would expect, although there is considerable variation around the regression line (the standard error of the estimate is 325 votes).

Is the result of the 1993 election unusual? A prediction interval can be formed for the value of DIFFABS based on the observed value of DIFFMACH, which is -564. Here is output for a 95% prediction interval:

```
PREDICTED VALUES OF DIFFABS

LOWER PREDICTED BOUND     -854.72
PREDICTED VALUE           -133.07
UPPER PREDICTED BOUND      588.58
SE (PREDICTED VALUE)       344.79

PERCENT COVERAGE              95.0

PREDICTOR VALUES: DIFFMACH = -564.00
```

The forecast difference in absentee ballot votes is about -133; that is, the Democrat would get 133 fewer absentee ballots than the Republican. The actual observed difference of 1025 more absentee ballots is well outside the 95% prediction interval, indicating that it was unusual. In fact, the observed value is at the upper limit of a 99.67% prediction interval:

```
PREDICTED VALUES OF DIFFABS

LOWER PREDICTED BOUND     -1291.0
PREDICTED VALUE           -133.07
UPPER PREDICTED BOUND      1024.9
SE (PREDICTED VALUE)       344.79

PERCENT COVERAGE              99.67

PREDICTOR VALUES: DIFFMACH = -564.00
```

That is, a difference in absentee ballot votes as far from the predicted value as is observed here, would be expected to occur only 0.33% of the time, or in about 1 out of every 300 elections where no fraud was alleged.

Inspection of the regression diagnostics and the standardized residuals versus fitted values plot indicates two potential problems: an unusual observation (case 18, the previously noted 1992 election in the First District), and a nonlinear pattern in the residuals:

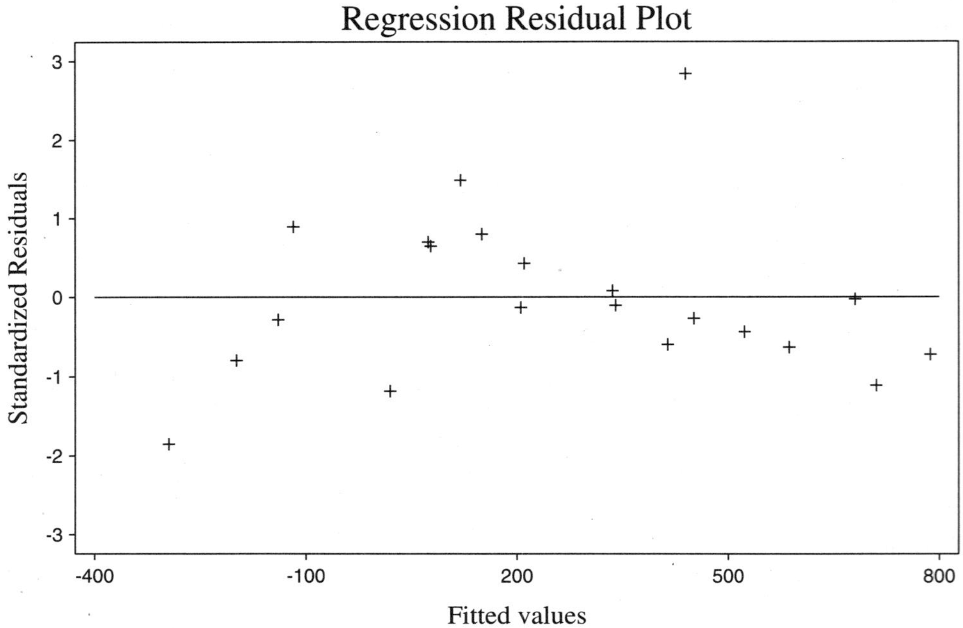

CASE	YEAR	DIFFABS	FITTED	RESIDUAL	LEVERAGE	STDRES	COOKSD
1	1982	346.00	209.81	136.19	0.0487	0.4299	0.0047
2	1982	282.00	76.132	205.87	0.0645	0.6552	0.0148
3	1982	223.00	413.33	-190.33	0.0605	-0.6045	0.0118
4	1984	593.00	121.10	471.90	0.0571	1.4961	0.0678
5	1984	572.00	786.17	-214.17	0.1942	-0.7345	0.0650
6	1984	-229.00	-138.82	-90.177	0.1289	-0.2974	0.0065
7	1984	671.00	679.57	-8.5719	0.1412	-0.0285	0.0001
8	1986	293.00	73.172	219.83	0.0651	0.6999	0.0170
9	1986	360.00	334.93	25.073	0.0509	0.0792	0.0002
10	1986	306.00	340.44	-34.440	0.0513	-0.1089	0.0003
11	1988	401.00	151.64	249.36	0.0533	0.7890	0.0175
12	1988	378.00	710.77	-332.77	0.1555	-1.1147	0.1144
13	1988	-829.00	-293.51	-535.49	0.2051	-1.8490	0.4412
14	1988	394.00	586.76	-192.76	0.1047	-0.6271	0.0230
15	1990	151.00	-117.01	268.01	0.1202	0.8796	0.0529
16	1990	-349.00	20.555	-369.55	0.0765	-1.1839	0.0581
17	1990	160.00	204.98	-44.983	0.0490	-0.1420	0.0005
18	1992	1329.0	438.45	890.55	0.0650	2.8352	0.2793
19	1992	368.00	452.26	-84.256	0.0677	-0.2686	0.0026
20	1992	-434.00	-198.31	-235.69	0.1553	-0.7894	0.0573
21	1992	391.00	524.59	-133.59	0.0852	-0.4300	0.0086

Let's address the latter issue first. The observed pattern in the residual plot occurs because the difference in absentee ballot votes levels off as the difference in machine votes increases. A way to address that effect is to fit a quadratic, rather than linear relationship, since the resultant parabola could fit the observed nonlinearity.

Here are the results of the fitted regression (DFMACHSQ represents the squared value of DIFFMACH):

```
LINEAR REGRESSION OF DIFFABS Democrat-Republican absentee vote

PREDICTOR
VARIABLES     COEFFICIENT    STD ERROR     STUDENT'S T       P        VIF
---------     -----------    ---------     -----------    ------     -----
CONSTANT        -219.007       105.355         -2.08      0.0522
DIFFMACH         0.02971       0.00689          4.31      0.0004       7.1
DFMACHSQ      -2.845E-07     1.068E-07         -2.67      0.0158       7.1

R-SQUARED               0.6335     RESID. MEAN SQUARE (MSE)     79859.5
ADJUSTED R-SQUARED      0.5928     STANDARD DEVIATION           282.594

SOURCE         DF        SS           MS          F         P
----------    ---    ----------   ----------    -----    ------
REGRESSION      2    2.485E+06    1.242E+06     15.56    0.0001
RESIDUAL       18    1.437E+06      79859.5
TOTAL          20    3.922E+06

CASES INCLUDED 21    MISSING CASES 0
```

Including the quadratic term increases the R^2 by .15, and both coefficients are statistically significantly different from zero. This model leads to a predicted value of about 236 more Republican absentee votes than Democratic absentee votes. This results in an assessment of the 1993 election that is even stronger than before; now, the observed difference in absentee ballot votes is at the limit of a 99.94% prediction interval. That is, a difference of absentee ballot votes as far as the observed one from the predicted value would only be expected to occur in about 1 in 1667 elections:

```
PREDICTED VALUES OF DIFFABS

LOWER PREDICTED BOUND     -1496.9
PREDICTED VALUE           -235.85
UPPER PREDICTED BOUND      1025.0
SE (PREDICTED VALUE)       302.42

PERCENT COVERAGE             99.94

PREDICTOR VALUES: DIFFMACH = -564.00, DFMACHSQ = 3.181E+05
```

These calculations certainly support the possibility of voter fraud in the 1993 election. They do not, however, address a subtler point: was the observed fraud (assuming it did, in fact, occur) large enough to change the outcome of the election? Put another way, could random chance account for a large enough difference in absentee ballot votes such that Mr. Stinson would have won the election anyway, even if no fraud had been committed? Since Mr. Marks obtained 564 more machine votes than Mr. Stinson, all that would have been required for Mr. Stinson to win was to obtain 565 more absentee ballot votes than Mr. Marks. How likely is that to happen by random chance? This can be determined by finding the appropriate confidence level that would correspond to the value 565 being at the upper limit of a prediction interval. The appropriate confidence level is easily determined to be 98.37%:

```
PREDICTED VALUES OF DIFFABS

LOWER PREDICTED BOUND     -1037.2
PREDICTED VALUE           -235.85
UPPER PREDICTED BOUND      565.45
SE (PREDICTED VALUE)       302.42

PERCENT COVERAGE             98.37

PREDICTOR VALUES: DIFFMACH = -564.00, DFMACHSQ = 3.181E+05
```

Thus, in about 1 in 120 elections (considering only the upper tail of the distribution), the absentee ballot vote difference would be expected to be far enough from the predicted value so as to be consistent with the result of the election being reversed due to the absentee ballots. The question then becomes whether this actually happening in one election out of 22 in the past decade is more often than the "1 in 120" chance derived, and is thus too unlikely to be allowed to stand.

Regression plots and diagnostics suggest that using the quadratic model has removed the observed pattern in the residuals, but the 1992 First Senatorial District election still shows up as an outlier:

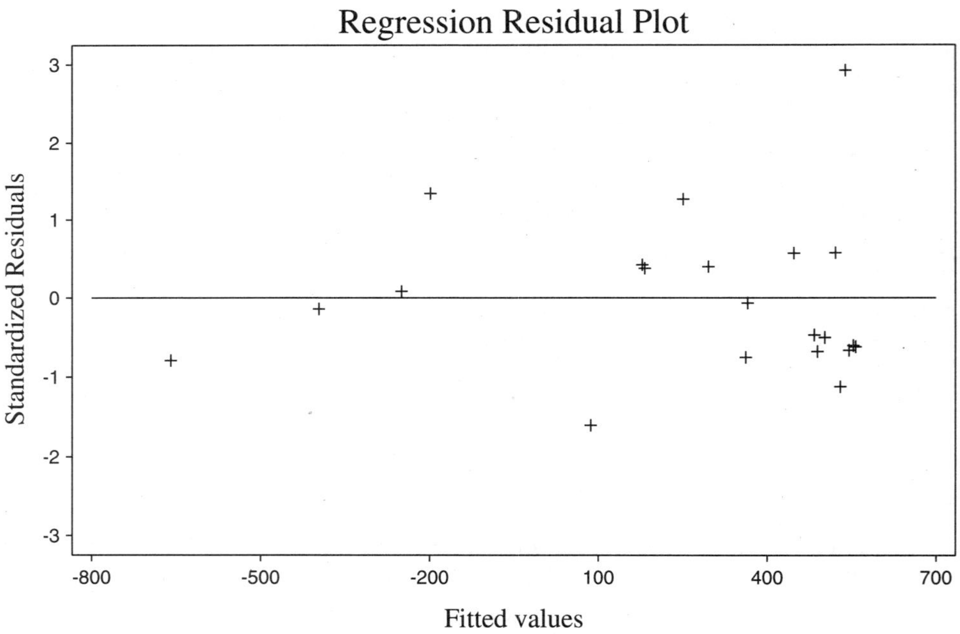

CASE	YEAR	DIFFABS	FITTED	RESIDUAL	LEVERAGE	STDRES	COOKSD
1	1982	346.00	367.40	-21.397	0.0925	-0.0795	0.0002
2	1982	282.00	181.52	100.48	0.0841	0.3715	0.0042
3	1982	223.00	529.39	-306.39	0.0843	-1.1330	0.0394
4	1984	593.00	251.08	341.92	0.0869	1.2662	0.0509
5	1984	572.00	447.25	124.75	0.3966	0.5683	0.0708
6	1984	-229.00	-249.52	20.516	0.1505	0.0788	0.0004
7	1984	671.00	520.77	150.23	0.1856	0.5891	0.0264
8	1986	293.00	176.69	116.31	0.0839	0.4300	0.0056
9	1986	360.00	484.28	-124.28	0.0902	-0.4611	0.0070
10	1986	306.00	488.16	-182.16	0.0898	-0.6756	0.0150
11	1988	401.00	294.26	106.74	0.0892	0.3958	0.0051
12	1988	378.00	503.40	-125.40	0.2313	-0.5061	0.0257
13	1988	-829.00	-660.53	-168.47	0.4425	-0.7985	0.1687
14	1988	394.00	552.16	-158.16	0.1068	-0.5922	0.0140
15	1990	151.00	-198.35	349.35	0.1319	1.3268	0.0891
16	1990	-349.00	85.690	-434.69	0.0840	-1.6072	0.0790
17	1990	160.00	361.78	-201.78	0.0923	-0.7495	0.0190
18	1992	1329.0	539.25	789.75	0.0829	2.9182	0.2565
19	1992	368.00	543.73	-175.73	0.0824	-0.6492	0.0126
20	1992	-434.00	-397.59	-36.405	0.2253	-0.1464	0.0021
21	1992	391.00	556.19	-165.19	0.0870	-0.6118	0.0119

Do the results change very much if this election is removed from the analysis?

```
LINEAR REGRESSION OF DIFFABS Democrat-Republican absentee vote

PREDICTOR
VARIABLES      COEFFICIENT    STD ERROR     STUDENT'S T       P        VIF
---------      -----------    ---------     -----------     ------    -----
CONSTANT         -214.477      78.7007         -2.73        0.0144
DIFFMACH          0.02607      0.00523          4.99        0.0001      7.2
DFMACHSQ       -2.410E-07    8.051E-08         -2.99        0.0082      7.2

R-SQUARED              0.7209      RESID. MEAN SQUARE (MSE)     44553.3
ADJUSTED R-SQUARED     0.6880      STANDARD DEVIATION           211.076

SOURCE         DF        SS            MS          F         P
----------     ---   ----------    ----------    -----    ------
REGRESSION      2    1.956E+06     9.779E+05     21.95    0.0000
RESIDUAL       17    7.574E+05       44553.3
TOTAL          19    2.713E+06

CASES INCLUDED 20    MISSING CASES 0
```

The strength of the regression has increased, and now the standardized residuals versus fitted values plot indicates no problems:

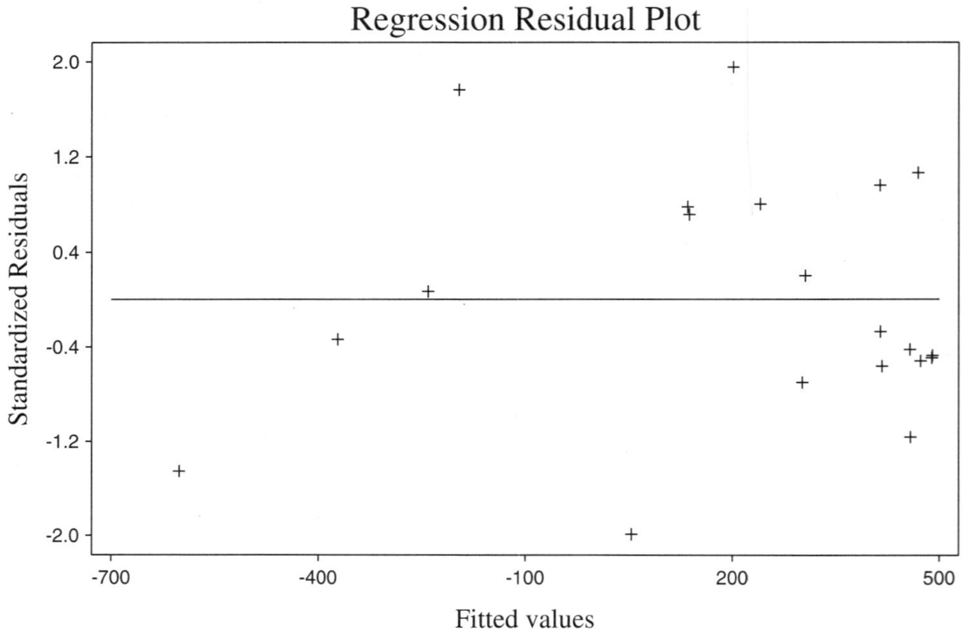

Once again we ask, how unusual is the 1993 election? First, how often would we observe a difference in machine and absentee ballots for the Democrat of this size or larger when the machine votes recorded a 564 vote Republican plurality? The observed difference in absentee ballot votes is at the upper limit of a 99.996% prediction interval, so the answer is about 1 in 25,000 elections:

```
PREDICTED VALUES OF DIFFABS

LOWER PREDICTED BOUND     -1483.5
PREDICTED VALUE           -229.26
UPPER PREDICTED BOUND      1025.0
SE (PREDICTED VALUE)       225.89

PERCENT COVERAGE              99.996

PREDICTOR VALUES: DIFFMACH = -564.00, DFMACHSQ = 3.181E+05
```

How likely would it have been for the difference in absentee votes to be large enough to reverse the result of the election? The required plurality of 565 votes for Mr. Stinson is at the upper limit of a 99.74% prediction interval, so the answer is about 1 in 770 elections:

```
PREDICTED VALUES OF DIFFABS

LOWER PREDICTED BOUND     -1025.5
PREDICTED VALUE           -229.26
UPPER PREDICTED BOUND      566.97
SE (PREDICTED VALUE)       225.89

PERCENT COVERAGE              99.74

PREDICTOR VALUES: DIFFMACH = -564.00, DFMACHSQ = 3.181E+05
```

We need to understand, however, that there is a logical inconsistency to the last calculations. We have determined that the 1993 election was very unusual (differing by 1254 votes from that expected), but we have removed an apparently equally unusual election in which no fraud was alleged. This casts a good deal of doubt on the probability calculations, since the existence of two elections of such unusual behavior out of 22 is extraordinarily unlikely (less than 1 in a million). Professor Ashenfelter chose to keep the 1992 election in the analyses he reported to Judge Newcomer, but this is also problematic, since the outlier invalidates the calculations somewhat. For these reasons, the probabilities determined here must be taken as very tentative.

Another factor that should be taken into account in interpreting these inferences is that while the 1993 election in question was a special election, all 21 elections used in the analyses here were regular elections. It is certainly plausible that the relationship between machine votes and absentee votes could be different for a special election, invalidating the inferences made here.

Summary

The results of a special election for the Second Senatorial District in Philadelphia in 1993 were overturned, because of the determination of voting fraud. Statistical analysis suggests that the observed plurality of Democratic absentee ballot votes, based on the observed plurality of Republican machine votes, is highly unusual. Based on a quadratic model relating the difference in absentee ballot votes to the difference in machine votes, it is estimated that the observed pattern would occur only 1 in 1667 times. Further, the estimated chance that the outcome of the election would be reversed due to absentee ballots based on the observed pattern of machine votes is about 1 in 120. The existence of another, unchallenged, election in which the absentee ballot voting pattern was also very unusual casts some doubt on the specific estimated probabilities determined, as does the fact that all of these calculations are based on regular (not special) elections.

Postscript

On April 27, 1994, Judge Newcomer made his final ruling on this case. He let stand his decision not to call a new election, and Mr. Marks was allowed to retain the seat that the judge had previously awarded to him. Mr. Stinson was acquitted of the charges against him on June 22, 1994.

Purchasing power parity and high inflation countries

Topics covered: Constant shift model. Full model. Pooled model. Prediction. Residual plots. Tests of hypotheses.

Key words: Indicator variables. t–statistic.

Data File: `ppp.dat`

Recall that in the earlier examination of the purchasing power parity data ("Purchasing power parity: is it true?"), there were six countries with unusually high difference in inflation rates (compared with that of the United States). In this case, we will investigate the appropriateness of the purchasing power parity hypothesis once again, now accounting for the two subgroups of high inflation and low–to–moderate inflation countries (all analyses are without the outlier, Iran). The previous regression model, based on predicting `CHNGEX75` from `CHNGIN75`, did not take this group membership into account. For this reason, it can be called a "pooled" model, in that it combines (or pools) the observations over the two subgroups, and can be represented mathematically as follows:

$$\texttt{CHNGEX75}_{ij} = \beta_0 + \beta_1 \times \texttt{CHNGIN75}_{ij} + \epsilon_{ij},$$

where i represents low–to–moderate or high inflation, and j is an index of the number of the country within its appropriate inflation group.

First, construct a variable to define the subgroups of low–to–moderate inflation and high inflation. The natural way to do this is using a 0/1, or *indicator* variable, where the variable takes on the value 1 if the difference in inflation variable has a value over 20%, and 0 otherwise. You might call this variable `IN`. Now, perform a regression of `CHNGEX75` on both `CHNGIN75` and `IN`. What do you find? Does this model seem to fit better than the pooled model?

In fact, it is possible to answer this question statistically. The regression model of `CHNGEX75` on both `CHNGIN75` and `IN` can be represented as follows:

$$\texttt{CHNGEX75}_{ij} = \beta_0 + \beta_1 \times \texttt{CHNGIN75}_{ij} + \beta_2 \times \texttt{IN}_{ij} + \epsilon_{ij}.$$

However, since `IN` is an indicator variable, this model is equivalent to the following two regression relationships:

$$\texttt{CHNGEX75}_{ij} = \beta_0 + \beta_1 \times \texttt{CHNGIN75}_{ij} + \epsilon_{ij}$$

for low–to–moderate inflation countries, and

$$\texttt{CHNGEX75}_{ij} = \beta_0 + \beta_1 \times \texttt{CHNGIN75}_{ij} + \beta_2 + \epsilon_{ij}$$

for high inflation countries. The two models describe lines with identical slopes, but they differ in their intercepts by β_2. For this reason, this model is called a "constant shift" model. The t–test for whether $\beta_2 = 0$ provides a test whether the constant shift model is an improvement over the pooled model. Note that here the test is not significant, showing that the constant shift model is not a significant improvement.

We might suppose that there is a difference in appropriate regression lines, but it is reflected in different slopes, rather than different intercepts. This also can be evaluated using regression methods. Construct a variable CHIN such that CHIN = IN × CHNGIN75, and then perform a regression of CHNGEX75 on CHNGIN75, IN and CHIN. This model is

$$\texttt{CHNGEX75}_{ij} = \beta_0 + \beta_1 \times \texttt{CHNGIN75}_{ij} + \beta_2 \times \texttt{IN}_{ij} + \beta_3 \times \texttt{CHIN}_{ij} + \epsilon_{ij}.$$

Once again, since IN is an indicator variable, this model is equivalent to the following two regression relationships:

$$\texttt{CHNGEX75}_{ij} = \beta_0 + \beta_1 \times \texttt{CHNGIN75}_{ij} + \epsilon_{ij}$$

for low–to–moderate inflation countries, and

$$\texttt{CHNGEX75}_{ij} = \beta_0 + \beta_1 \times \texttt{CHNGIN75}_{ij} + \beta_2 + \beta_3 \times \texttt{CHNGIN75}_{ij} + \epsilon_{ij}$$

for high inflation countries (this is because CHIN equals either 0, if $\texttt{IN} = 0$, or CHNGIN75, if $\texttt{IN} = 1$). This model is completely general, since it allows different intercepts for the two groups (differing by β_2) and different slopes for the two groups (differing by β_3). For this reason, this is called the "full" model. Examination of t–statistics testing whether these coefficients are different from zero allows us to see if different lines for the two groups provides statistically significantly better fit than a common line for the two groups. Note that for these data the coefficient for CHIN is significant, indicating that different slopes for the two groups are appropriate.

Fit a final regression model of CHNGEX75 on CHNGIN75 and CHIN (dropping the insignificant IN). Check the assumptions for this regression model. The model implies two regression lines:

$$\texttt{CHNGEX75} = -.0762 + 1.1205 \times \texttt{CHNGIN75}$$

for the low–to–moderate inflation countries, and

$$\texttt{CHNGEX75} = -.0762 + 1.0147 \times \texttt{CHNGIN75}$$

for the high inflation countries. Test the purchasing power parity hypotheses ($\beta_0 = 0$, $\beta_1 = 1$) for the low–to–moderate inflation countries. Does PPP seem to hold? In fact, the slope coefficient is (marginally) different from 1, suggesting a mild violation of PPP. In contrast, PPP seems to hold for the high inflation countries (this can be tested by repeating the previous analysis, interchanging the values of 0 and 1 in the IN variable and making the corresponding change to CHIN).

Comments to the Instructor

The six countries with CHNGIN75 greater than 20% are Chile, Ghana, Israel, Mexico, Turkey and Uruguay. The regression corresponding to the constant shift model has a coefficient for CHNGIN75 significantly different from zero, but the coefficient for IN is not ($t = -1.62$, $p = .114$); thus, the constant shift model is not an improvement over the pooled model.

The regression corresponding to the full model has significant coefficients for CHNGIN75 and CHIN ($t = -2.09$, $p = .043$), indicating improvement over the pooled model. The coefficient for different intercepts (i.e., that of IN) is not significantly different from zero, however ($t = 0.94$, $p = .354$).

The final model has the following structure:

PREDICTOR VARIABLES	COEFFICIENT	STD ERROR	STUDENT'S T	P
CONSTANT	-0.07616	0.29159	-0.26	0.7953
CHNGIN75	1.12052	0.04281	26.18	0.0000
CHIN	-0.10585	0.04202	-2.52	0.0159

For the low–to–moderate inflation countries, the test of $\beta_0 = 0$ is $t = -0.26$, with $p = .795$, so the hypothesis is not rejected. The test of $\beta_1 = 1$ is

$$t = \frac{1.12052 - 1}{.04281} = 2.815,$$

which has tail probability .0075; thus, overshooting is apparently occurring.

The model that reverses the 0/1 nature of IN has the following structure:

PREDICTOR VARIABLES	COEFFICIENT	STD ERROR	STUDENT'S T	P
CONSTANT	-0.07616	0.29159	-0.26	0.7953
CHNGIN75	1.01467	0.01700	59.70	0.0000
CHIN	0.10585	0.04202	2.52	0.0159

Therefore, for the high inflation countries, the test of $\beta_0 = 0$ is again $t = -0.26$, with $p = .795$, so the hypothesis is not rejected. The test of $\beta_1 = 1$ is

$$t = \frac{1.01467 - 1}{.017} = 0.863,$$

which has tail probability .393; thus, PPP is apparently appropriate.

There is still some slight structure apparent in the residuals versus fitted values plot.

To maintain consistency with the unannotated edition,
the next case begins on the following page.

Prediction of the time interval between "Old Faithful" eruptions

Topics covered: Building regression models. Examining subgroups. Prediction. Removing autocorrelation.

Key words: Autocorrelation. Durbin–Watson statistic. Lagged variable. Time series plot.

Data File: `geyser1.dat`, `geyser2.dat`

This case builds on the earlier investigations of "Old Faithful" eruptions. In the analyses presented earlier, based on 16 days of data, it was found that the "intereruption times of the 'Old Faithful' geyser in Yellowstone National Park are apparently bimodal, centering at around 55 minutes and 75–80 minutes, respectively. These times are directly related to the duration of the previous eruption, with longer eruptions followed by longer intereruption times (and shorter eruptions followed by shorter intereruption times)." We will explore now some aspects of the data that were not investigated previously, using `geyser1.dat` for this analysis.

Data were provided for 16 different days. We would like to know if the pattern of the duration of eruption and the interval between eruptions was the same for each of the sixteen days. If we found that the patterns for the days were very dissimilar, we could not generalize our conclusions to apply to all days. On the other hand, if we found the pattern to be similar for each day, we then could speak about the prevailing pattern for all days. A quick graphical way to do this is to look at boxplots for the time of duration for each day. We should do this also for the intereruption times.

Examine the side–by–side boxplots for these two variables. What do you see? There is some variation between the days but it is not substantial enough to treat the days separately. The data are homogeneous with respect to the days so we do not have to analyze the data separately for each day. This is an important concept to keep in mind while analyzing data that can be viewed as having natural subgroups. When analyzing such data, the homogeneity of the groups should be examined first. If the groups are homogeneous, the subgroups need not be taken into account and the data can be analyzed as a whole. On the other hand, if the groups are heterogeneous with respect to the characteristic studied, then the data should be analyzed separately for each group (or, alternatively, the model being considered for the data should incorporate group membership as a predictive component). A situation where this problem often arises is in studying data that contain values for both males and females. If the variable studied differs between the sexes then an analysis should be carried out that accounts for the differences between males and females. Analyzing the pooled (combined) data will give a misleading picture.

The conclusion that the duration and intereruption time do not vary significantly across the days can be arrived at formally by doing a statistical test of significance. The graphical analysis is sufficient here, however.

Let us now turn to another aspect of the data. The data were collected to see if the time to next eruption could be predicted accurately, to give the visitors to the Park more information to organize their visit. Our current prediction rule (based on the analysis carried out earlier) is quite simple: if the last eruption was short (less than 3 minutes) the next eruption is predicted to occur after 55 minutes, while if the last eruption was a long one (longer than 3 minutes) the next eruption is predicted to occur after 80 minutes. Can we do better by finding a more precise relationship between the duration of eruption and the interval between the eruptions?

Fit a linear regression line connecting the interval between eruptions and the duration of eruptions. Although the coefficients are significant, the fitted model is not satisfactory. This can be seen in two different ways. Graphically, if we plot the standardized residuals against the order of the observations (a time series plot), two kinds of time dependency can be seen. On the one hand, there is a general long–term cyclical pattern to the residuals, but there is also a pattern of positive residuals being followed by negative ones, and vice versa (negative autocorrelation). The Durbin–Watson statistic, which is high (2.50), also highlights the latter effect, showing that the residuals are negatively autocorrelated.

We can attempt to remedy this type of model deficiency by performing a regression using *lagged variables*. A lagged variable is one where the value for a particular case is the value of the original variable corresponding to the previous case. The lagged variables will often remove the time dependencies. Fit a linear regression model predicting the interval between eruptions from the duration of the current eruption, duration of the previous eruption, and the interval between the current and previous eruptions. Note that observations 1, 14, 27, 40, 54, 68, 82, 95, 108, 122, 136, 150, 164, 180, 194 and 208 should be treated as missing in the transformed data set, since they correspond to the first observations of new days (and the values for the lagged variables are not known).

All of the coefficients are significant, and the residuals no longer appear autocorrelated. This can be seen from a plot of the standardized residuals against the order of the observations (a time series plot).

Now, let us apply the regression model derived to predict future intereruption time (that is, the 1985 data discussed in the earlier case, given in `geyser2.dat`).

Construct a histogram of the errors made when using the regression model to predict the 1985 values (note that the existence of cases with unknown duration value leads to 87 observations for which it is impossible to make a forecast). This histogram is similar to that formed (and given in the earlier case "Eruptions of the 'Old Faithful' geyser") when using the earlier simple prediction rule. This, combined with the simplicity of that rule (and its applicability even if durations are only noted as "Short" or "Long"), implies that the simpler rule is the one of choice.

One factor that we have not considered here is that the 16 days in `geyser1.dat` come from two different years (days 1–8 are from August 1978, while days 16–23 are from August 1979). Does that matter here?

Technical terms

Autocorrelation: the correlation between the values of a variable and its lagged values. For example, the first order autocorrelation of a variable X is the correlation between values of the form X_i and X_{i-1}.

Durbin–Watson statistic: a statistic calculated to examine the independence of successive residuals from a fitted regression model. Values of the statistic close to 2 indicate independence of the residuals. Values of the Durbin–Watson statistic far removed from 2 indicate autocorrelated residuals, with values above 2 indicating negative autocorrelation and values below 2 indicating positive autocorrelation. Application of the

Durbin–Watson statistic requires that the regression model satisfy various properties, including normality and homoscedasticity of the errors. Several methods are available for dealing with autocorrelated residuals. These include introduction of new variables, introduction of lagged variables, taking differences, or transforming variables.

Lagged variable: a variable derived from an original variable, where the value for a particular case is the value for the original variable corresponding to an earlier case (usually, the previous case). For example, a variable that has been lagged once has as its ith value the $(i-1)^{st}$ value of the original variable.

Comments to the Instructor

Side–by–side boxplots confirm that there are few appreciable differences in the distributions of `INTERVAL` or `DURATION` when separated by day. Note, by the way, that this can be confirmed using a one–way analysis of variance model, if so desired.

The fitted regression line connecting the two variables is

$$\widehat{\texttt{INTERVAL}} = 33.967 + 10.358 \times \texttt{DURATION},$$

with $F = 734.56$ and $R^2 = 76.95\%$. As noted in the case, however, the Durbin–Watson statistic equals 2.50.

A regression using `DURATION`, the lagged value of `DURATION`, and the lagged value of `INTERVAL` is highly significant ($F = 265.22$, $R^2 = 79.75\%$), with each variable's coefficient being statistically significantly different from zero. The fitted regression line is

$$\begin{aligned}\widehat{\texttt{INTERVAL}}_t = 55.386 &+ 9.032 \times \texttt{DURATION}_t \\ &+ 2.089 \times \texttt{DURATION}_{t-1} \\ &- 0.344 \times \texttt{INTERVAL}_{t-1}.\end{aligned}$$

Application of this derived model to the 1985 data yields errors with mean 2.67 minutes and standard deviation 6.15 minutes, with values in the range $(-13.0, 33.2)$. The simple rule derived in the earlier case yields errors in the range $(-25, 28)$, with mean 1.02 minutes and standard deviation 6.74 minutes. Thus, the performances of the two rules are quite comparable.

A constant shift model that includes a variable separating on year provides statistically significant improvement ($R^2 = 81.03\%$). Note, however, that this additional factor is not useful in predicting the 1985 values.

The success of teams in the National Hockey League

Topics covered: Prediction. Regression. Simple versus complicated models.

Key words: Forecasting. Scatter plot.

Data File: `hockey1.dat, hockey2.dat`

What determines the success of a hockey team? The following data set represents the total points, goals scored (`GF`) and goals given up (`GA`), for the 24 teams in the National Hockey League (NHL) during the 1992–1993 season. The season during this year consisted of 84 games for each team. A team is awarded two points for a win, one point for a tie, and no points for a loss (so, dividing points by 168 gives the winning percentage, a more common measure of success of a sports team).

Team	Points	GF	GA
Pittsburgh Penguins	119	367	268
Boston Bruins	109	332	268
Chicago Blackhawks	106	279	230
Quebec Nordiques	104	351	300
Detroit Red Wings	103	369	280
Montreal Canadiens	102	326	280
Vancouver Canucks	101	346	278
Toronto Maple Leafs	99	288	241
Calgary Flames	97	322	282
Washington Capitals	93	325	286
Los Angeles Kings	88	338	340
New York Islanders	87	335	298
Winnipeg Jets	87	322	320
New Jersey Devils	87	309	299
Buffalo Sabres	86	335	297
St. Louis Blues	85	282	278
Philadelphia Flyers	83	319	319
Minnesota North Stars	82	272	293
New York Rangers	79	304	308
Edmonton Oilers	60	242	337
Hartford Whalers	58	284	369
Tampa Bay Lightning	53	245	332
San Jose Sharks	24	218	414
Ottawa Senators	24	202	395

Can a model be built relating success (`POINTS`) to offense (`GF`) and defense (`GA`)? Is there a simpler model that works just as well?

First, look at the descriptive statistics for the data. The average number of points is 84 (in an 84 game season), as it must be, and the average number of goals scored and given up are the same. Now, construct scatter plots with `POINTS` on the vertical axis, and `GF` and `GA`, respectively, on the horizontal axis. Is there a relationship between team success and offensive and/or defensive success? Are the relationships what you would expect?

A natural model to use to try to predict POINTS would be based on both GF and GA. Fit this model to the data. Note that the regression is very strong. Knowing the total goals scored and given up tells you almost everything about how successful a team is. That might seem obvious, but it does say something of interest: all of the other things that fans think lead to a team's success (home ice advantage, penalty killing, power play success, toughness, etc.) simply get absorbed into the only things that actually matter — put the puck in the other guy's net, and keep it out of yours!

Look again at the fitted regression model. Do the assumptions seem to hold? What is the interpretation of the estimated coefficients? Are those interpretations sensible, given the context of the data?

It is reasonable to wonder if this model can be simplified. Perhaps it is only the *difference* in goals scored and given up that matters. That is, if, for example, a team gives up 30 more goals than it scores, it does poorly, whether that's GF = 310, GA = 340, or GF = 250, GA = 280. This can be investigated by simply creating the variable DIFF = GF - GA, and doing the regression on that variable. Construct this difference variable, and form a scatter plot of POINTS on DIFF. Does this plot suggest that this variable alone would be a good predictor of team success?

Now fit the simple regression model of POINTS on DIFF. Note that the model fits very well — almost as well as the more complicated model based on both GF and GA. Thus, it is apparently only the difference in goals scored and goals given up that matters. A team can choose to play a defensive style by, for example, keeping close physical contact with the opposition players (called "tight checking"). By this strategy they are trying to minimize the goals scored against them; however it often results in fewer opportunities for the team to score and leads to games in which the number of goals scored by either team is low. There is no evidence that playing a defensive style is any better or worse than playing an offensively oriented game, which would lead to both more goals scored and more goals given up.

The benefits of a simpler model are not merely that a simplifying description of the data you have might be obtained. It is generally the case that simpler models tend to forecast more accurately than more complicated models, because fewer aspects of the underlying process are required to remain stable over time. Over the last 50 years, the NHL, and the sport of hockey in general, changed dramatically, from a low–scoring, "bump and grind" sort of game, played in a six–team league (before 1967), to a much more wide–open, high scoring game after expansion of the league (1967 through 1984 or 1985), to a game incorporating aspects of European hockey (after 1985). Thus, it would be quite impressive if a model based on 1993 data could accurately forecast performance under such different circumstances.

This can be investigated by examining how well the two alternative models discussed here do in predicting other years of NHL hockey. Using the data provided for the 1987–1988, 1981–1982, 1969–1970, 1965–1966, 1952–1953 and 1942–1943 seasons in hockey2.dat, apply each of the two models to the appropriate data, and see how accurate the predictions based on the models are. You will have to adjust for changing season lengths; seasons lasted 50 games in 1942–1943, 70 games in 1952–1953 and 1965–1966, 76 games in 1969–1970, and 80 games in 1981–1982 and 1987–1988. One way to do this is to convert each of the models, and the new data, so that they refer to winning percentage (WINNING = POINTS/GAMES), rather than POINTS. An alternative way is to adjust only the predictions so as to be proportionally aligned with the actual number of games for each season (that is, take the predictions from the model based on the 1992-93 data, and multiply them by GAMES/84 for any other year's data).

Which model predicts better?

One way to answer this is to look at the prediction errors. Use the model based on goal differences to predict the winning percentage for each team for each season (PREDDIF). Then calculate the error of the prediction, that is WINNING - PREDDIF. This measures how far the prediction is from the actual winning percentage. Now do the same thing for the model based on both goals scored and goals given up.

Based on the prediction errors, here are a few of the possible ways to summarize the quality of the prediction. First, for 61.1% of the predictions, the error based on the model using goal differences alone is smaller than that based on both goals scored and goals given up. Second, the average absolute value of the prediction error for the simpler model is 3.64 points, compared with 4.49 points for the more complex model. This summary measure focuses on the absolute size of the prediction error. Alternatively, the absolute proportional error is 5.19% versus 6.74%. This summary measure focuses on the size of the relative prediction error.

By any of these measures, the model based on goal differences outperforms that using both goals scored and goals given up. Thus, goal difference is apparently all that is needed to predict team success, even over a time period of 50 years when the style of the game has changed considerably.

Comments to the Instructor

Scatter plots indicate that, as would be expected, there is a fairly strong direct relationship between POINTS and GF, and a fairly strong inverse relationship between POINTS and GA. The regression on both of these variables is very strong ($R^2 = 96.6\%$, $F = 300.7$). The fitted regression is

$$\widehat{\texttt{POINTS}} = 102.67 + .275 \times \texttt{GF} - .336 \times \texttt{GA}.$$

That is, an additional goal scored, holding GA fixed, is associated with an expected increase of .275 in POINTS, while an additional goal given up, holding GF fixed, is associated with an expected decrease of .336 in POINTS. There is a hint of heteroscedasticity in the residuals versus fitted plot.

The scatter plot of POINTS on DIFF indicates a very strong relationship. This is supported by the regression analysis, as the model based on DIFF fits almost as well as the model on GF and GA ($R^2 = 96.4\%$). The residuals versus fitted plot is virtually identical to that for the more complicated model. The fitted regression model has the form

$$\widehat{\texttt{POINTS}} = 84 + .305 \times (\texttt{GF} - \texttt{GA}).$$

Using these models on the other years' data gives the prediction error figures given in the case.

The effectiveness of NBA guards: assists per minute

Topics covered: Finding structure and unusual observations. Multiple regression modeling.

Key words: Descriptive statistics. Midhinge. Prediction. Regression. Residual plots. Scatter plot.

Data File: `nba.dat`

In the case "The effectiveness of National Basketball Association guards" we only considered the single dimension of basketball success represented by the scoring rate, as measured by points scored per minute played. Another measure of success is the number of assists per minute played. A player is credited with an assist when he passes the ball to another player who scores immediately, and as a consequence of the pass. This is often used to measure how effectively the player distributes the ball to other players. Assisting the scoring is the primary goal for many players, so their contributions may not appear in the scoring rate as measured directly by the points per minute.

Can we build a model, using the same guidelines as for `PPM`, that can be used to predict the assist rate?

The natural measure is the assists per minute played (`APM`). What do the descriptive statistics indicate about the assist rate? The median and mean of `APM` are close and the midpoint of the quartiles (a location estimate sometimes called the *midhinge*) is similar to the mean, so the distribution appears to be symmetric. The mean rate of 0.16 assists per minute works out to about 7.7 per full game (i.e., 48 minutes) played. Construct a stem–and–leaf display of `APM`. What additionally can you see that is interesting? There is one unusual player — John Stockton. Look him up in the data–table and comment on his characteristics. We will need to keep an eye on him throughout the analysis.

As in the previous case, we will omit the players playing very few minutes and focus on those players playing 10 or more minutes per game and appearing in 10 or more games. This excludes 10 players.

Construct a scatter plot with `APM` on the vertical axis and `MPG` on the horizontal axis.

At first glance there does not appear to be a clear relationship with `MPG`. There is an upper outlier at about 35 minutes per game. This is John Stockton ($\texttt{APM} = 0.34$). Interestingly, there could be a hint of a bifurcation in guards. Focusing on those playing more than 25 minutes per game, there is a tier at about $\texttt{APM} = 0.23$ and a tier at about $\texttt{APM} = 0.12$. These tiers are not clear cut, but are noticeable. Staring at it a little longer, it could appear that there is a decreasing relationship for the bottom tier as `MPG` increases. Intriguing (although not definitive) differences! Looking at the part–time players, we also see a hint of a tiered plot. Can you suggest a reason for such a tiered pattern? Do you think these patterns are real, based on what you see, or might they just be a result of random chance? In the upper tier, `APM` appears to be increasing with `MPG`. We could interpret the upper tier as the "play making" guards. The lower tier presumably contributes in other ways — scoring or defense, for example.

Construct a scatter plot with APM on the vertical axis and HEIGHT on the horizontal axis.

What do you immediately see? There are a few players that are quite different from the rest. The two very short players are Muggsy Bogues and Spud Webb (5'3" and 5'7", respectively). They are different, but still seem to be somewhat consistent with the general pattern. We will check this more closely later. John Stockton is still an outlier, even when placed in the perspective of height. We will omit him (for the moment) from the data used on this trail.

As before, we can build a model for APM based on MPG, HEIGHT and FTP, as well as other possible factors.

Fit the model with only MPG. Is the coefficient of MPG statistically significantly different from zero? Construct a residual plot for this model. It looks quite reasonable. Is there much evidence that APM is related to MPG?

But wait, don't give up! We know that APM is related to HEIGHT. Fit the model for APM based on HEIGHT. How does it look? Construct a residual plot for this model. The two observations with high leverage values are, as we would expect, Muggsy Bogues and Spud Webb. The influence of the latter on the model is noteworthy, although not extreme (Cook's $D = 0.31$). The estimate of the rate of decrease of APM with HEIGHT is 0.00518 assists per minute per centimeter. It is over 7 standard errors away from zero, so there is ample evidence to indicate the two variables are linearly related. Why do you suppose it might be that taller players are less effective passers than shorter players?

Try adding MPG to this model. How does the numerical summary of the model look? The coefficient of MPG is statistically significantly different from zero ($t = 2.28$, $p = 0.025$), even though it was not before. That is, while minutes per game does not provide significant predictive power of assists per minute by itself, given a player's height it does add to predictive power. The estimate suggests that APM increases at a rate of 0.00123 assists per minute per additional minute per game played, given a player's height. Construct a residual plot for this model. Do Bogues and Webb still have a lot of leverage and influence on the model?

Fit the model for APM based on HEIGHT, MPG and FTP. How important is FTP? Construct a residual plot for this model. Has the additional variable reduced the leverage and influence of Bogues and Webb?

Finally, predict the value of APM of a player with John Stockton's characteristics, based on whichever of these models you think is best. Follow the procedure used to predict Michael Jordan's scoring average in the previous case. How different is Stockton from the typical player with his characteristics?

Comments to the Instructor

The mean number of assists per minute is .160; the median number is .158. The midhinge equals $(.107 + .211)/2 = .159$.

John Stockton is the prototype of a point guard — below average in height, a good (but not great) scorer (15.1 points per game), but extraordinary in his assist rate (12 assists per game).

If the noted bifurcation in the plot really exists, it could easily correspond to the distinction between shooting guards (bottom tier) and point guards (top tier). The evidence for such a bifurcation, however, is weak at best.

The coefficient of `MPG` in the simple regression of `APM` on `MPG` is not statistically significantly different from zero ($t = 1.51$). On the other hand, `HEIGHT` is strongly (inversely) related to `APM`. Presumably smaller players are better able to maneuver around the court, thereby creating scoring opportunities for their teammates.

Spud Webb and Muggsy Bogues still show up as leverage points with respect to the model on `HEIGHT` and `MPG` (with leverage values 0.120 and 0.244, respectively). Michael Adams, at 5'10", is a borderline leverage point as well (leverage value 0.064). The influence of each of these points is still rather modest, however (Cook's D equaling 0.12, 0.38 and 0.01, respectively).

The addition of `FTP` to the model does not improve the fit very much, and the coefficient of this variable is not statistically significantly different from zero ($t = -1.78$, $p = .078$). Spud Webb and Muggsy Bogues are still leverage points, and now Sherman Douglas shows up as a potential leverage point, due to his 56% free throw percentage. Given the little gain apparent from using `FTP`, a reasonable choice for the final model is the one based on only `HEIGHT` and `MPG`.

A 99.82% prediction interval has the observed value of `APM` for John Stockton at its upper limit:

```
PREDICTED VALUE of APM

LOWER PREDICTED BOUND      0.0509
PREDICTED VALUE            0.1973
UPPER PREDICTED BOUND      0.3438
SE (PREDICTED VALUE)       0.0455

PERCENT COVERAGE             99.82

PREDICTOR VALUES: HEIGHT = 185.00, MPG = 34.915
```

A player with Stockton's characteristics is predicted to average 0.197 assists per minute; fewer than 1 in 1100 players with his characteristics would be expected to have assists per minute as large, or larger, than that of Stockton.

To maintain consistency with the unannotated edition,
the next case begins on the following page.

The effectiveness of NBA guards: beyond points and assists

Topics covered: Finding structure and unusual observations. Multiple regression modeling.

Data File: `nba.dat`

In our consideration of the performance of guards in the NBA, we noted that many factors determine an individual player's contribution to the team. In each case we considered a single dimension; scoring rate as measured by points scored per minute played, originally, and later the assists rate as measured by the assists made per minute played. Here we will look at other dimensions of success.

There are several paths to take. Up to now we have measured success as production per minute. Do you think that measuring success as production per game would provide an interesting perspective? How about production per individual opportunity to score?

Perform similar analyses on other measures of effectiveness beside `PPM` and `APM`.

Do the same players stand out? What relationships exist with `MPG`, `HEIGHT` and `AGE`?

Two other dimensions of player performance are rebounds made (i.e., the number of times the player controlled the ball after a missed shot) and field goal percentage (i.e., the percentage of the shots that the player made). Each rebound made gives the player's team another opportunity to score. Thus rebounds could be regarded as another dimension of scoring. Field goal percentage is another measure of the rate of scoring, using the opportunities to score as a basis.

Is Michael Jordan still extremely unusual? Or does he appear earth–bound from these perspectives? What about John Stockton?

Comments to the Instructor

Production measures in a per game, rather than per minute, scale would seem to be a useful approach. Indeed, a regression of `PPG` on `HEIGHT`, `MPG` and `FTP` (omitting Michael Jordan) has an R^2 of 86.7%, which is much higher than the corresponding measures in the per minute scale. This advantage is illusory, however. The main contributor to the improved fit is `MPG`, and this is just reflecting a scale effect (greater minutes per game naturally leads to greater points per game, even if scoring rate per minute is fixed). This sort of misleading apparent regression strength arises in many other contexts as well — for example, when examining absolute, rather than per capita, measures for different states, one will find strong regressions merely due to a size effect.

It is not surprising to see that the predictability of rebounds per minute (`RPM`) and `FGP` are considerably less than those of `PPM` and `APM`. Rebounds are not a guard's primary responsibility, and even high scorers sometimes have low shooting percentages. A regression of `RPM` on `HEIGHT` and `MPG` has $R^2 = 12\%$, after removing Clyde Drexler as a marginal outlier. This choice of variables is intuitive, since being taller would clearly aid in getting rebounds. The fitted model is as follows:

$$\widehat{\texttt{RPM}} = -.108 + .000983 \times \texttt{HEIGHT} + .000645 \times \texttt{MPG}.$$

A regression of FGP on MPG and FTP yields an R^2 of 19.5%. Again, the choice of predictors is intuitive, since FTP and FGP are both measures of shooting ability. The fitted model is as follows:

$$\widehat{\texttt{FGP}} = 31.89 + .1333 \times \texttt{FTP} + .1131 \times \texttt{MPG}.$$

To maintain consistency with the unannotated edition,
the next case begins on the following page.

Another look at emergency calls to the New York Auto Club

Topics covered: Model selection. Prediction. Residual plots. Multiple regression. Tests of hypotheses.

Key words: Lagged variable. Transformation.

Data File: `ers.dat`

In the case "Emergency calls to the New York Auto Club" we built a simple linear regression model for the daily percentage of the monthly number of calls to the Emergency Road Service (ERS) of the New York Auto Club in terms of the daily low temperature forecasted the previous day. Based on the model, we predicted the percentage monthly calls for the next day based on the previous day's forecast.

Besides the forecast low, there are several easily recorded variables that could help predict the number of calls (CALLS) answered by the ERS. The data set includes the following possibilities:

```
FLOW      Forecasted daily low temperature
FHIGH     Forecasted daily high temperature
RAIN      Indicator of rain or snow forecasted for that day
SNOW      Indicator of snow forecasted for that day
WEEKDAY   Indicator of the day being a regular workday
SUNDAY    Indicator of the day being a Sunday
```

Explore the relationship between CALLS and these variables. Assume that your objective is to predict CALLS for the day. Build a multiple regression model for CALLS in terms of some or all of these variables.

Use diagnostic plots to check the assumptions of the model.

Are there transformations of the variables that will improve the predictive power of the model? Will lagged versions of some of the variables help?

Are there transformations of CALLS that lead to a better model?

Predict January 27, 1994 based on your final model. Is the prediction accurate?

Repeat the modeling process using the percentage of the monthly calls answered (PCALLS) as the target variable. How does the accuracy of the prediction of January 27, 1994 based on your final model compare with the prediction made in the original case?

Comments to the Instructor

A model to predict CALLS that includes FLOW and RAIN seems to fit, although a bit weakly ($R^2 = 41\%$, $F = 8.67$). However, examination of residuals shows that there is strong autocorrelation in the errors (the Durbin–Watson statistic is 1.05). This suggests using the previous day's number of calls as a predictor as well. This leads to **two** cases being considered as missing — the first, and also the fifteenth, since it corresponds to January 16, 1994, while the fourteenth case corresponds to January 29, 1993.

A regression on FLOW and LAGCALLS is strong; $R^2 = 76.6\%$ and $F = 37.71$. The autocorrelation has disappeared. However, there is a clear outlier, which corresponds to January 27, 1994. The fitted model with this day omitted is

$$\widehat{\texttt{CALLS}}_t = 1889.83 + .7954 \times \texttt{CALLS}_{t-1} - 49.558 \times \texttt{FLOW}_t,$$

with $R^2 = 86.2\%$ and $F = 68.81$. Unfortunately, this model does a poor job of predicting January 27, 1994. The predicted number of calls is only 4089.7, and the observed value of 8947 is at the upper limit of a 99.983% prediction interval. Thus, the previous day's calls is apparently not enough to account for a particularly busy day.

The number of calls is somewhat long–tailed, so a log transformation could be considered. This does not, however, improve the prediction.

The monthly percentage of calls (PCALLS) can also be used as a target variable in a multiple regression. The variables FLOW, RAIN, SNOW, WEEKDAY and SUNDAY, taken together, are all significant in a model with $R^2 = 58.7\%$. This complicated model does not predict January 27, 1994 any better than the model on just FLOW in the original case; the prediction is 3.99% of the monthly total of calls, with the observed value of 5.3% at the upper limit of a 95% prediction interval. Thus, the simple model of PCALLS on FLOW seems best for prediction of the unusual January 27, 1994.

To maintain consistency with the unannotated edition,
the next case begins on the following page.

Subgroups in the electricity consumption data

Topics covered: Indicator variables. Multiple regression.

Data File: `elusage.dat`

It was noted in the previous analysis of electrical consumption data ("Electricity usage, temperature and occupancy") that the house in question was partially or completely unoccupied from December 1992 through June 1993 (cases 41 through 47). It would be reasonable to expect that consumption might be lower for those months, in a systematic way.

Investigate this question more closely by evaluating the usefulness of the pooled, constant shift and full models for this data set, defining the groups by being in this time period or not.

Does it seem that the best model for these data does treat this time period differently?

If so, what is the apparent effect of the house being partially or completely unoccupied on the electrical consumption?

Note: The discussion given in the case "Purchasing power parity and high inflation countries" is particularly relevant here. There the concepts of pooled, constant shift, and full models are applied in a situation similar to the situation here.

Comments to the Instructor

The pooled model was fit in the earlier case, and was clearly inappropriate; the months where the house was partially or completely unoccupied show up as clear outliers. Thus, it is no surprise that models that account for these months fit better.

Let GONE be the indicator variable taking on the value 1 for the months December 1992 through June 1993, and 0 otherwise. The constant shift model fits significantly better than the pooled model, with the t-statistic for GONE equaling -4.17. Residual plots and diagnostics indicate no problems with the model. The R^2 equals 85.86%, and the implied models are

$$\widehat{\texttt{LOGKWH}} = 4.4354 - 1.6614 \times \texttt{LOGTEMP}$$

for months with full occupancy, and

$$\widehat{\texttt{LOGKWH}} = 4.2783 - 1.6614 \times \texttt{LOGTEMP}$$

for months with partial occupancy. Transforming back to the original scale, the models are

$$\widehat{\texttt{KWH}} = 27248.96 \times \texttt{TEMPERATU}^{-1.6614}$$

and

$$\widehat{\texttt{KWH}} = 18978.42 \times \texttt{TEMPERATU}^{-1.6614},$$

respectively. Thus, the months with partial occupancy are expected to have electricity usage that is 30.4% lower than would be expected for full occupancy months for a given temperature.

The full model is not a significant improvement over the constant shift model ($R^2 = 86.66\%$). Interestingly, however, a model with equal intercepts, but different slopes, fits about as well as the constant shift model:

$$\widehat{\texttt{LOGKWH}} = 4.4174 - 1.6507 \times \texttt{LOGTEMP}$$

for months with full occupancy, and

$$\widehat{\texttt{LOGKWH}} = 4.4174 - 1.7476 \times \texttt{LOGTEMP}$$

for months with partial occupancy ($R^2 = 86.14\%$). Still, the simple multiplicative form of the constant shift model makes it the preferred model.

To maintain consistency with the unannotated edition,
the next case begins on the following page.

A closer look at productivity and quality in the assembly plant

Topics covered: Indicator variables. Illusory correlation.

Data File: `prdq.dat`

Recall that in our previous examination of productivity and quality in automotive assembly plants ("Productivity versus quality in the assembly plant"), we saw evidence of an illusory correlation, in that the association in the sample as a whole was different from that in the subgroups defined by Japanese and non–Japanese ownership.

Investigate this question more closely by evaluating the usefulness of the pooled, constant shift and full models for this data set, defining the groups by whether ownership is Japanese or not.

Does it seem that the best model that accounts for group membership suggests the presence of illusory correlation, when compared with the pooled model (which does not account for group membership)?

Note: The discussion given in the case "Purchasing power parity and high inflation countries" is particularly relevant here. There the concepts of pooled, constant shift, and full models are applied in a situation similar to the situation here.

Comments to the Instructor

The pooled model, fitting `PRODUCTIV` on `QUALITY`, provides virtually no predictive power ($R^2 = 1.1\%$, $F = 0.29$), although the slope of `QUALITY` is positive (as a scatter plot indicates).

The constant shift model fits much better than the pooled model ($R^2 = 36.6\%$, $F = 6.94$), and the sign of the coefficient of `QUALITY` is now negative (the two fitted models are

$$\widehat{\texttt{PRODUCTIV}} = 34.406 - .0579 \times \texttt{QUALITY}$$

for non-Japanese plants, and

$$\widehat{\texttt{PRODUCTIV}} = 22.199 - .0579 \times \texttt{QUALITY}$$

for Japanese plants). This is the illusory correlation noted earlier. However, the coefficient of `QUALITY` in this model is not significantly different from zero ($t = -1.07$), meaning that the observed evidence of illusory correlation is not strong enough to not attribute it to random chance.

The full model does not improve significantly on the constant shift model ($R^2 = 36.82\%$). Thus, the simplest effective model just removes `QUALITY` altogether, and simply separates based on group membership. This is exactly equivalent to a two–sample t–test; $\overline{\texttt{PRODUCTIV}}_{NONJ} = 29.688$, while $\overline{\texttt{PRODUCTIV}}_{JAPAN} = 19.091$, with the difference in means statistically significant, as indicated by $t = 3.56$ ($p = .0015$).

All of these regression models exhibit heteroscedasticity, in that the productivity levels for non–Japanese firms are considerably more variable than those for Japanese firms ($s_{PRODUCTIV} = 9.112$ for non–Japanese forms, while $s_{PRODUCTIV} = 4.4823$ for Japanese firms). These data can thus be used as an example of the use of weighted linear regression. A weighting variable can be constructed that takes on the appropriate value of $1/s^2_{PRODUCTIV}$ for each firm, with weighted least squares analysis then following. If this is done, the strength of each of the observed relationships is considered a bit higher, but the qualitative results of the analysis are unchanged. In particular, a weighted version of the two–sample t–test comparing productivity values in Japanese and non–Japanese firms is $t = 4.00$.

To maintain consistency with the unannotated edition,
the next case begins on the following page.

Predicting mortgage rates for different types of mortgages

Topics covered: Indicator variables. Multiple regression.

Data File: `mort.dat`

Recall that in the previous examination of the data set about mortgages in the New York metropolitan area ("Mortgage rates for different types of mortgages"), it was apparent that there was a significant difference in mortgage rate for fixed versus adjustable rate mortgages. The data set includes another variable: the number of *points* charged to the borrower by the lender (the points are a fee charged as a percentage of the amount borrowed; for example, a fee of one point corresponds to 1% of the amount borrowed).

Construct a model to predict mortgage rates using the type of mortgage and points charged.

Interpret the coefficients of your final model. Do they make sense?

Is this model better than the one fit earlier, which did not include `POINTS` as a predictor?

Hint: Note that the type of mortgage, fixed and adjustable, is a qualitative variable and cannot be entered into the regression model directly. A qualitative variable has to be translated into a numerical form before it can be used in the model. The data set includes a variable that takes the value 0 when the mortgage is fixed rate and the value 1 when the mortgage is adjustable rate. Carry out the regression analysis using `POINTS` and the new ("indicator") variable for mortgage type as predictors.

Comments to the Instructor

The fitted regression of mortgage rate (`RATE`) on type of mortgage (`TYPE`) and `POINTS` is very strong ($R^2 = 90.86\%$, $F = 84.51$), but an outlier is apparent. It is case 10, EAB, which offers a mortgage with a 6% rate (when the predicted rate is 4.88%). It is possible that EAB feels that it must offer such a mortgage, but is not very interested in getting much mortgage business. Omitting this lender strengthens the results ($R^2 = 95.95\%$, $F = 189.03$). The final model, which is of the constant shift form, is

$$\widehat{\texttt{RATE}} = 7.802 - .283 \times \texttt{POINTS}$$

for fixed rate mortgages, and

$$\widehat{\texttt{RATE}} = 5.153 - .283 \times \texttt{POINTS}$$

for adjustable rate mortgages. Thus, given `POINTS`, adjustable rate mortgages have an opening rate that is, on average, 2.65% lower than fixed rate mortgages; given the type of mortgage, an increase of one point is associated with a decrease in rate of 0.28%. Both of these results make sense, given the relative risk of fixed and adjustable rate mortgages, and the increased cost to the borrower if `POINTS` are higher.

There is no evidence that the full model improves on the constant shift model.

Better prediction of the yield in a small vineyard

Topics covered: Forming hypotheses and models. Non–constant variance. Simple versus complicated models. Variation.

Key words: Hypothesis testing. Prediction interval. Residual plots. Scatter plot. Simple regression. t–test. Transformation. Weighted linear regression.

Data File: `vine3.dat`

Let's consider again the vineyard from the case "Volume and weight from a vineyard harvest." You should reread the introduction to that case for a description of the vineyard and the process by which the grapes are harvested. In that case we looked at the average weight of the lugs and used it to predict the total weight of the yield. Separately we used lug counts from the first few rows to do the same thing. In doing so we studied the variables separately. Here we shall try to understand in more depth the relationship between the lug counts and the total yield.

To explore this relationship, let's start by looking at a scatter plot of the total yield and number of lugs for the years 1976–91:

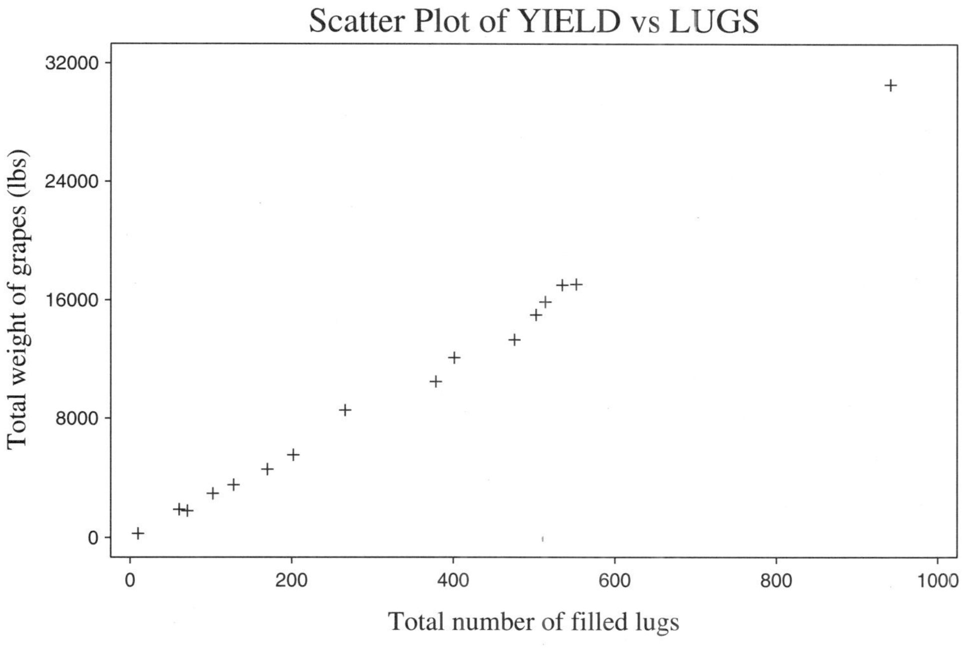

As we might expect, they appear to have a strong linear relationship. In the earlier case the yield was modeled by the equation

$$\text{Total yield} \quad = \quad \text{Average weight per lug} \times \text{Number of lugs}, \qquad (1)$$

where the average weight of the lugs was assumed to be the same from year to year.

By modeling the relationship between the total yield and the number of lugs we can address the assumption that the average weight of the lugs is the same from year to year. Based on the scatter plot, let's consider a simple regression model for yield based on the number of lugs. That is,

$$\text{Expected total yield} = \beta_0 + \beta_1 \times \text{Number of lugs}. \tag{2}$$

Comparing equation (2) to model (1), we see that the claim that the average weight of the lugs is constant corresponds to the particular model with $\beta_0 = 0$. Then we can, in addition, interpret β_1 as the average weight of the lugs. We can fit the model to the data and test if this claim is true. The numerical summary for this model is as follows:

```
LINEAR REGRESSION OF YIELD  Total weight of grapes (pounds)

PREDICTOR
VARIABLES     COEFFICIENT     STD ERROR     STUDENT'S T       P
---------     -----------     ---------     -----------     ------
CONSTANT        -612.240       266.181         -2.30        0.0373
LUGS              32.0391      0.64819         49.43        0.0000

R-SQUARED               0.9943   RESID. MEAN SQUARE (MSE)  3.936E+05
ADJUSTED R-SQUARED      0.9939   STANDARD DEVIATION          627.353

SOURCE          DF       SS            MS           F         P
----------     ---   ----------    ----------    -----    ------
REGRESSION       1   9.616E+08     9.616E+08   2443.18    0.0000
RESIDUAL        14   5.510E+06     3.936E+05
TOTAL           15   9.671E+08

CASES INCLUDED 16    MISSING CASES
```

The plot of the standardized residuals versus the number of lugs shows some interesting structure (note that for this simple regression, this plot has the same appearance as one of the standardized residuals versus fitted values):

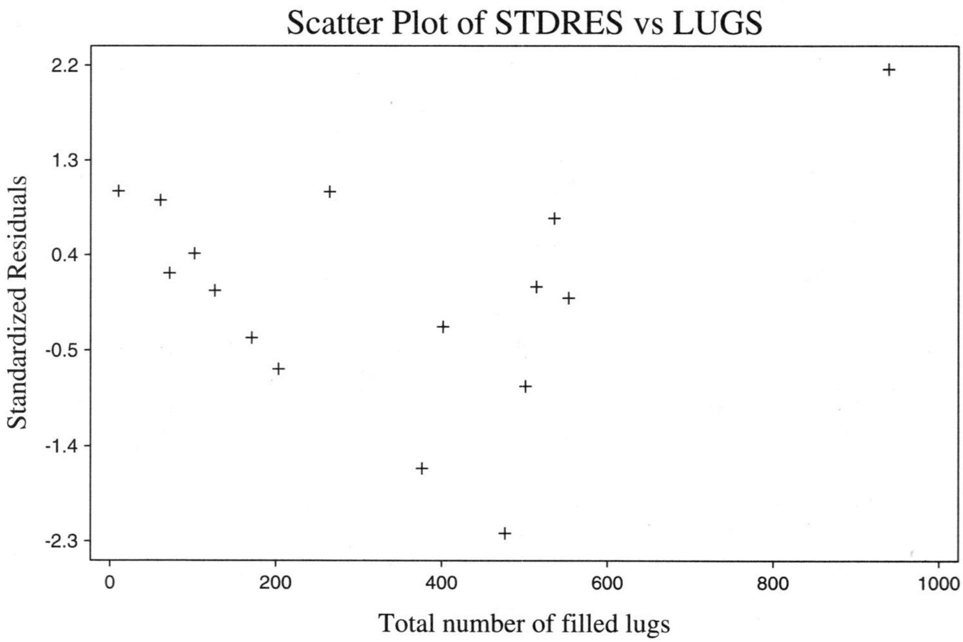

The standardized residuals are not random and there appears to be some structure to them. First, the final year has a large positive standardized residual. It is not extreme, but it is disturbing. Second, the variation of the standardized residuals for a fixed number of lugs is not constant, but increases as the number of lugs increases. Another way of saying this is that there is a "fanning out" of the residuals as the number of lugs increases. Within this pattern the large standardized residual of the final year is not so odd. This suggests that the variance of the weights of the yields increases with the number of lugs. Why would this be so? We can think of the total yield of the harvest as being composed of the sum of the weights of each individual lug. In addition, we expect the individual weights of the lugs to have the same variance and be approximately independent of each other. Hence the variance of the total yield should be approximately equal to the sum of the variances of each lug weight, that is, proportional to the number of lugs that comprised it. This would be consistent with the pattern in the standardized residuals.

As the assumptions of the regression model do not appear to hold, we cannot use the usual machinery of hypothesis testing to check the claim that the average weight of the lugs is constant from year to year.

We can incorporate this non–constant variance feature into the regression model by adjusting the influence (i.e., the "weighting") of each case inversely relative to its variance. This fitting procedure is called **weighted least squares.** Conceptually it is the same as ordinary linear regression, except that the different variances of the cases are taken into account.

Here is the weighted linear regression:

```
WEIGHTED LINEAR REGRESSION OF YIELD  Total weight of grapes (pounds)

WEIGHTING VARIABLE: RECLUGS

PREDICTOR
VARIABLES     COEFFICIENT    STD ERROR     STUDENT'S T       P
---------     -----------    ---------     -----------     ------
CONSTANT        -122.327      100.227         -1.22        0.2424
LUGS              30.5625     0.56686         53.92        0.0000

R-SQUARED             0.9952    RESID. MEAN SQUARE (MSE)     1221.44
ADJUSTED R-SQUARED    0.9949    STANDARD DEVIATION           34.9491

SOURCE        DF       SS            MS           F         P
----------    ---   ----------    ----------    -----    ------
REGRESSION     1    3.551E+06     3.551E+06   2906.88    0.0000
RESIDUAL      14      17100.2       1221.44
TOTAL         15    3.568E+06

CASES INCLUDED 16    MISSING CASES 0
```

Here RECLUGS is the reciprocal of LUGS.

The plot of the standardized residuals versus the number of lugs is now as follows:

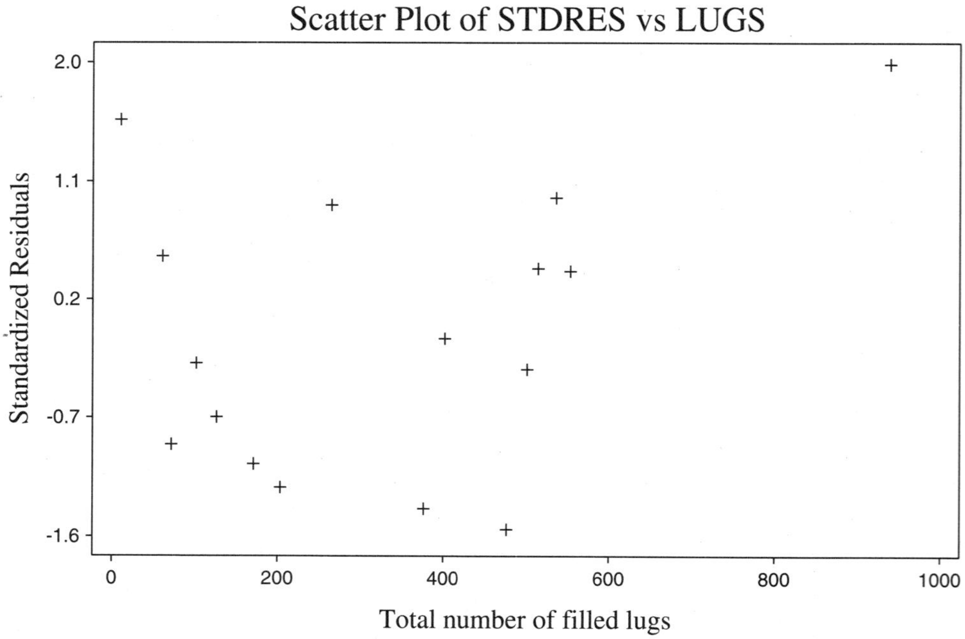

The residual plot looks much better in terms of variance, suggesting that our hunch about the variance was right. The final year still has a reasonably large standardized residual. Let's look at the leverages and influence of each case. The latter is measured by Cook's distance:

CASE	YEAR	LUGS	YIELD	STDRES	LEVERAGE	COOKSD
1	1976	11.00	313.0	1.56	0.70	2.87
2	1977	476.00	13299.0	-1.55	0.09	0.12
3	1978	60.00	1848.0	0.53	0.10	0.02
4	1979	102.00	2894.0	-0.29	0.06	0.00
5	1980	71.00	1792.0	-0.91	0.08	0.04
6	1981	203.00	5483.0	-1.23	0.04	0.04
7	1982	127.00	3490.0	-0.70	0.05	0.01
8	1983	534.75	16970.0	0.98	0.11	0.06
9	1984	552.50	17083.0	0.41	0.11	0.01
10	1985	401.00	12067.5	-0.10	0.08	0.00
11	1986	266.00	8519.0	0.92	0.05	0.02
12	1987	514.00	15912.0	0.43	0.10	0.01
13	1988	377.50	10507.0	-1.39	0.07	0.07
14	1989	170.50	4618.0	-1.05	0.04	0.03
15	1990	502.00	14975.0	-0.33	0.10	0.01
16	1991	940.50	30521.0	1.99	0.21	0.52

The harvest of 1976 has higher leverage on the model (0.70, over five times the average leverage of 2/16 = 0.125). Its influence on the model is also quite high (2.87), indicating that removing it would substantially change the fitted coefficients. The harvest in 1976 was the first ever in the vineyard and produced a miniscule yield. Given that we wish to use the model to predict future yields, the harvest in 1976 is different enough so that we will omit it from determining the model. Note that the standardized residual plot does not provide this information, since the small number of lugs in 1976 has caused it to be weighted very strongly in the weighted least squares regression.

Refitting the model with 1976 omitted produces the following standardized residuals versus fitted values plot:

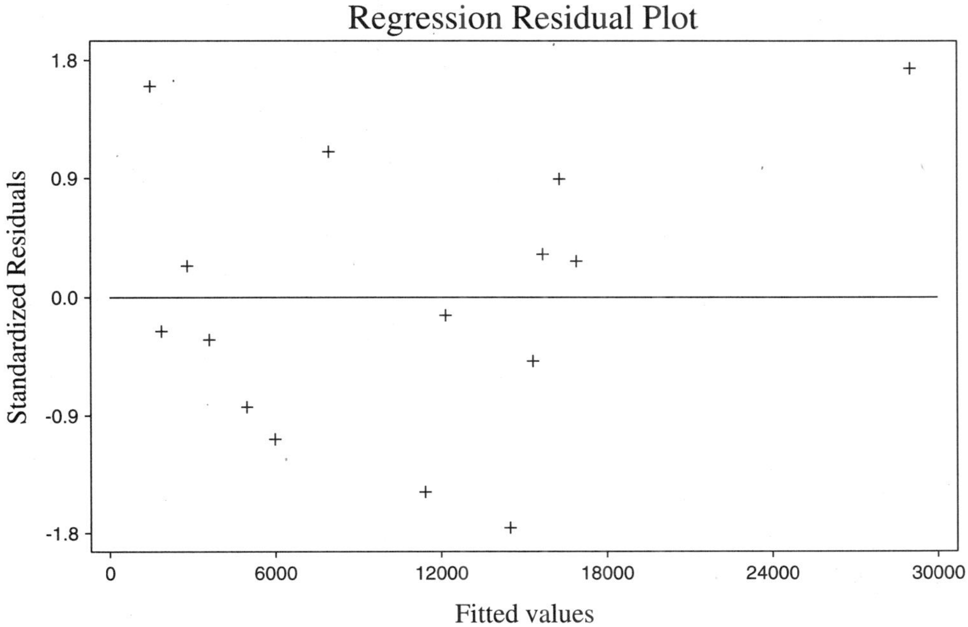

The leverages and influences of the remaining harvests are modest, and are not reported here. The 1991 harvest is less aberrant than the plot would suggest (since it is weighted less, because of the large number of lugs for that year). Thus the assumptions of this (weighted) model look reasonable. As this is so, let's look at the numerical summary:

```
WEIGHTED LINEAR REGRESSION OF YIELD  Total weight of grapes (pounds)

WEIGHTING VARIABLE: RECLUGS

PREDICTOR
VARIABLES     COEFFICIENT    STD ERROR     STUDENT'S T        P
---------     -----------    ---------     -----------     ------
CONSTANT        -361.997      172.613         -2.10        0.0561
LUGS             31.2224      0.66616         46.87        0.0000

R-SQUARED             0.9941    RESID. MEAN SQUARE (MSE)     1085.54
ADJUSTED R-SQUARED    0.9937    STANDARD DEVIATION           32.9475

SOURCE        DF       SS            MS          F         P
----------    ---   ----------    ----------   -----    ------
REGRESSION     1    2.385E+06     2.385E+06   2196.73   0.0000
RESIDUAL      13      14112.0       1085.54
TOTAL         14    2.399E+06

CASES INCLUDED 15   MISSING CASES 0
```

The removal of the 1976 harvest has altered the model; the coefficient of LUGS has decreased by about two standard errors and the intercept has increased by over one standard error. The $R^2 = 99.41\%$ is excellent, and the p–value for the hypothesis test that the coefficient of LUGS is zero is very small. Thus the relationship is very strong.

The real value of this model is that it gives us a way to address the question in the earlier case about the average weight of the lugs. To test the claim all we need to do is to test the hypothesis that the constant term is zero. The t–statistic for the hypothesis that $\beta_0 = 0$ is given as –2.10 (listed in the line labeled CONSTANT), which is barely not significant (that is, statistically significantly different from zero) using the 5% Type I error rule. Thus there is insufficient evidence in the data to reject the hypothesis that the yield is, *on average,* proportional to the number of lugs. A more liberal interpretation of the evidence provided by the p–value (0.0561) states that the evidence is borderline statistically significant. In addition, our best estimate of the average weight per lug is $\widehat{\beta}_1 = 31.22$ pounds per lug.

A couple of technical notes. Suppose we had not done a residual analysis on the original model and focused on its numerical summary given before. The same hypothesis test has a t–statistic of –2.30 and we would reject the model ($p = 0.0373$). Thus, the evidence against the simpler model (1) appears stronger, based on the (incorrect) unweighted analysis. Second, the estimate of 31.22 pounds per lug from the model is superior to the mean of the averages given in the descriptive statistics (29.26 pounds per lug), because it takes into account the differing variances.

In the earlier case we predicted the yield based on the model given in (1). We can use the more sophisticated model (2) to predict the yield of another harvest given the total lug count. In 1992 the lug count was 1057.

Let's look at the 95% prediction interval for yield given this lug count:

```
PREDICTED VALUES OF YIELD

LOWER PREDICTED BOUND  3.137E+04
PREDICTED VALUE        3.264E+04
UPPER PREDICTED BOUND  3.392E+04
SE (PREDICTED VALUE)      590.20

PREDICTOR VALUES: LUGS = 1057.0
```

The 95% prediction interval for a harvest comprising of 1057 lugs is (31370, 33920) pounds. The observed yield in 1992 was 31,229 pounds, so the interval does not, in fact, include the observed value for that year (remember — 5% of all 95% prediction intervals constructed will not cover the true value).

In this case we have a reasonable model for the relationship between the number of lugs and the variance of the yield. This enabled us to specify the weights for the regression model. If the weights are unknown, we can often adjust for non–constant variance in the data by considering transformations of the underlying relationship. For example, we can re–express equation (1) by transforming each side of the equality by the logarithmic function:

$$\log(\text{Total yield}) = \log(\text{Average weight per lug}) + \log(\text{Number of lugs}). \qquad (3)$$

This suggests that we consider a simple regression model for the logarithm of the total yield as a linear function of the logarithm of the total number of lugs. That is,

$$\text{Expected } \log(\text{Total yield}) = \beta_0 + \beta_1 \times \log(\text{Number of lugs}). \qquad (4)$$

Comparing equation (4) to the model (3), we see that the claim that the average weight of the lugs is constant corresponds to the model with $\beta_1 = 1$. If the claim is correct, we can additionally interpret 10^{β_0} as the average weight of the lugs. As before, we can fit the model to the data using regression.

The numerical summary for this model is as follows:

```
LINEAR REGRESSION OF LOGYIELD  Logarithm of the total yield

PREDICTOR
VARIABLES     COEFFICIENT    STD ERROR     STUDENT'S T       P
---------     -----------    ---------     -----------     ------
CONSTANT         1.39542      0.03637         38.37        0.0000
LOGLUGS          1.02983      0.01522         67.66        0.0000

R-SQUARED              0.9970    RESID. MEAN SQUARE (MSE)  8.449E-04
ADJUSTED R-SQUARED     0.9967    STANDARD DEVIATION          0.02907

SOURCE          DF        SS           MS          F         P
----------     ---    ----------   ----------    -----    ------
REGRESSION       1     3.86807      3.86807    4578.32    0.0000
RESIDUAL        14     0.01183    8.449E-04
TOTAL           15     3.87990

CASES INCLUDED 16    MISSING CASES 0
```

The standardized residuals versus fitted values plot has the following form:

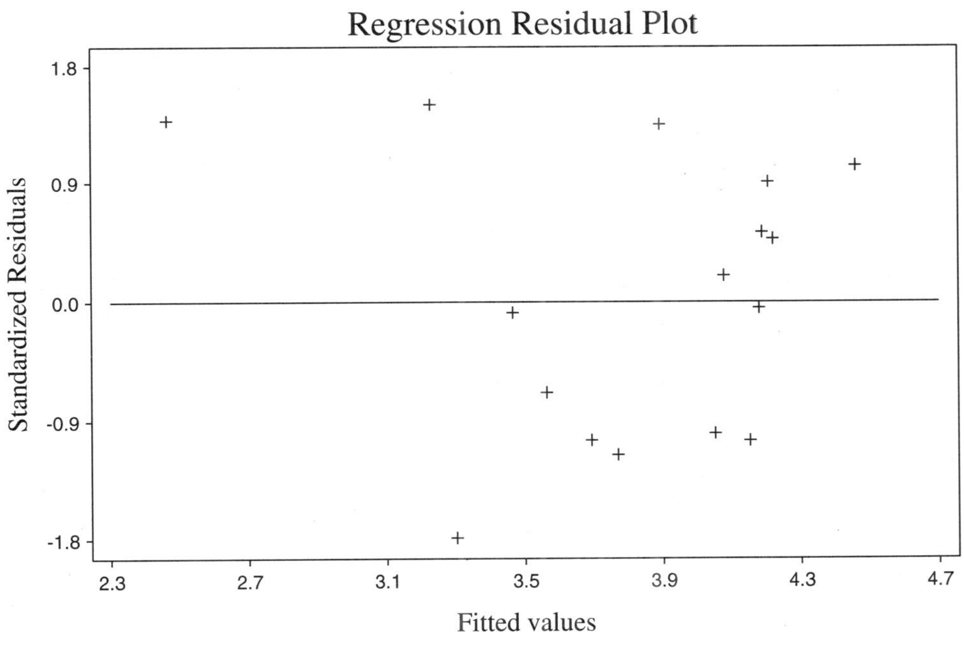

The residual plot does not look ideal. The very low lug counts in 1976, 1978 and 1980 do not appear to follow the pattern of the later years. They may not be representative of the future harvests and, in addition, have higher leverage and influence on the model. Let's consider the model with those three years removed. Refitting the model produces the following standardized residuals versus fitted values plot:

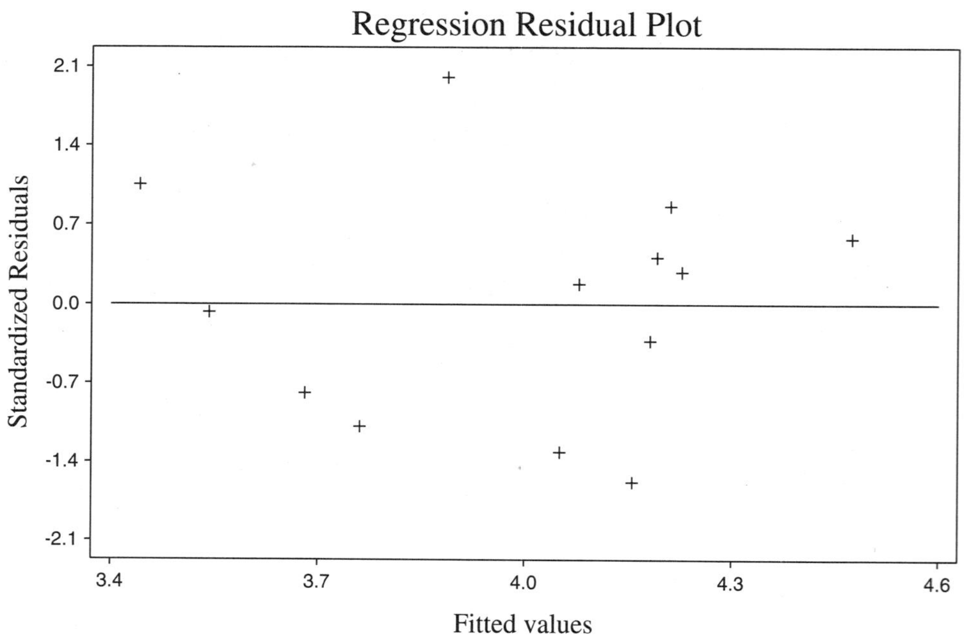

The residual plot looks a lot better, with a hint of a decreasing variance for the larger harvests. The leverage and influences of the remaining harvests are modest, and are not reported here. As the assumptions for this reduced data set appear to be roughly satisfied, let's look at the numerical summary:

```
LINEAR REGRESSION OF LOGYIELD  Logarithm of the total yield

PREDICTOR
VARIABLES     COEFFICIENT    STD ERROR     STUDENT'S T        P
---------     -----------    ---------     -----------     ------
CONSTANT         1.29502      0.05784         22.39        0.0000
LOGLUGS          1.06905      0.02280         46.90        0.0000

R-SQUARED              0.9950      RESID. MEAN SQUARE (MSE)  5.064E-04
ADJUSTED R-SQUARED     0.9946      STANDARD DEVIATION          0.02250

SOURCE         DF       SS            MS          F          P
----------    ---   ----------    ----------    -----     ------
REGRESSION      1    1.11360        1.11360    2199.15    0.0000
RESIDUAL       11    0.00557      5.064E-04
TOTAL          12    1.11917

CASES INCLUDED 13   MISSING CASES 0
```

The removal of the smaller harvests has altered the model; the coefficient of LOGLUGS has increased by about two standard errors and the intercept has decreased by about two standard errors. The $R^2 = 99.50\%$ and the $p-$ value for the hypothesis test that the coefficient of LOGLUGS is zero shows that the regression is very strong. However the real interest lies in testing the hypothesis that $\beta_1 = 1$. The t–statistic for this can be calculated manually:

$$t = \frac{1.06905 - 1}{0.0228} = 3.03.$$

This is significant at any reasonable Type I error level, so we reject this hypothesis. That is, for harvests that are not too small, the model that assumes constant average lug weight from year to year is rejected. Even though the estimate is quite close to one, it is (statistically significantly) different from one. This suggests that we may need to model how the average weight of the lugs varies with the size of the harvest to improve our model.

In the earlier case we predicted the yield based on the model given in (1). We can use this more sophisticated model to predict the yield of another harvest given the total lug count. Since the 1992 lug count was 1057 (`LOGLUGS` = 3.0241), the 95% prediction interval is as follows:

```
PREDICTED VALUES OF LOGYIELD

LOWER PREDICTED BOUND      4.4707
PREDICTED VALUE            4.5279
UPPER PREDICTED BOUND      4.5851
SE (PREDICTED VALUE)       0.0260

PREDICTOR VALUES: LOGLUGS = 3.0241
```

The 95% prediction interval for a logarithm of the yield comprised of 1057 lugs is (4.4707, 4.5851) log–pounds. If the logarithm of the yield falls in this interval then the yield itself falls in the interval

$$(10^{4.4707},\ 10^{4.5851}) \quad = \quad (29560,\ 38468)\text{ pounds}$$

(and vice versa). That is, we are 95% confident that the yield of a harvest comprised of 1057 lugs is between 29,560 and 38,468 pounds. The observed yield in 1992 was 31,229 pounds, so the interval includes the observed value for that year.

Summary

In the vineyard the yield of the vines is measured by the total weight of the harvested grape bunches. Here we investigate the relationship between the harvest yield and the number of lugs, with the objective of using the number of lugs as a measure of yield.

In the previous case study of this vineyard ("Volume and weight from a vineyard harvest"), it was assumed that the average lug weight is the same from year to year. By considering models that incorporate this claim as a particular case, we are able to show that there is evidence in the data that the claim is false.

Based on a weighted linear regression model for the total yield in terms of the number of filled lugs for each harvest we find that the deviation from the claim is borderline statistically significant. Our best point estimate of the average weight per lug is 31.22 pounds per lug.

An alternative simple regression model for the logarithm of the total yield in terms of the logarithm of the total number of lugs can be considered. This model strongly suggests that the claim of constant average lug weight is false. Based on this model we are able to predict the total yield of a harvest based on the observed total lug count. In particular, we are 95% confident that the yield of the 1992 harvest (comprised of 1053 lugs) is between 29,437 and 38,309 pounds.

Technical terms

Weighted linear regression: fitting a regression model to data where the errors about the regression line have different variances ("heteroscedasticity"). Each observation is weighted by an estimate of (a constant times) the inverse of the variance of the error for that observation. Ordinary linear regression is a special case of weighted regression, with all weights equaling one.

Incomes of Long Island communities

Topics covered: Transformation. Weighted linear regression.

Key words: Histogram. Mean. Median. Outliers. Regression. Residual plots. Scatter plot. Skewness.

Data File: `liinc.dat`

Demography is the science of social and vital statistics. In practical terms, demographers attempt to understand and describe society from patterns in the populations, including birth, death, migration and income patterns. This case examines income patterns for Long Island, New York. The data set gives the median income in 1989 dollars for 257 Long Island communities in 1979 and 1989, as reported in *Newsday* on August 31, 1992. Values over $150,000 are coded as $150,001. Note that changes in income between 1979 and 1989 are in **real** dollars, with inflation effects removed. Was there a consistent pattern in median incomes between 1979 and 1989 that can be identified statistically?

First, examine each of the income variables (`INCOME79` and `INCOME89`, say) using both summary statistics and graphs. Both variables are long right–tailed; note, for example, that for each variable the mean income is about $7,000 higher than the median income.

Now, construct a scatter plot of 1989 income versus 1979 income. Three facts are apparent from the plot: there could be a linear relationship connecting the two variables; there is a "fanning out" effect off the regression line, suggesting heteroscedasticity; and there is a clear outlier in the lower right of the plot. This is Saltaire, which went from a 1979 median income of $133,502 to a 1989 median income of $25,714. How could this happen? Well, Saltaire is a small community on Fire Island in Suffolk County, with virtually no year–round residents. This drop could be the result of just one or two wealthy people moving out!

Remove this point and fit a regression model. The regression is quite strong, and the coefficient of 1979 income being greater than one coincides with our belief that 1989 income is, in general, greater than 1979 income. But do the regression assumptions hold? A plot of residuals versus fitted values shows a widening of the plot from left to right, implying non–constant variance.

A natural approach, given the long tails of the original income variables and the fact that variables that deal with money often operate multiplicatively rather than additively, is to reanalyze in a log scale. Create transformed versions of the 1979 and 1989 income variables based on taking logarithms (`LOGINC79` and `LOGINC89`, say). The long tails have now disappeared.

Now, construct a scatter plot of logged 1989 income versus logged 1979 income. The resultant plot is interesting — the relationship looks reasonably linear, and the spreading out effect is lessened, but it now appears that the variability off the regression line is higher for both lower income and higher income communities. Actually, that makes sense — communities with lower income could be ones that are genuinely poor, or ones with few permanent residents (as we saw with Saltaire before), while higher income communities could provide evidence of differing gains because of the expanding economy in the 1980's.

A regression of logged 1989 income on logged 1979 income indicates a strong fit, but there's definitely nonconstant variance. Save the standardized residuals from this regression.

We don't have any theoretical reasons to hypothesize any specific functional form for the variances of the errors, but we can use the patterns observed in the data to derive a reasonable form. The scatter plot of logged 1989 income versus logged 1979 income suggested that there were three groups in the data: communities with logged 1979 income less than about 4.4 (1979 income less than $25,000), with large variability off the regression line; communities with logged 1979 income between about 4.4 and 4.8 ($25,000 $<$ 1979 income $\leq$ $63,250), with small variability off the regression line, and communities with logged 1979 income greater than about 4.8 (1979 income $>$ $63,250), with large variability off the regression line.

Once these groups are identified, they can be used to form a weighting variable. Calculate the standard deviations of the standardized residuals for each of the three subgroups; they turn out to be $s_1 = 1.5974$, $s_2 = .8008$ and $s_3 = 1.2298$, respectively. Now, create a weight variable, which takes on the values

$$\texttt{WEIGHT}_i = \frac{1}{s^2_{j(i)}},$$

where the notation $s_{j(i)}$ means the standard deviation estimate s_1, s_2 or s_3, being determined by the subgroup j to which case i belongs. Perform a weighted least squares regression of logged 1989 income on logged 1979 income, using this weight variable.

Note that the coefficients have changed very little, although the fit is assessed as being slightly weaker than in the earlier fit. Examine residual plots and diagnostics from this fit. There are five potentially unusual communities:

	COMMUNITY	INCOME79	INCOME89
39	Cove Neck	65621	112114
80	Greenport West	34025	30926
112	Laurel	40050	36953
167	Ocean Beach	28480	26250
183	Poquott	38359	62216

The income in Cove Neck increased by over 70%, while that in Poquott increased by over 60%, which are unusually high increases. On the other hand, the incomes in Greenport West, Laurel and Ocean Beach *dropped* by more than 7%, which is a surprise. We can try removing these points, although (based on measures like the Cook's distance) it's unlikely it will change things much.

The final model can be used to describe the general pattern in income change over the 1980's on Long Island. The model is

$$\texttt{LOGINC89} = .556 + .899 \times \texttt{LOGINC79};$$

transforming back to original units gives

$$\begin{aligned}\texttt{INCOME89} &= 10^{.556} \times (\texttt{INCOME79})^{.899}\\ &= 3.597 \times (\texttt{INCOME79})^{.899}.\end{aligned}$$

The multiplier of almost 3.6 would seem to imply that real income in 1989 was higher than it was in 1979 by a large amount, but we have to be careful in this interpretation. The median value of the 1979 incomes is $44,470, so the median predicted 1989 income is $3.597 \times (44470)^{.899} = \$54,269.12$, an increase of 22% (remember, this figure includes correcting for inflation). The 1980's were very good for Long Island economically; the 1990's have been pretty bad. The exponent of .899, being significantly less than one, implies that the proportional increases for wealthier communities were smaller than they were for poorer communities (that is, a community that had 1979 income that was 2 times that of another community is predicted to have 1989 income that is only $2^{.899} = 1.865$ times as high). In that sense, the poorer communities were "catching up," but clearly there's a lot more catching up to do!

Comments to the Instructor

The fitted regression of INCOME89 on INCOME79 (after omitting Saltaire) has $R^2 = 91.73\%$, with $F = 2805.94$, but heteroscedasticity is quite apparent.

The fitted regression of LOGINC89 on LOGINC79 has $R^2 = 91.4\%$, with $F = 2688.98$, but heteroscedasticity is still apparent. After performing the weighted regression suggested in the case, the fit has weakened slightly ($R^2 = 89.49\%$, $F = 2153.25$), but is still very strong. The five communities mentioned in the case all have absolute standardized residuals greater than 2.7; omitting them gives the final model summarized in the case, which fits very well ($R^2 = 91.08\%$, $F = 2531.45$), with no apparent violations of regression assumptions.

Looking more deeply into the average weight per lug from a vineyard

Topics covered: Forming hypotheses and models. Non–constant variance. Variation.

Key words: Confidence interval. Hypothesis testing. Prediction interval. Residual plots. Scatter plot. Simple regression. t–test. Weighted linear regression.

Data File: `vine3.dat`

Let's consider again the vineyard from the cases "Volume and weight from a vineyard harvest" and "Better prediction of the yield in a small vineyard." In the latter case we found that the average weight of the lugs varied from harvest to harvest. In this case we will try to build a model for the average weight of the lugs in terms of the total number of lugs collected.

Let's think some more about the factors that affect the average weight of the lugs. A more detailed breakdown of the total weight is:

$$\text{Total weight} = \text{Density of grapes per volume filled in the lug} \\ \times \text{Average percentage of the lug filled} \times \text{Volume of each lug} \times \text{Number of lugs.}$$

The first factor is the weight per volume of the dumped bunches. The volume of the bunches changes from year–to–year, although we might expect the density to be less variable. As we discussed in the first case, the proportion of each lug that is filled depends on the pickers. Carrying the heavy filled baskets far is tiring for the pickers, so they tend to dump in the closest non–filled lug. We could hypothesize that in years that there is a heavy crop, the lugs close to the vines tend to be filled more heavily. In poor years some lugs could be left partially filled as the picker moves to the next vine. If this hypothesis is correct, we might expect that the average weight per lug is related to the total yield, that is, the total weight. We then would hypothesize a relationship with the number of lugs, which, as we saw in the second case, is highly correlated with the total yield.

To explore this hypothesis, first create a scatter plot of the average weight per lug versus the number of lugs. What do you see?

There does appear to be some relationship, although it is not strong. Fit a simple linear regression model and look at a scatter plot of the standardized residuals versus the number of lugs.

We see a pattern in the variation of the standardized residuals as the number of lugs increases. However this time it is *decreasing* as the number of lugs increases. Let's think about this. The average weight per lug is the average of the weights of each individual lug. In addition, for a given number of lugs, we expect the weights to have about the same variance and be approximately independent of each other. Thus we can think of the average weight per lug as the sample average, so its variance will be *inversely* proportional to the number of lugs. This is the same principle we used before, but using the Central Limit Theorem now. So, we should use weighted linear regression again.

Fit the weighted linear regression to these data using `LUGS` as the weight variable.

Create a plot of the standardized residuals versus the number of lugs.

The residual plot looks a lot better, although there is a hint that the variance is increasing with the number of lugs. The leverage of the 1991 harvest (940.5 lugs) is six times the average leverage (0.75, where the average leverage = 2/16 = 0.125). This is due to the weighting by LUGS that the standardized residuals versus fitted values plot does not reflect. However, the influence of the point is not great (Cook's D = 0.28). Thus the harvest appears to be consistent with the pattern of the other harvests, and we will retain it.

How well does this model fit? The $R^2 = 51.54\%$ is moderate. The p–value for the hypothesis test that there is no (linear) relationship between the average weight and the number of lugs is 0.0017. This shows that there is a definite relationship here and the hypothesis that the average weight per lug is constant from year–to–year is rejected. This conclusion is consistent with the conclusion from the previous case. In fact, we can interpret $\beta_0 + \beta_1 \times$ (number of lugs) as the expected average weight per lug for a given number of filled lugs.

Calculate a 95% confidence interval for β_1.

We are 95% confident that the expected increase in average weight for the lugs when the number of lugs increases by 100 lugs is

$$100 \times (0.00251,\ \ 0.00876) \ = \ (0.251,\ \ 0.876) \text{ pounds per lug.}$$

What do you think these results suggest about the relationship between the average weight of the lugs and the total number of lugs collected? Are you surprised that the lugs would be filled to higher capacity when the harvest is larger?

Construct a 95% prediction interval for the average weight of the lugs if the number of lugs equals 1057 (the value for 1992). Now, multiplying the upper and lower limits of this interval by 1057 gives a prediction interval for the yield for 1992, or (-22863, 93237). The observed yield in 1992 was 31,229 pounds, so the interval does contain the true value. Still, the interval is very wide, and even includes negative values, so it cannot be considered very useful here.

Comments to the Instructor

The unweighted regression relationship between average weight per lug (AVEWEIGHT) and LUGS is fairly weak ($R^2 = 38.61\%$, $F = 8.81$), and heteroscedasticity of the residuals exists. The weighted regression is noticeably stronger (although still not tremendously strong), and has the form

$$\widehat{\texttt{AVEWEIGHT}} = 27.327 + .00564 \times \texttt{LUGS}.$$

A 95% confidence interval for β_1 then has the form

$$\begin{aligned} &\hat{\beta}_1 \pm t^{14}_{.025} \times \widehat{\mathrm{SE}}(\hat{\beta}_1) \\ &= .00564 \pm (2.14)(.00146) \\ &= (.00251, .00876). \end{aligned}$$

It is not surprising that harvests with greater `LUGS` are characterized by having lugs that are filled to a greater capacity, since pickers will have more grapes to load into nearby lugs. In a smaller harvest, pickers will move on to an unfilled lug while a given lug is not very full.

A 95% prediction interval for `AVEWEIGHT` for a harvest with `LUGS` equaling 1057 is $(-21.630, 88.209)$ (the predicted value is 33.290 pounds per lug). Multiplying the limits of the prediction interval by 1057 gives the stated prediction interval (the prediction itself is $1057 \times 33.290 = 35,188$ pounds).

To maintain consistency with the unannotated edition,
the next case begins on the following page.

Predicting incomes of Long Island communities

Topics covered: Weighted linear regression.

Data File: `liinc.dat`

Recall that in the earlier case "Incomes of Long Island communities," the final model derived was based on using logged income variables (i.e., variables that are the logarithmic transforms of the original variables). A scatter plot of the untransformed 1989 and 1979 income variables, however, also suggests a linear relationship connecting them, albeit with nonconstant variance.

Construct a (weighted) regression for predicting 1989 income from 1979 income.

You might base your weights on the apparent change in variation off the regression line that occurs for 1979 income greater than or less than \$60,000. To do this first only consider communities with values less than \$60,000. For these data, regress the untransformed 1989 values on the 1979 values. Record the standard deviation around the regression line (that is, the standard error of the estimate). Repeat the process for communities with values greater than \$60,000. The relative variation about the regression line for the two groups is then approximately measured by the square of the ratio of the first standard deviation to the second. Create a new variable of weights that takes the value 1 if the 1979 income is less than \$60,000 and the value of the relative variation if the 1979 income is greater than \$60,000.

Which model do you prefer — the model based on the original variables, or the one based on the logged variables?

Comments to the Instructor

First, we must remember to omit Saltaire from any analyses from the start. The standard error of the estimate from a regression of INCOME89 on INCOME79 for the 204 communities with 1979 income less than \$60,000 is \$4399.45. The standard error of the estimate from the regression for the 51 communities with 1979 income greater than \$60,000 is \$13279.5. Thus, the weight variable has the value 1 for the former set of communities, and $(4399.45/13279.5)^2 = 0.1027$ for the latter set of communities.

The fitted weighted regression is then

$$\widehat{\texttt{INCOME89}} = 6521.99 + 1.0687 \times \texttt{INCOME79},$$

with $R^2 = 87.85\%$ and $F = 1828.55$. Four of the five outliers noted for the log model now have absolute standardized residual less than 2.8 (Poquott is still an outlier); in addition, Lido Beach, with its increase in income from \$53,672 to \$77,279, is an outlier. Omitting these two communities gives a fitted model

$$\widehat{\texttt{INCOME89}} = 6516.86 + 1.0656 \times \texttt{INCOME79},$$

with $R^2 = 88.6\%$ and $F = 1951.3$. There are no apparent violations of regression assumptions.

There aren't any compelling reasons to choose one of the two fitted models for these data (original scale versus log scale) over the other, based on the statistics. The log scale has the potential advantage of assessing changes in income in a multiplicative, rather than additive fashion, which seems intuitive. In fact, the two models are likely to give similar predictions over the range of the data. For example, the predicted median 1989 income for the model in the original scale is $6516.86 + (1.0656)(44470) = \$53,894.09$, which can be compared to the value using the log model ($54,269.12).

To maintain consistency with the unannotated edition,
the next case begins on the following page.

Prediction of yields from partial harvesting in a small vineyard

Topics covered: Forming hypotheses and models. Non–constant variance. Simple versus complicated models. Variation.

Key words: Hypothesis testing. Normal plot. Prediction. Prediction interval. Regression through the origin. Residual plots. Scatter plot. Simple regression. t–test.

Data File: `vine3.dat`

Let's consider again the vineyard from the cases "Volume and weight from a vineyard harvest," "Better prediction of the yield in a small vineyard," and "Looking more deeply into the average weight per lug from a vineyard." In the first case we explored a simple model for the relationship between the total yield of the vineyard and the sum of the lug counts from rows 3 through 6. We then used this model to predict the total yield for other harvests. It is worthwhile to reread this part of the first case before you continue here.

In this case we will try to improve the quality of the prediction of the yield by looking at more sophisticated regression models for the total yield in terms of the total row count from the initial rows (that is, rows 3 through 6).

To start, create a scatter plot of the total harvest weight versus the total row count from the initial rows. What do you see? There is a positive relationship here, and the final year (1991) has a much greater count and yield than the other years.

Fit a simple linear regression model for the total harvest weight based on the total row count from the initial rows.

Create a scatter plot of the standardized residuals versus the fitted values.

With only nine cases the plot is quite sparse; it is hard to say if there is a pattern.

Create a normal plot of the standardized residuals.

This appears to suggest that the distribution of the standardized residuals is roughly normal.

The $R^2 = 82.31\%$ is reasonably high. The slope estimate is 5.71 standard errors away from zero; strong evidence that there is a positive linear relationship between the variables. If β_0 is zero then the expected total weight of the harvest would be zero if the total row count from the initial rows was zero. This is reasonable (although a zero lug count is unlikely to occur). The p–value for the hypothesis test that the intercept term β_0 is zero is 0.3569, and the estimate is about one standard error away from zero. We could therefore consider the model with the constant term removed. Note that this model specifies that the total yield is proportional to the total row count from the initial rows.

Fit a simple linear regression model for the total harvest weight based on the total row count from the initial rows that does not include a constant term (i.e., the intercept is taken to be zero).

Create a plot of the standardized residuals versus the fitted values. What do you see?

It is important to note that the measures of fit generated by most regression packages when performing regression through the origin are *not* directly comparable to those produced when a constant term is included. For example, the R^2 value of 0.96 here does not mean that this model provides a better fit than the previous model that included a constant term. The reason for this is that the usual measures are mean–corrected, which is not sensible for the model without an intercept.

Based on either of these models we can predict the yield for a new harvest.

What is the point prediction of the total weight in the model with a constant term, given a total row count from the initial rows?

In 1992, the lug count from rows 3 through 6 was 19, 20, 28 and 24 for a total of 91 lugs. Predict the yield for the 1992 harvest based on both models.

These predictions are not of much use unless we indicate how close we can expect them to be to the true harvest yields. To do this let's look at the prediction intervals.

For the model with a constant term, calculate a 95% prediction interval for the total weight of the 1992 harvest. Also calculate a 50% prediction interval.

Calculate 95% and 50% prediction intervals based on the simpler model of proportionality.

These data were collected during the 1992 harvest before the total weight of the harvest was known. Subsequently, the harvest weight was found to be 31,229 pounds.

Which of the above intervals include the observed yield? Do the predictions from the models appear to be consistent with the actual yield? Which prediction method seems most reasonable, given the models that underlie them?

Evaluate these predictions and summarize the relative performance of the various models. Which model would you recommend for predicting the total weight of the harvest from the total lug count from rows 3 through 6?

Additional technical note

As we noted in the case, the R^2 measure produced by most statistical packages for models without an intercept is not comparable to the usual R^2 for models without one. Edward L. Korn and Richard Simon, in a 1991 paper in *American Statistician*, and Roy T. St. Laurent, in a presentation at the 1994 Joint Statistical Meetings, suggested that a better R^2 measure to use is as follows:

$$R_C^2 = 1 - \frac{\sum_{i=1}^n (y_i - \hat{y}_i)^2}{\sum_{i=1}^n y_i^2 - n(\overline{\hat{y}})^2},$$

where $\sum_{i=1}^n (y_i - \hat{y}_i)^2$ is the residual sum of squares, $\sum_{i=1}^n y_i^2$ is the sum of squared target values, and $\overline{\hat{y}}$ is the sample mean of the fitted values. The value R_C^2 equals the usual R^2 measure for regression with an intercept term. For these data, $R_C^2 = 84.46\%$, which is much closer to the R^2 value with an intercept term, reflecting that the fit of the two models is similar.

Using a computer to produce the prediction intervals is easy for us. The Barnhills may need a formula that can be done on a hand–calculator. In some books there may not be a technical exposition of the formula. Let's work from first principles to see where these numbers come from for the simpler model (without a constant term). The standard error of the prediction error for a new observation (i.e., the standard error of the estimate) is $s = 3279.28$. The estimated standard error of the fitted value $\widehat{\texttt{YIELD}}$ itself is then

$$\widehat{\text{SE}}(\widehat{\texttt{YIELD}}) = \widehat{\text{SE}}(\widehat{\beta}_1) \times \texttt{INITIAL} = 27.2592 \times \texttt{INITIAL}.$$

Thus the estimated standard error for the prediction of the total weight for a harvest is:

$$\sqrt{\widehat{\text{SE}}^2(\widehat{\texttt{YIELD}}) + s^2}.$$

The $100 \times p\,\%$ prediction interval is then

$$\widehat{\texttt{YIELD}} \pm t^8_{(1-p)/2} \times \sqrt{\widehat{\text{SE}}^2(\widehat{\texttt{YIELD}}) + s^2}.$$

Using this formula, you can calculate the 50% and the 95% prediction intervals.

Comments to the Instructor

There is a positive relationship in the scatter plot of the total harvest weight versus the initial row count, and the final year (1991) has a much greater count and yield than the other years.

Let's fit a simple linear regression model and look at a plot of the standardized residuals. With only nine cases the standardized residuals plot is quite sparse; it is hard to say if there is a pattern. The normal probability plot appears to indicate that the distribution of the standardized residuals is approximately normal.

Based on either of the models we can predict the yield for a new harvest. The point prediction of the total weight in the model with a constant term is:

$$\widehat{\texttt{YIELD}} = 2369.59 + 341.934 \times \texttt{INITIAL} \quad \text{pounds,}$$

and in the model of proportionality,

$$\widehat{\texttt{YIELD}} = 394.534 \times \texttt{INITIAL} \quad \text{pounds.}$$

This slope term can be compared to the average ratio of YIELD to INITIAL that was calculated earlier, 432.08.

The prediction for the 1992 yield based on the model with a constant term is:

$$\widehat{\texttt{YIELD}} = 2369.59 + 341.934 \times 91 = 33,486 \quad \text{pounds,}$$

and in the model of proportionality

$$\widehat{\texttt{YIELD}} = 394.534 \times 91 = 35,903 \quad \text{pounds.}$$

For the model with a constant term we predict that 95% of the future harvests which produce 91 lugs from rows 3 through 6 will have total weights in the interval

$$(22152, \quad 44819) \text{ pounds.}$$

We also predict that 50% of the future harvests which produce 91 lugs from rows 3 through 6 will have total weights in the interval

$$(30077, \quad 36894) \text{ pounds.}$$

Prediction intervals for the simpler model of proportionality are as follows: we predict that 95% of the future harvests which produce 91 lugs from rows 3 through 6 will have total weights in the interval

$$(26421, \quad 45384) \text{ pounds.}$$

Further, we predict that 50% of the future harvests which produce 91 lugs from rows 3 through 6 will have total weights in the interval

$$(32998, \quad 38807) \text{ pounds.}$$

The actual value for the 1992 yield fell in the 95% prediction interval for both models. It only fell inside the 50% confidence interval for the model with a constant term. If the models were correct, we would expect it to fall inside about half of the 50% prediction intervals. Thus the predictions from the models appear to be consistent with the actual total.

Given that dropping the intercept term makes it more difficult to evaluate the strength of the fit, and imposes a strict proportionality of `YIELD` and `INITIAL`, the model with an intercept term is likely to perform better in predicting yields from new harvests.

To maintain consistency with the unannotated edition,
the next case begins on the following page.

The birth of a beluga whale

Topics covered: Autocorrelation. Multiple regression.

Key words: Autocorrelation. Differencing a time series. Durbin–Watson statistic. Histogram. Lagged variable. Residual plots. Scatter plot. Time series.

Data File: `whale.dat`

The New York Aquarium's exhibition of beluga whales (*Delphinapterus leucas*) dates back to June 5, 1897, in Battery Park, Manhattan. Since 1961 the Aquarium has continuously maintained a population of beluga whales in its facilities at Coney Island, Brooklyn. As of August 1, 1991, there were five belugas housed in the Aquarium's Marine Mammal Holding Facility.

It is estimated that there are between 65,000 and 88,000 beluga whales in the wild. At present, there are approximately 30 of this species in captivity in the United States and Canada. It is the goal of the New York Aquarium to contribute to the establishment of a self–sustaining captive population of beluga whales (thereby avoiding their capture). To this end, a commitment to establish a breeding group was put in place in the Winter/Spring of 1989 – 1990. Before August 1991, there had been six captive beluga whale births — three at the New York Aquarium, two at Sea World, and one at the Vancouver Aquarium. Four of the calves lived less than one week; one lived two months, while one lived four months. In none of these cases was any detailed information gathered about the birth or post–birth period, so there was very little known about beluga whale birth (in fact, there is little quantitative information about the births of any cetaceans, the species that also includes other whales, dolphins and porpoises).

On August 7, 1991, the 11–year–old beluga whale Natasha gave birth to a male calf, subsequently named Hudson, at the New York Aquarium. Seven days later, the 20–year–old beluga Kathy gave birth to a male calf, subsequently named Casey. The births of these two calves provided the opportunity for marine biologists to study in detail the birth process and early months of the calves' lives. In particular, it was believed that nursing behavior of the calf would be highly informative regarding its general health. To this end, the calves and mothers were watched 24 hours a day for approximately the first two weeks after birth, and approximately 9 hours a day after that. The following analysis is based on observations of the calf Hudson. The variables are as follows:

```
PERIOD1  : the time period since birth, which corresponds to the case
           number.
```

Each period corresponds to a six hour consecutive time period. Since some periods were only partially observed, or were unobserved, the variables had to be adjusted to account for the missingness. This was done by proportionally adjusting values for periods that were partially observed, and smoothing values for periods that were unobserved. Subsequent analysis indicated no biases introduced using this procedure.

```
BOUTS1   : the number of nursing bouts that occurred in the period.
```

A nursing bout is defined as a successful nursing episode where milk was obtained.

```
LOCKONS1 : the number of lockons that occurred in the period.
```

A lockon occurs when the calf attaches itself to the mother while suckling milk from a mammary gland.

```
DAYNIGHT1 : an indicator variable indicating whether the period was
            during the day (1) or night (0).
```

Day is defined here as 8AM–8PM, while night is 8PM–8AM.

```
NURSING1 : the total number of seconds spent successfully nursing
           during the period.
```

As the observations are recorded over time, these data form a time series. If beluga whales are to be bred successfully in captivity, or if intervention in the wild is to be productive, it is imperative that marine biologists understand the nursing process. To this end, a regression model can be built to try to model the nursing time as a function of the other variables. The data source is the New York Aquarium's study of the birth of the beluga whales, organized by John Nightingale, Associate Director of the New York Aquarium. The database was constructed, modified and analyzed by Jeanne McLaughlin–Russell, under the supervision of one of the authors of this book.

First let's take a look at the variables. The data set consists of 228 time periods, or the first 57 days of Hudson's life.

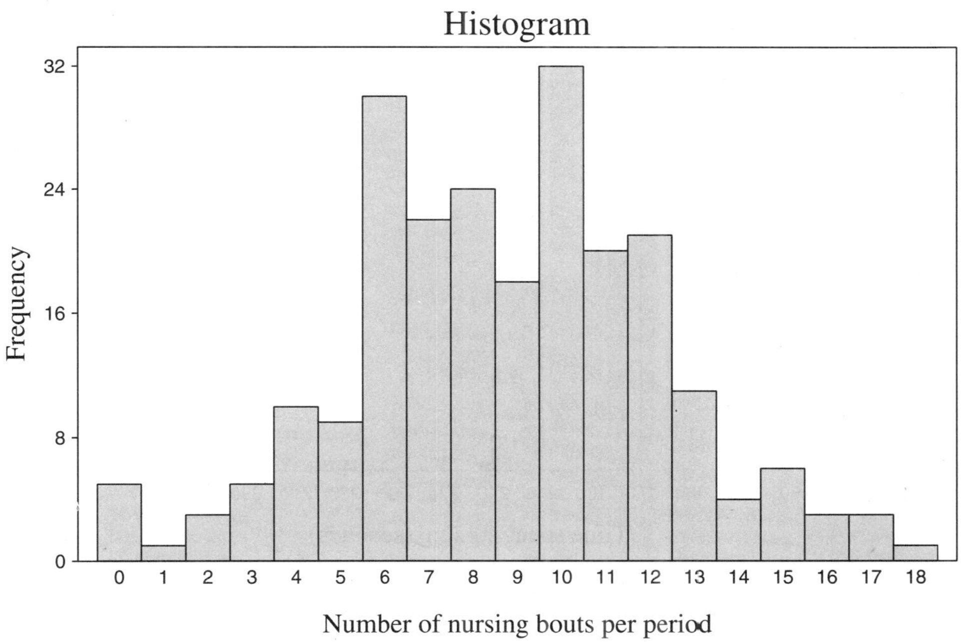

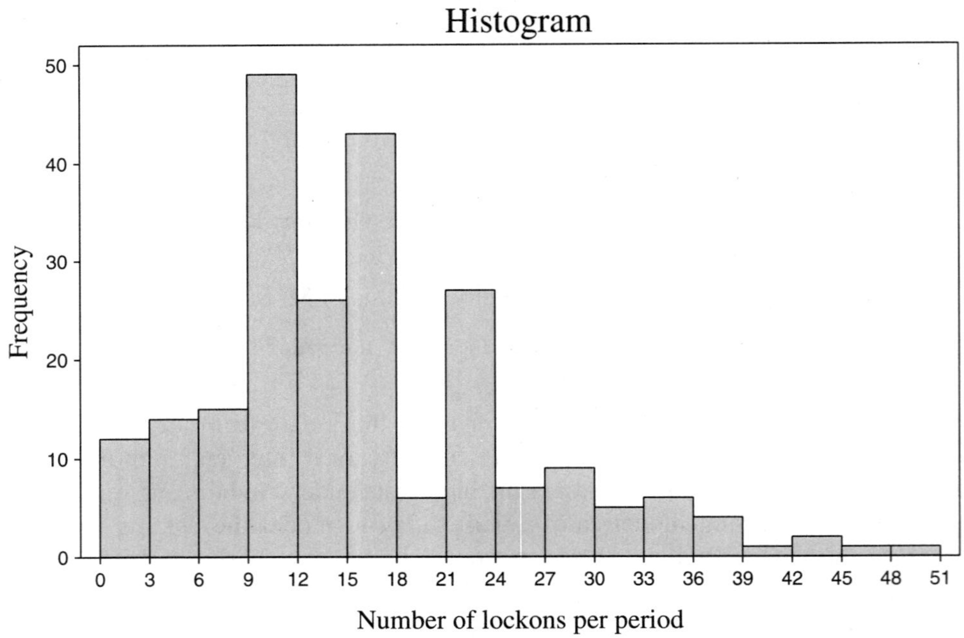

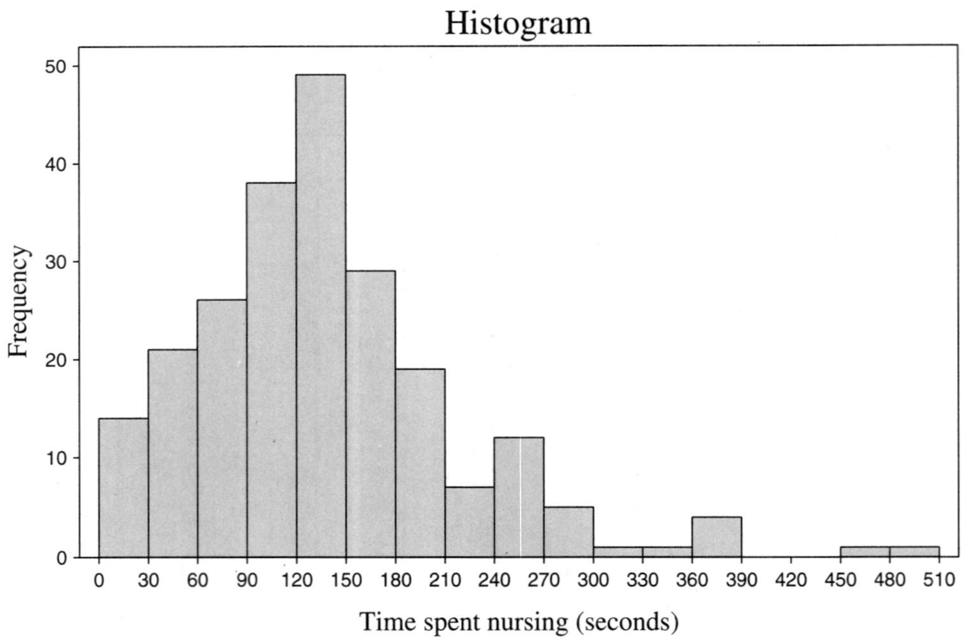

The variables look reasonably well–behaved, although perhaps a bit long–tailed. Now let's look at scatter plots. First a plot against `PERIOD1` (which is simply a time series plot).

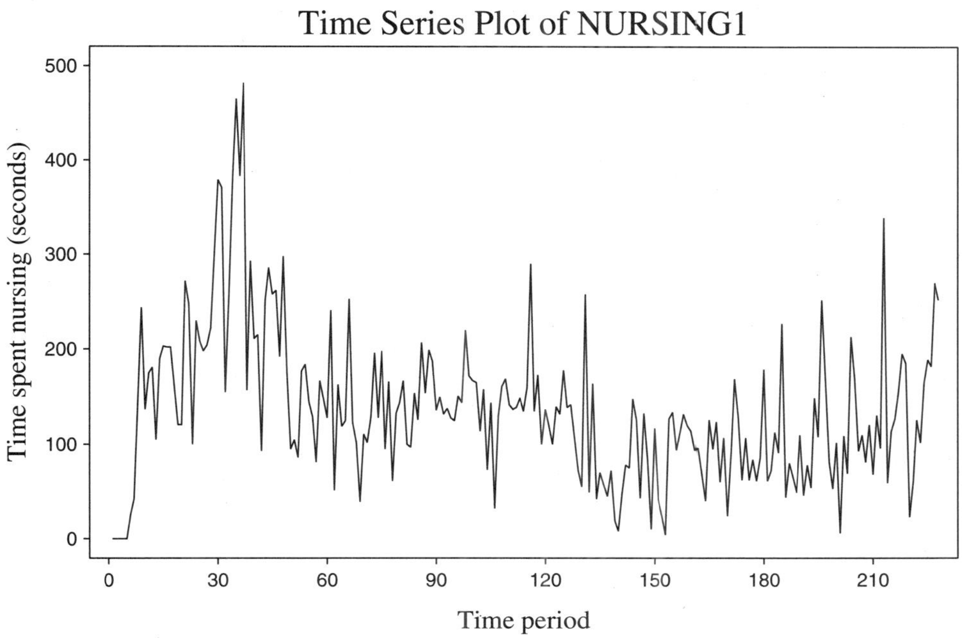

There definitely appears to be a cyclical pattern in the data. Nursing time increases until about 7–10 days post–partum, after which it begins to decline steadily. This same pattern has been observed in the nursing of killer whales and dolphins, and has been attributed to the need for the calf to produce rapidly a layer of blubber, to protect against the cold and provide a supplemental energy source.

Note the dip in nursing starting at around time period 140 (35 days). This was about ten days before Hudson was diagnosed as having a bacterial infection. Antibiotics were administered for 24 days, and the calf recovered. That nursing dipped 10 days **before** any distress was noted in the calf has important implications in future births, since the observation of such a dip could be an "early warning sign" of potential problems.

Here are scatter plots of nursing versus the other variables:

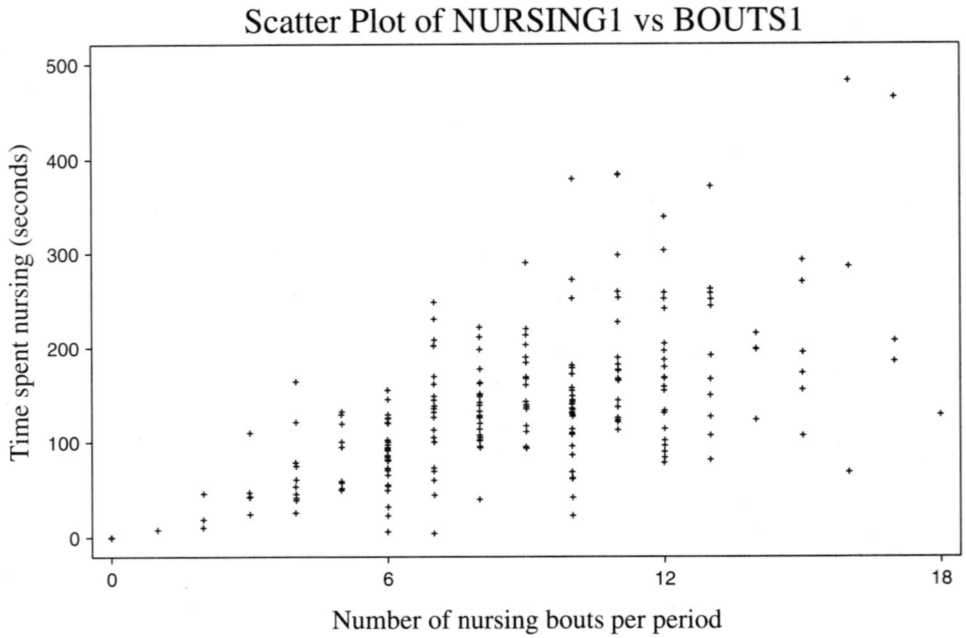

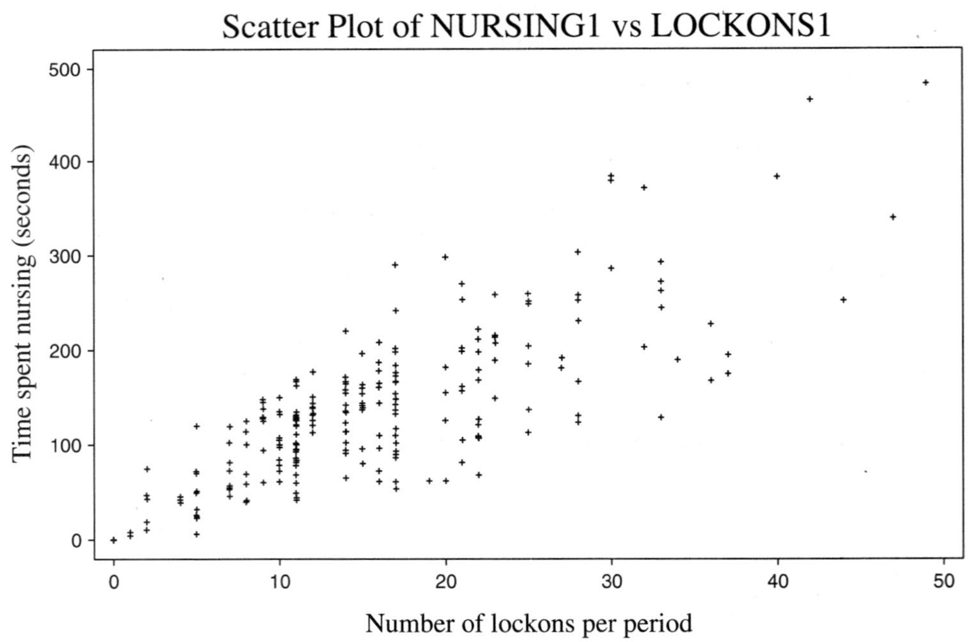

Both variables show the expected direct relationship with nursing time. Note, however, a pattern by which the variation in nursing time increases as the level of nursing increases (i.e., the "clouds widening"). There's no apparent circadian (i.e., day versus night) effect:

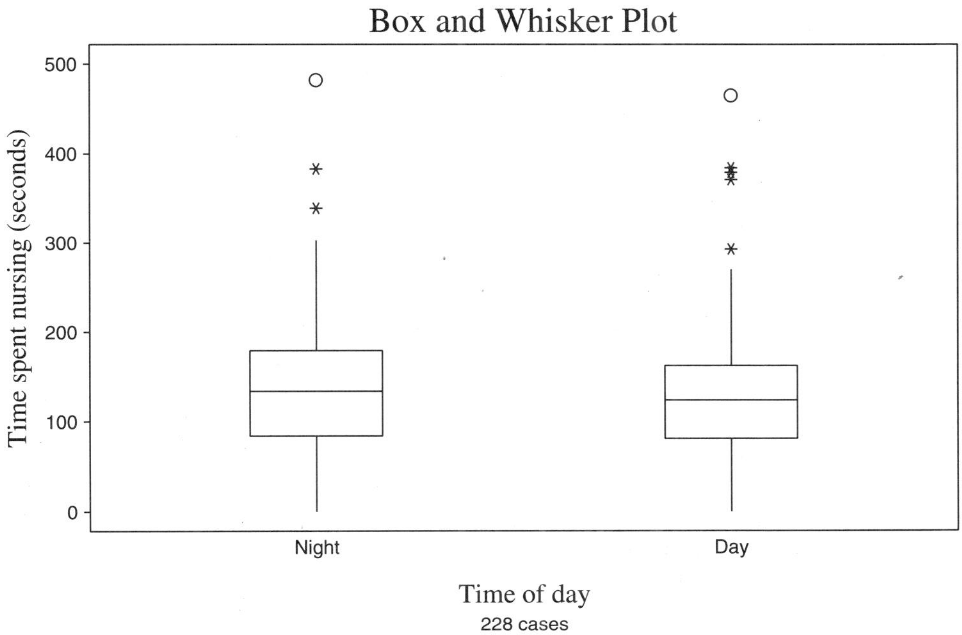

Here is the result of a regression analysis:

```
LINEAR REGRESSION OF NURSING1  Time spent nursing

PREDICTOR
VARIABLES    COEFFICIENT    STD ERROR     STUDENT'S T       P        VIF
---------    -----------    ---------     -----------    ------    -----
CONSTANT        58.4824      9.38888          6.23       0.0000
PERIOD1        -0.37179      0.04507         -8.25       0.0000      1.0
BOUTS1          3.73538      1.23170          3.03       0.0027      2.2
LOCKONS1        5.85240      0.46930         12.47       0.0000      2.1
DAYNIGHT1      -8.10288      5.84793         -1.39       0.1673      1.0

R-SQUARED             0.7108      RESID. MEAN SQUARE (MSE)     1936.21
ADJUSTED R-SQUARED    0.7056      STANDARD DEVIATION           44.0024

SOURCE        DF       SS           MS           F        P
----------   ---   ----------   ----------   -----    ------
REGRESSION     4   1.061E+06    2.654E+05    137.05   0.0000
RESIDUAL     223   4.318E+05      1936.21
TOTAL        227   1.493E+06

CASES INCLUDED 228   MISSING CASES 0
```

The day/night indicator variable isn't needed, so we fit a model without it:

```
LINEAR REGRESSION OF NURSING1   Time spent nursing

PREDICTOR
VARIABLES    COEFFICIENT    STD ERROR     STUDENT'S T       P        VIF
---------    -----------    ---------     -----------    ------    -----
CONSTANT         54.7897      9.02116            6.07    0.0000
PERIOD1         -0.37084      0.04516           -8.21    0.0000      1.0
BOUTS1           3.59885      1.23027            2.93    0.0038      2.2
LOCKONS1         5.89830      0.46909           12.57    0.0000      2.1

R-SQUARED              0.7083     RESID. MEAN SQUARE (MSE)      1944.17
ADJUSTED R-SQUARED     0.7044     STANDARD DEVIATION            44.0927

SOURCE          DF        SS            MS           F          P
----------     ---    ----------    ----------     -----     ------
REGRESSION       3    1.058E+06     3.526E+05     181.35     0.0000
RESIDUAL       224    4.355E+05       1944.17
TOTAL          227    1.493E+06

CASES INCLUDED 228    MISSING CASES 0
```

The fit is pretty good ($R^2 \approx .7$), with all variables significant, and there is no collinearity problem. Do the assumptions seem to hold?

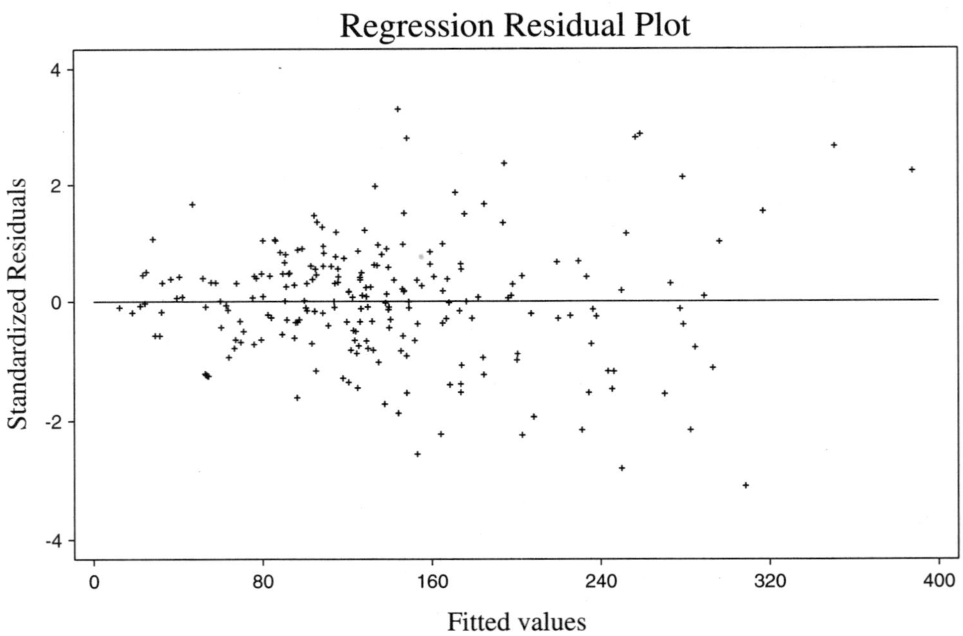

This plot suggests a problem — the variability of the residuals increases as the fitted values increase. That is, we have heteroscedasticity. In addition, the residuals are positively autocorrelated, as can be seen in the following plot, and as is confirmed by the Durbin–Watson statistic ($DW = 1.00$):

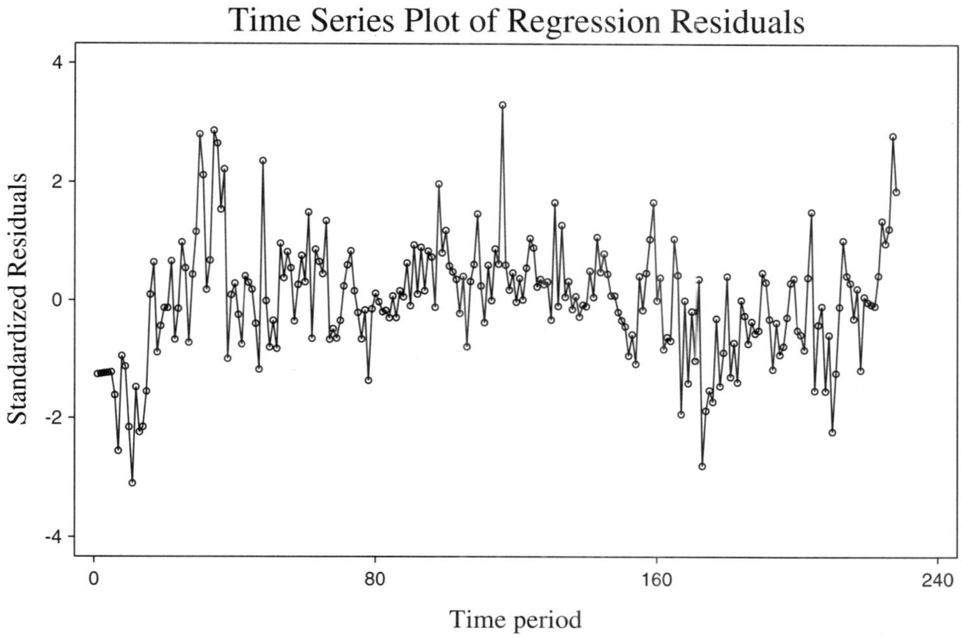

What should we do about this? There are several approaches that can be taken to tackle the problem of autocorrelation. One simple transformation that can sometimes make the analysis of a time series variable easier is to work with the *differenced variable*; i.e., instead of looking at the original series, we analyze the difference between two values adjacent in time. We will use as the value for the ith time period the change in the variable from the previous [i.e., $(i-1)^{st}$] time period. If the process is stable, this differenced variable (which we will call `CHNGNURS1`) will exhibit less dependence between successive cases. In addition it will often not have trends with time and usually varies about a constant (often zero).

Here is a time series plot of the variable:

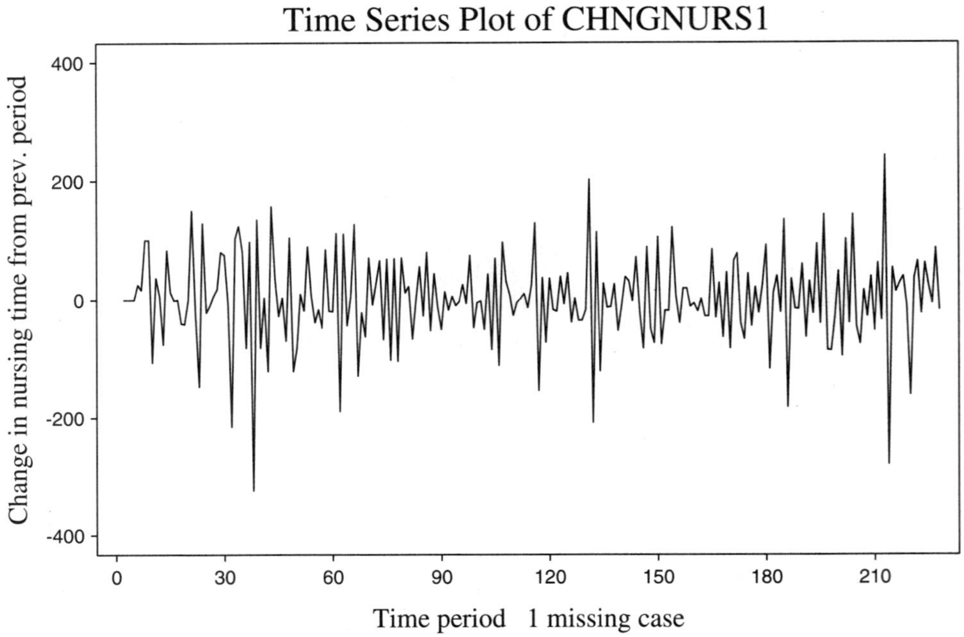

Note that the changes in nursing time are, indeed, relatively flat around zero. Interestingly, the variability around zero seems to change smoothly, first increasing, and then decreasing, and then increasing, and so on. Several of the changes are much larger than the others, as is evident from the spikes in the plot.

It is reasonable to suppose that the change in nursing time of Hudson from the previous time period could be related to the other variables of lockons and bouts, and, in fact, it is:

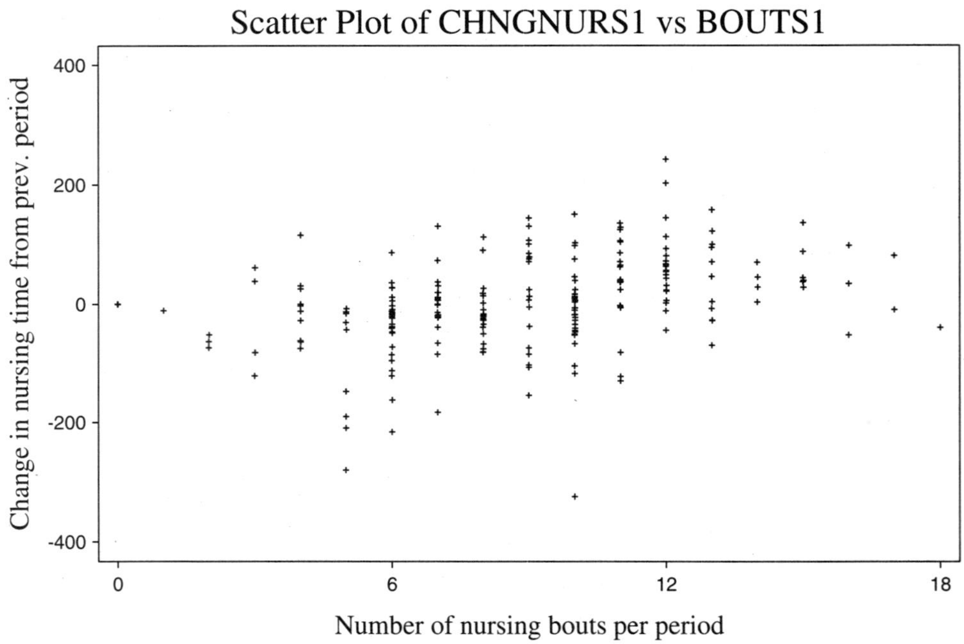

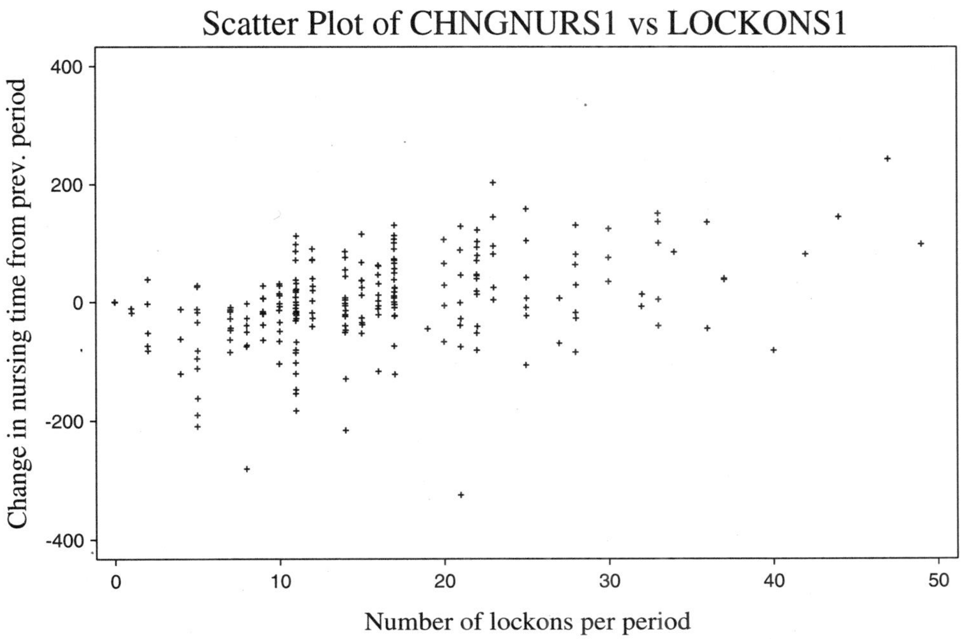

Each variable exhibits a direct relationship with change in nursing time. This makes sense, since we would expect that more nursing activity (as measured by the number of bouts or lockons) would be associated with an increase in nursing time. We also could hypothesize the existence of carryover effects, where the level of nursing activity in the previous time period (as measured by number of lockons or bouts, or nursing time itself) could be a predictor of change in activity in the current time period. This would show up as a relationship between change in nursing time and the lagged versions of lockons, bouts or nursing time. This is the case, as can be seen from these scatter plots:

Scatter Plot of CHNGNURS1 vs LAGBOUT1

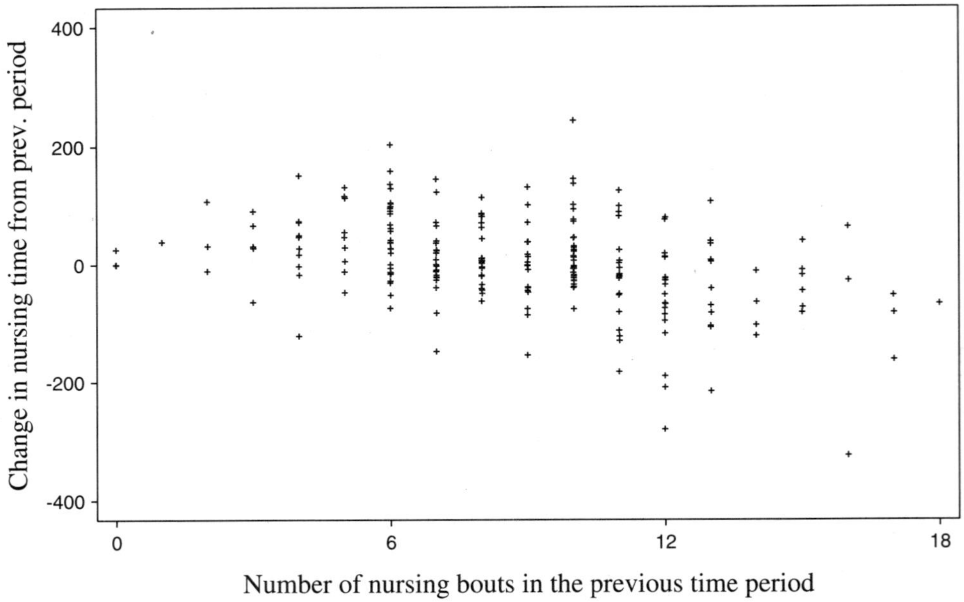

Scatter Plot of CHNGNURS1 vs LAGLOCK1

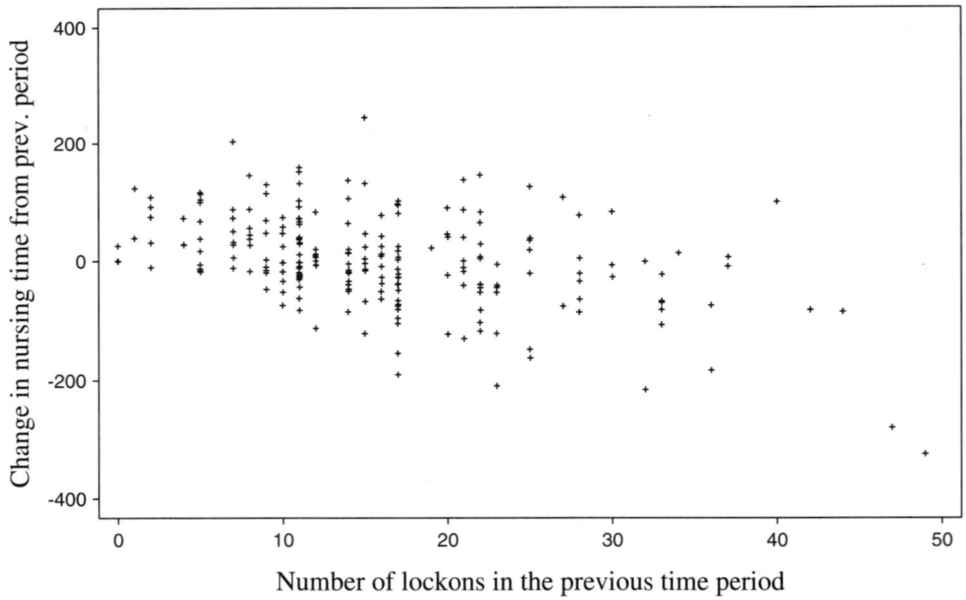

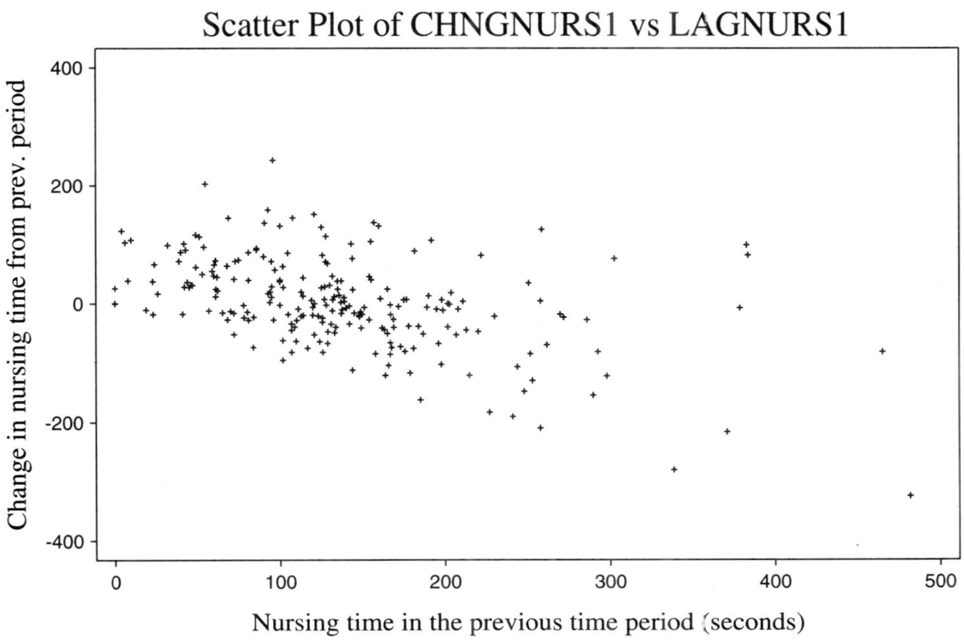

The negative associations in these plots are also intuitive, since they imply that increased activity in the previous time period is associated with a reduction in activity in the current time period (perhaps due to fatigue, or a fuller stomach).

Here is the regression of CHNGNURS1 on the possible predictors:

```
LINEAR REGRESSION OF CHNGNURS1  Change in nursing time

PREDICTOR
VARIABLES     COEFFICIENT    STD ERROR      STUDENT'S T        P        VIF
---------     -----------    ---------      -----------     ------     -----
CONSTANT        29.7287        9.55064          3.11        0.0021
PERIOD1         -0.18595       0.04611         -4.03        0.0001       1.4
BOUTS1           2.34183       1.13160          2.07        0.0397       2.3
LOCKONS1         6.14611       0.46067         13.34        0.0000       2.7
LAGBOUT1        -0.83983       1.13780         -0.74        0.4612       2.4
LAGLOCK1        -3.16795       0.57802         -5.48        0.0000       4.2
LAGNURS1        -0.49908       0.05896         -8.47        0.0000       3.5

R-SQUARED             0.7604     RESID. MEAN SQUARE (MSE)     1477.06
ADJUSTED R-SQUARED    0.7538     STANDARD DEVIATION           38.4325

SOURCE        DF       SS            MS           F         P
----------    ---   ----------    ----------    -----    ------
REGRESSION      6   1.031E+06     1.719E+05    116.35    0.0000
RESIDUAL      220   3.250E+05       1477.06
TOTAL         226   1.356E+06

CASES INCLUDED 227    MISSING CASES 1
```

The lagged bouts variable is not needed, and can be removed:

LINEAR REGRESSION OF CHNGNURS1 Change in nursing time

PREDICTOR VARIABLES	COEFFICIENT	STD ERROR	STUDENT'S T	P	VIF
CONSTANT	28.0187	9.25586	3.03	0.0028	
PERIOD1	-0.19318	0.04501	-4.29	0.0000	1.3
LOCKONS1	6.22081	0.44895	13.86	0.0000	2.5
BOUTS1	2.12921	1.09320	1.95	0.0527	2.2
LAGLOCK1	-3.36424	0.51269	-6.56	0.0000	3.3
LAGNURS1	-0.50658	0.05802	-8.73	0.0000	3.4

R-SQUARED 0.7598 RESID. MEAN SQUARE (MSE) 1474.02
ADJUSTED R-SQUARED 0.7544 STANDARD DEVIATION 38.3929

SOURCE	DF	SS	MS	F	P
REGRESSION	5	1.030E+06	2.061E+05	139.80	0.0000
RESIDUAL	221	3.258E+05	1474.02		
TOTAL	226	1.356E+06			

CASES INCLUDED 227 MISSING CASES 1

The current bouts variable is only marginally statistically significant, and can be removed with little loss in fit:

LINEAR REGRESSION OF CHNGNURS1 Change in nursing time

PREDICTOR VARIABLES	COEFFICIENT	STD ERROR	STUDENT'S T	P	VIF
CONSTANT	35.3164	8.51668	4.15	0.0000	
PERIOD1	-0.17433	0.04424	-3.94	0.0001	1.3
LOCKONS1	6.83428	0.32193	21.23	0.0000	1.3
LAGLOCK1	-3.53594	0.50822	-6.96	0.0000	3.2
LAGNURS1	-0.49049	0.05778	-8.49	0.0000	3.3

R-SQUARED 0.7557 RESID. MEAN SQUARE (MSE) 1492.56
ADJUSTED R-SQUARED 0.7513 STANDARD DEVIATION 38.6337

SOURCE	DF	SS	MS	F	P
REGRESSION	4	1.025E+06	2.562E+05	171.65	0.0000
RESIDUAL	222	3.313E+05	1492.56		
TOTAL	226	1.356E+06			

CASES INCLUDED 227 MISSING CASES 1

The model fits well. Residual plots indicate a reasonably good fit, with most of the time series problems removed (the Durbin–Watson statistic is $DW = 2.31$; the time series plot does, however, suggest a possible [long memory] time structure):

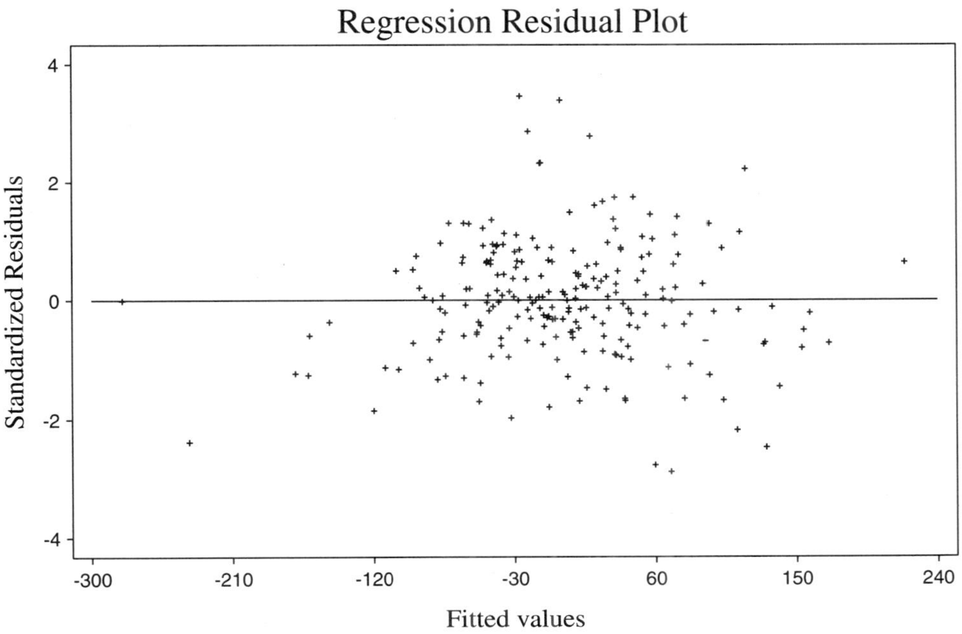
Regression Residual Plot
Standardized Residuals
4
2
0
-2
-4
-300
-210
-120
-30
60
150
240
Fitted values

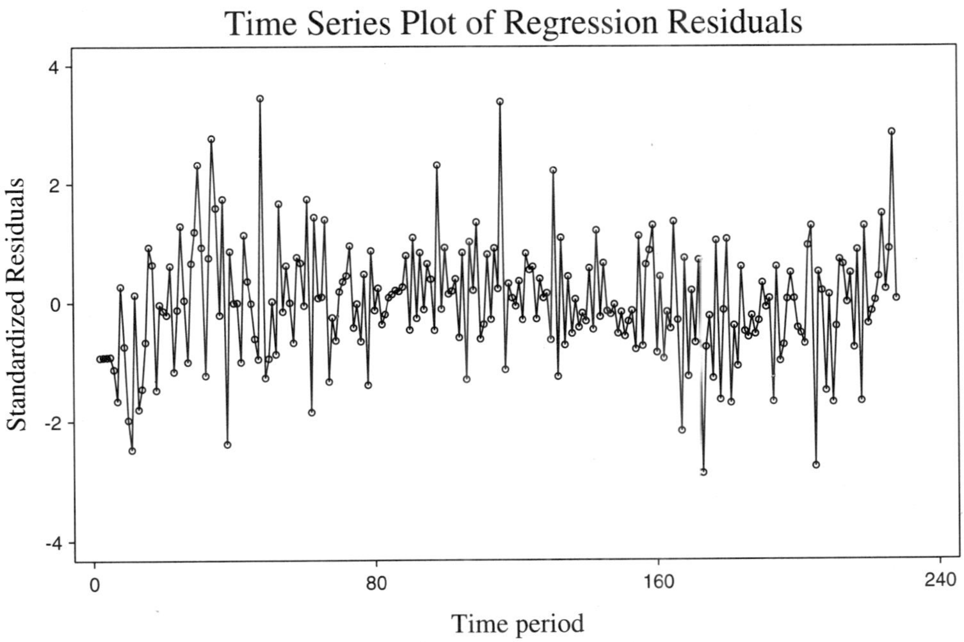
Time Series Plot of Regression Residuals
Standardized Residuals
4
2
0
-2
-4
0
80
160
240
Time period

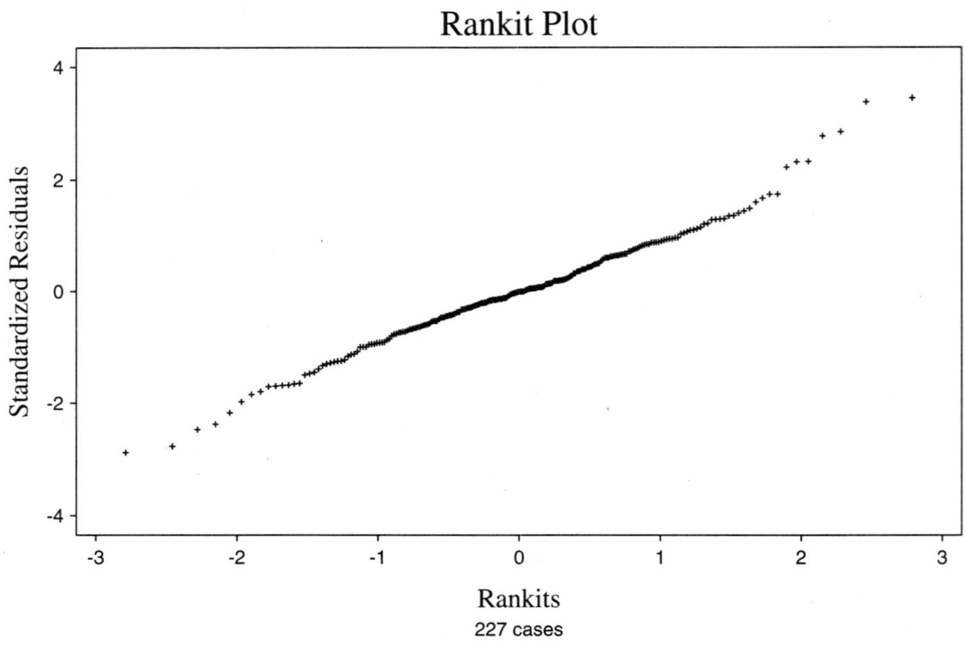

Here are the regression diagnostics for some of the cases, including those with large standardized residuals:

PERIOD1	NURSING1	CHNGNURS1	FITTED	STDRES	LEVERAGE	COOKSD
1	0	M	M	M	M	M
2	0	0	34.968	-0.9276	0.0479	0.0087
3	0	0	34.793	-0.9228	0.0475	0.0085
4	0	0	34.619	-0.9180	0.0472	0.0083
5	0	0	34.445	-0.9132	0.0468	0.0082
6	26	26	68.442	-1.1225	0.0422	0.0111
7	42	16	78.841	-1.6567	0.0360	0.0205
8	143	101	90.609	0.2746	0.0408	0.0006
9	244	101	129.035	-0.7396	0.0377	0.0043
10	137	-107	-31.935	-1.9785	0.0356	0.0289
.	.		.	.	.	.
34	384	125	18.982	2.7768	0.0234	0.0369
.	.		.	.	.	.
48	298	106	-26.010	3.4517	0.0200	0.0486
.	.		.	.	.	.
116	290	130	-0.241	3.3821	0.0065	0.0149
.	.		.	.	.	.
205	168	-45	59.812	-2.7741	0.0436	0.0702
.	.		.	.	.	.
227	270	88	-20.725	2.8523	0.0265	0.0442
228	252	-18	-19.758	0.0471	0.0685	0.0000

The low Cook's distances associated with the flagged values (cases 34, 48, 116, 205 and 227) suggest that removing them will have little effect on the fit.

One group of notable observations is the first five, which all had zero nursing times (Hudson did not start nursing until approximately 33 hours after birth). Each of these time periods has a fitted value around 35 seconds (except the first, which has no defined fitted value due to the lagged and differencing variables), which is not particularly sensible. It is reasonable to remove these five cases, and have our model refer to time periods after nursing has begun:

```
LINEAR REGRESSION OF CHNGNURS1  Change in nursing time

PREDICTOR
VARIABLES      COEFFICIENT    STD ERROR     STUDENT'S T        P        VIF
---------      -----------    ---------     -----------     ------     -----
CONSTANT         43.5309       9.45967          4.60        0.0000
PERIOD1         -0.20451       0.04675         -4.37        0.0000       1.4
LOCKONS1         6.73075       0.32610         20.64        0.0000       1.3
LAGLOCK1        -3.46035       0.50965         -6.79        0.0000       3.1
LAGNURS1        -0.51640       0.05923         -8.72        0.0000       3.3

R-SQUARED            0.7600      RESID. MEAN SQUARE (MSE)     1492.69
ADJUSTED R-SQUARED   0.7556      STANDARD DEVIATION           38.6353

SOURCE          DF       SS           MS           F          P
----------     ---    ----------   ----------    -----     ------
REGRESSION       4    1.031E+06    2.577E+05    172.63     0.0000
RESIDUAL       218    3.254E+05      1492.69
TOTAL          222    1.356E+06

CASES INCLUDED 223   MISSING CASES 0
```

The estimated regression coefficients have changed very little, but the residual plots look a little better:

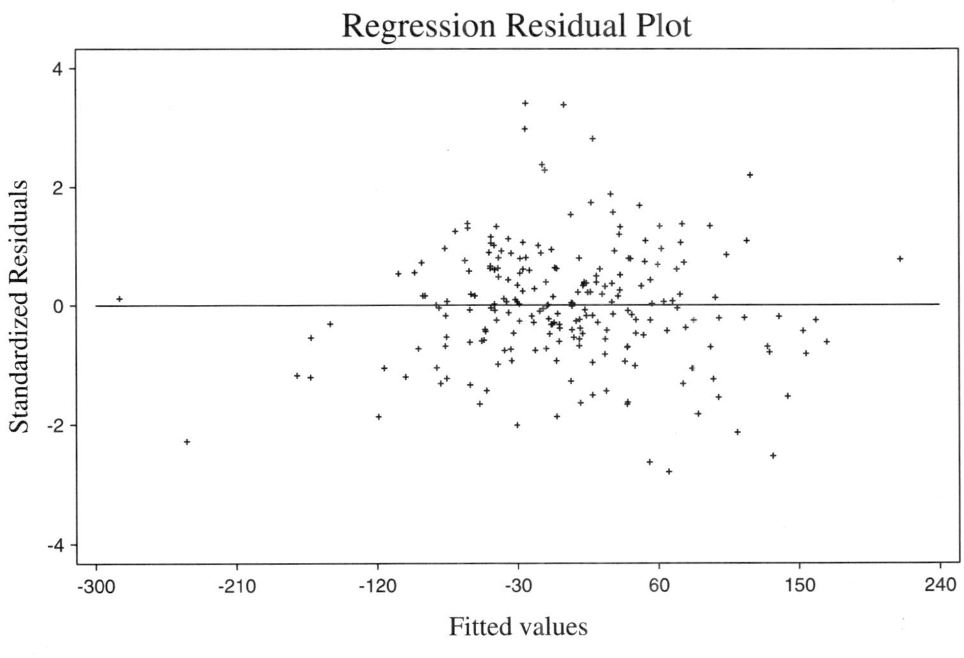

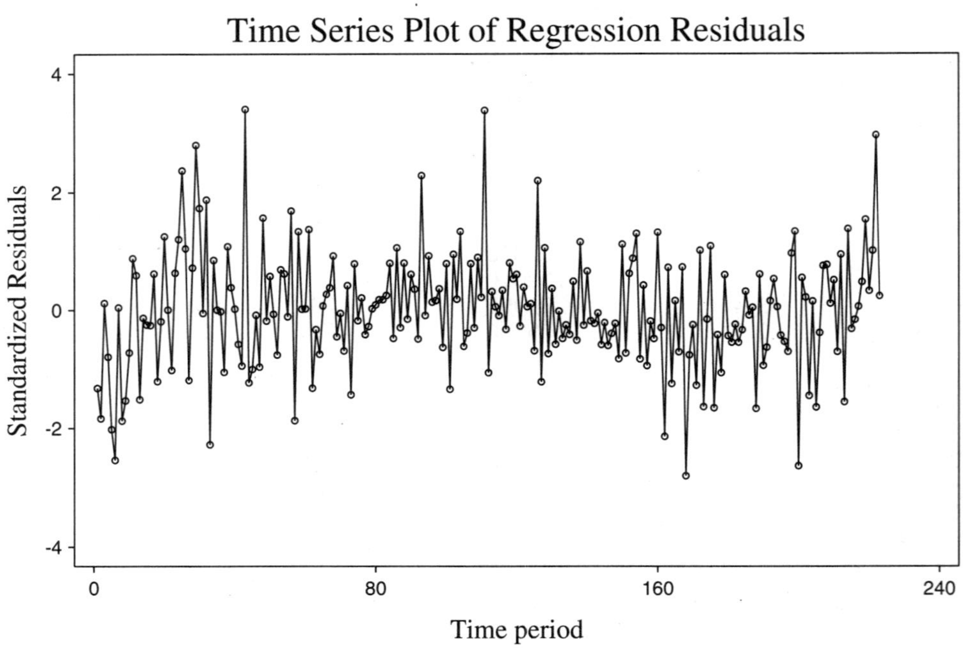

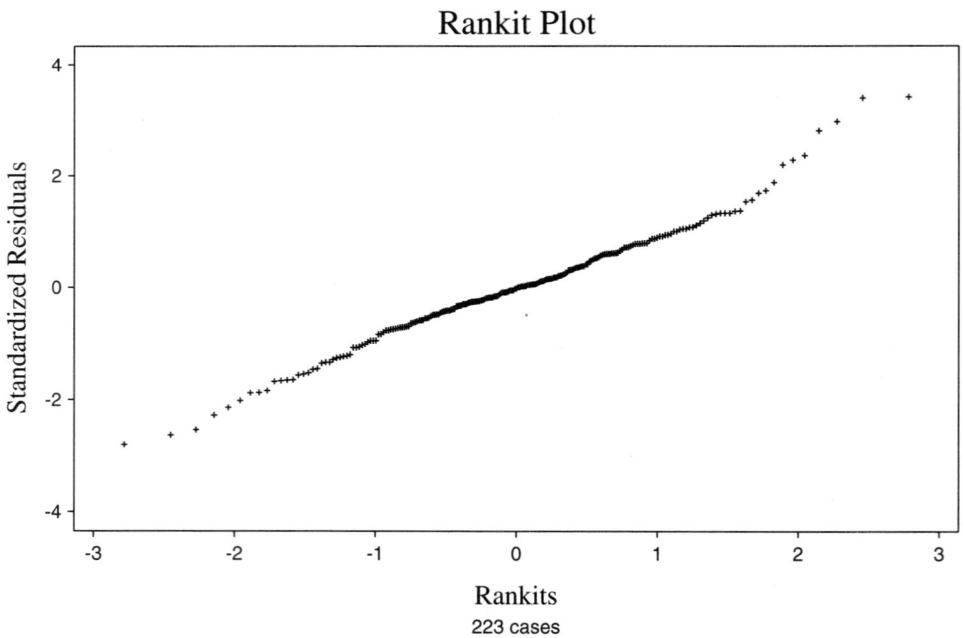

How can we use this model? We've seen that it is possible to predict the amount of nursing from a few readily available variables; that given the other variables, the change in nursing has a negative association with time since birth; that given the other variables, the change in nursing has a positive association with the number of lockons and a negative association with the number of lockons in the last time period; and that given the other variables, the change in nursing has a negative association with the amount of nursing in the last time period.

Note that even though this model is presented in terms of the change in nursing time, it can be used to predict directly the actual nursing time in any time period t. Since the fitted model is

$$\widehat{\mathtt{NURSING}_t - \mathtt{NURSING}_{t-1}} =$$
$$43.53 - .205 \times t + 6.731 \times \mathtt{LOCKONS}_t - 3.460 \times \mathtt{LOCKONS}_{t-1} - .516 \times \mathtt{NURSING}_{t-1},$$

we can predict the nursing time by:

$$\widehat{\mathtt{NURSING}_t} =$$
$$43.53 + .484 \times \mathtt{NURSING}_{t-1} - .205 \times t + 6.731 \times \mathtt{LOCKONS}_t - 3.460 \times \mathtt{LOCKONS}_{t-1}.$$

In fact, since $\mathtt{NURSING}_{t-1}$ is one of the predictors, by mathematical identity these coefficients are identical to the least squares regression coefficients for direct prediction of $\mathtt{NURSING}_t$. How does this model do in predicting Hudson's nursing times? We can check this by looking at the errors of prediction: $\mathtt{NURSING}_t - \widehat{\mathtt{NURSING}_t}$. If the model fits well we expect the errors to be close to zero. The absolute size of the error is a measure of how well the model fits.

Here is a histogram of the errors made by the model:

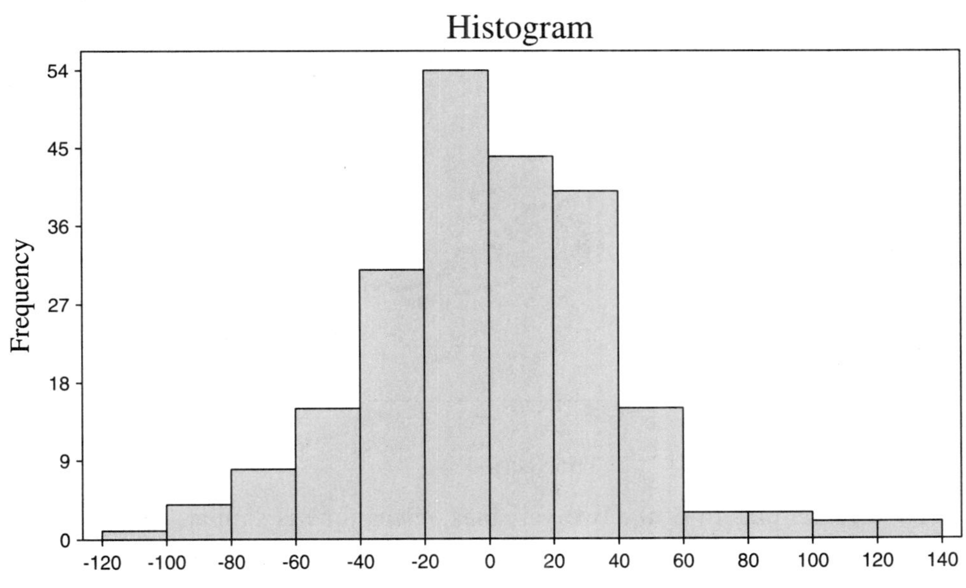

Descriptive statistics show that the errors average out to zero, with a standard deviation of about 38.3 seconds (if the second through fifth time periods are also included, the standard deviation is 38.4 seconds). Thus the model fits very well. Note that the estimated standard deviation of the errors from the original regression on (the undifferenced) nursing time was 44.1 seconds, illustrating that differencing the nursing variable was a useful thing to do.

```
                 ERROR1
N                   223
MEAN                  0
SD               38.286
MINIMUM         -106.87
MEDIAN          -0.7790
MAXIMUM          130.30
```

Presumably this model could now be used for other calves, and a calf whose nursing behavior fell outside what was predicted here might be in trouble. But is that actually true? Let's see. Remember, we have data for another calf (Casey). How well does the model derived from Hudson's data fit Casey's data? We can use the model given above to try to predict Casey's nursing times from his lockons, lagged lockons and lagged nursing time (and time period since birth).

Here is a scatter plot of both the actual nursing times (marked with a ∘) and the predicted nursing times (marked with a +) versus time:

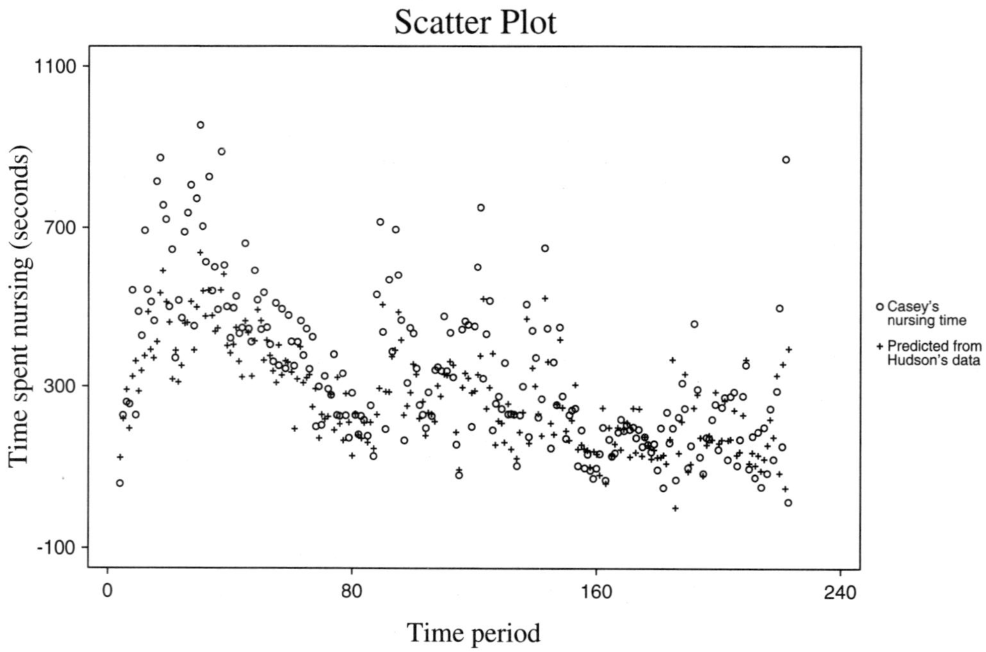

The general pattern of the two variables versus time is similar, except for one important point — the predictions are consistently too low. Here are summary statistics for the difference between the true value and the prediction:

```
DESCRIPTIVE STATISTICS

                 ERROR2
N                   220
MEAN             63.025
SD               125.99
MINIMUM         -380.30
MEDIAN           40.636
MAXIMUM          821.74
```

It can be seen that the predictions turn out to be too low, with median error of about 41 seconds per six hour period. Why did this happen? The reason is that right from birth, Casey was much bigger, stronger and more robust than Hudson, even though both calves were healthy. This illustrates a fact of which the marine biologists should be aware — that is, calf–to–calf variation is high, and only general lessons can be learned from any one birth.

Summary

The nursing behavior of a newborn beluga calf was studied, to see if it is possible to model and predict nursing levels from other, generally available, information. The nursing time in a given six–hour time period can be predicted using the time period number, the number of lockons in that time period, the number of lockons in the previous time period, and the nursing time in the previous time period. This provides a model that can usually predict the nursing time to within roughly 80 seconds per time period, for the calf being used to derive the model. When the model is validated using the data for another newborn calf, its predictions are consistently too low, with a median error of 41 seconds per time period, reflecting that there is considerable calf–to–calf variation in nursing behavior.

Estimating a demand function — it's about time

Topics covered: Autocorrelation. Multiple regression. Time series.

Key words: Autocorrelation. Durbin–Watson statistic. Lagged variable. Normal plot. Outliers. Regression diagnostics. Residual plots.

Data File: `demand.dat`

The earlier case "Estimating a demand function" showed how multiple regression could be used to estimate the demand for gasoline as a function of various predictors, including its price. The simple model recommended at the end of the case was based on the price index of gasoline (PG), per capita real disposable income (I) and the price index of used cars (PUC):

```
LINEAR REGRESSION OF G Total gasoline consumption

PREDICTOR
VARIABLES     COEFFICIENT    STD ERROR     STUDENT'S T       P        VIF
---------     -----------    ---------     -----------     ------    -----
CONSTANT        -103.506      9.68365         -10.69      0.0000
PG              -10.9454      2.47158          -4.43      0.0002      6.2
I                0.04058      0.00147          27.52      0.0000      3.4
PUC             -8.25978      3.11278          -2.65      0.0142      6.9

R-SQUARED             0.9839     RESID. MEAN SQUARE (MSE)     34.8246
ADJUSTED R-SQUARED    0.9818     STANDARD DEVIATION           5.90124

SOURCE        DF        SS            MS          F         P
----------   ---    ----------    ----------    -----    ------
REGRESSION     3      49076.0       16358.7    469.75    0.0000
RESIDUAL      23      800.966       34.8246
TOTAL         26      49877.0

CASES INCLUDED 27    MISSING CASES 0
```

Although this model fits the data well, it does suffer from a difficulty — it does not address the time ordering of the data. The residuals from this model exhibit autocorrelation, as can be seen from this time series plot:

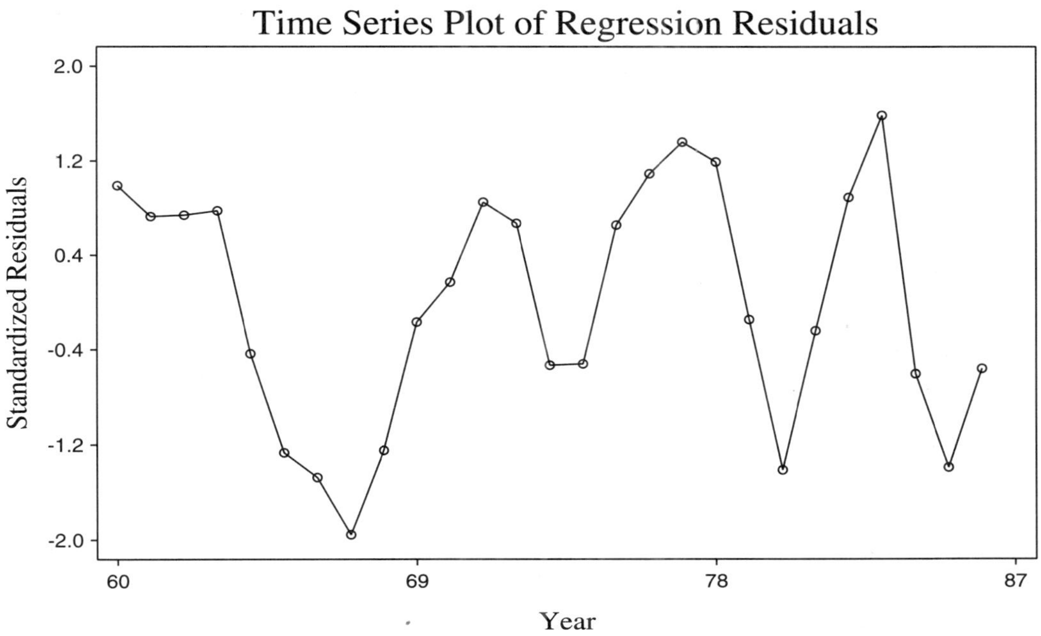

The Durbin–Watson statistic supports this, as it equals 0.74. This is an important problem to try to correct, because autocorrelation of this type can inflate the measures of fit for the model (F , R^2), so that the strength of the model is overstated.

One approach to handling this problem is to use as predictors of the ith observation of the target variable values from previous time periods, thereby accounting for possible "carry over" effects. This is called *lagging* variables. For these data, two obvious candidates for lagging are the gasoline consumption itself, `GLAG`, (saying that the previous year's gasoline consumption goes a long way to predicting this year's consumption, because of basic stability in the process), and the price index of gasoline, `PGLAG` (saying that consumption might be affected by the perception of people that prices are increasing or decreasing).

Here is output for a regression using these variables, along with PG, as predictors:

```
LINEAR REGRESSION OF G Total gasoline consumption

PREDICTOR
VARIABLES     COEFFICIENT     STD ERROR     STUDENT'S T      P        VIF
---------     -----------     ---------     -----------    ------    -----
CONSTANT          3.20191       4.48477            0.71    0.4828
GLAG              1.04612       0.02681           39.02    0.0000      1.9
PG               -27.8336       2.85093           -9.76    0.0000     16.4
PGLAG             25.0800       2.66826            9.40    0.0000     14.3

R-SQUARED              0.9914    RESID. MEAN SQUARE (MSE)     17.1295
ADJUSTED R-SQUARED     0.9902    STANDARD DEVIATION           4.13878

SOURCE         DF        SS            MS          F          P
----------     ---   ----------    ----------    -----     ------
REGRESSION      3      43289.7       14429.9     842.40    0.0000
RESIDUAL       22      376.849       17.1295
TOTAL          25      43666.5

CASES INCLUDED 26    MISSING CASES 1
```

The model fits very well, and according to the Durbin–Watson statistic, the autocorrelation has been removed ($DW = 2.29$). A time series plot of the residuals, however, shows that there is a clear outlier:

This outlier corresponds to 1974:

CASE	YR	G	STDRES	LEVERAGE	COOKSD
1	60	129.7	M	M	M
2	61	131.3	-1.4162	0.1691	0.1020
3	62	137.1	-0.2117	0.1635	0.0022
4	63	141.6	-0.6577	0.1416	0.0178
5	64	148.8	-0.0300	0.1266	0.0000
6	65	155.9	0.1347	0.1077	0.0005
7	66	164.9	0.4575	0.0907	0.0052
8	67	171.0	-0.3017	0.0752	0.0018
9	68	183.4	1.1149	0.0680	0.0227
10	69	195.8	1.1101	0.0617	0.0202
11	70	207.4	0.6265	0.0695	0.0073
12	71	218.3	0.3126	0.0879	0.0024
13	72	226.8	-0.3801	0.1138	0.0046
14	73	237.9	0.8578	0.1236	0.0259
15	74	225.8	-2.9762	0.1454	0.3767
16	75	232.4	0.1055	0.0695	0.0002
17	76	241.7	0.5192	0.0815	0.0060
18	77	249.2	0.2334	0.0995	0.0015
19	78	261.3	1.2467	0.1194	0.0527
20	79	248.9	-0.8134	0.2275	0.0487
21	80	226.8	-0.2163	0.4569	0.0098
22	81	225.6	1.9442	0.2570	0.3268
23	82	228.8	-1.3905	0.2220	0.1380
24	83	239.6	1.1052	0.1769	0.0656
25	84	244.7	-0.0979	0.1466	0.0004
26	85	245.8	-0.6029	0.1378	0.0145
27	86	269.4	-1.1145	0.4614	0.2661

This result is not a surprise. The first of the great oil shocks occurred in 1974, due to the oil embargo instituted by Arab oil producing countries. Prices increased by about 35%, resulting in a dramatic drop in gasoline consumption. The relationship between gasoline consumption and the price index of gasoline in 1974 was undoubtedly unusual. We could contemplate removing the entire case for 1974 and then reanalyzing the remaining data. Unfortunately, if we do that, we will disturb the natural time ordering in the data. An alternative approach is to substitute a "reasonable" value, such as the average of the two neighboring values, for the outlying value, and then reanalyze the entire adjusted data set. This is admittedly an *ad hoc* solution, and more complex (and theoretically justified) substitution methods are possible. Still, very simple techniques like this can often work quite adequately.

For these data, the gasoline consumption of 225.8 is too low, compared with the values of 237.9 for 1973 and 232.4 for 1975, so the averaged value of 235.1 is substituted (of course, when discussing our results, we must note that they no longer apply to 1974, or future years that might be like 1974). Here is the resultant regression output:

```
LINEAR REGRESSION OF G Total gasoline consumption

PREDICTOR
VARIABLES     COEFFICIENT    STD ERROR    STUDENT'S T       P      VIF
---------     -----------    ---------    -----------    ------   -----
CONSTANT          1.56435      3.57126           0.44    0.6656
GLAG              1.05999      0.02135          49.65    0.0000     1.9
PG               -27.2260      2.27022         -11.99    0.0000    16.4
PGLAG             23.9950      2.12476          11.29    0.0000    14.3

R-SQUARED            0.9946    RESID. MEAN SQUARE (MSE)    10.8619
ADJUSTED R-SQUARED   0.9938    STANDARD DEVIATION          3.29574

SOURCE        DF       SS           MS          F         P
----------    ---   ----------   ----------   -----    ------
REGRESSION     3     43804.5      14601.5   1344.28    0.0000
RESIDUAL      22     238.962      10.8619
TOTAL         25     44043.4

CASES INCLUDED 26   MISSING CASES 1
```

The model fits slightly better, although the coefficients have changed little. Let's look at the time series plot of the standardized residuals and a normal plot of the standardized residuals:

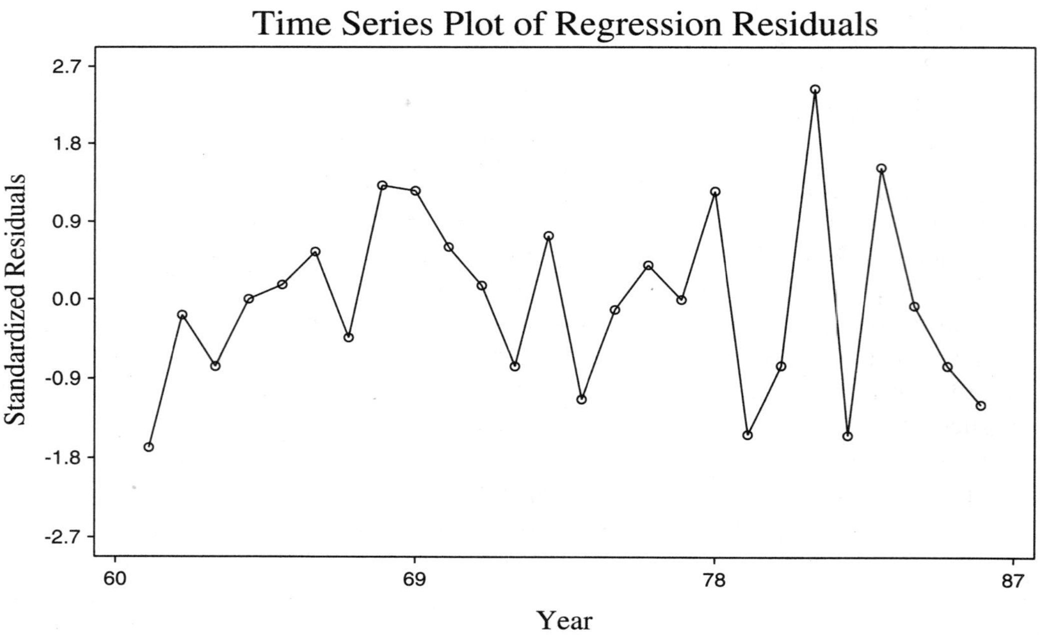

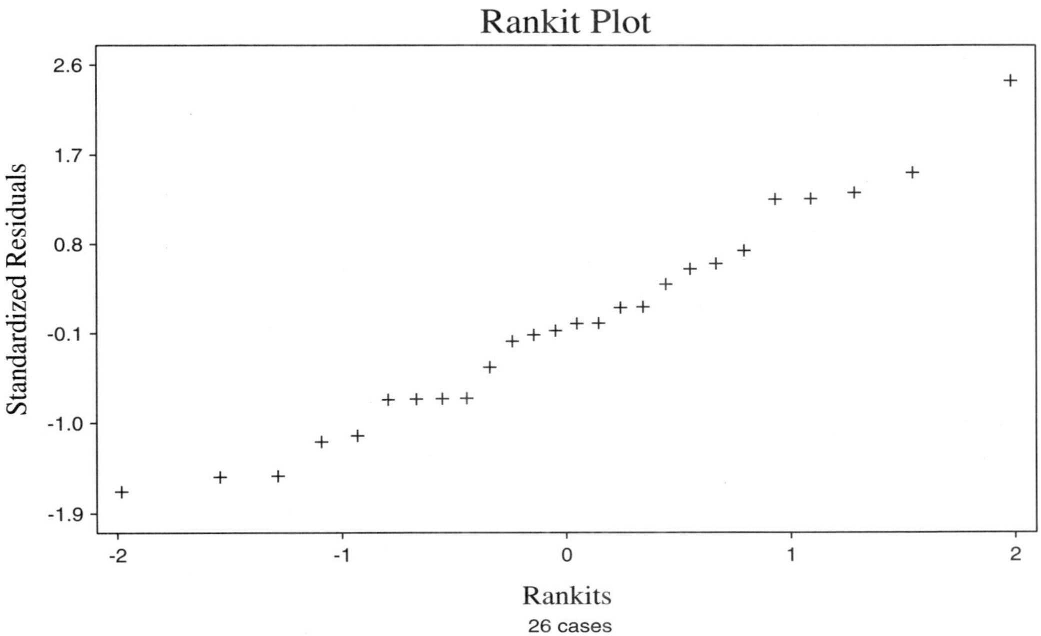

We see that the model is also more appropriate; there is little evidence for autocorrelation ($DW = 2.35$), and no outliers are apparent.

Let's look at the standardized residuals versus fitted values plot:

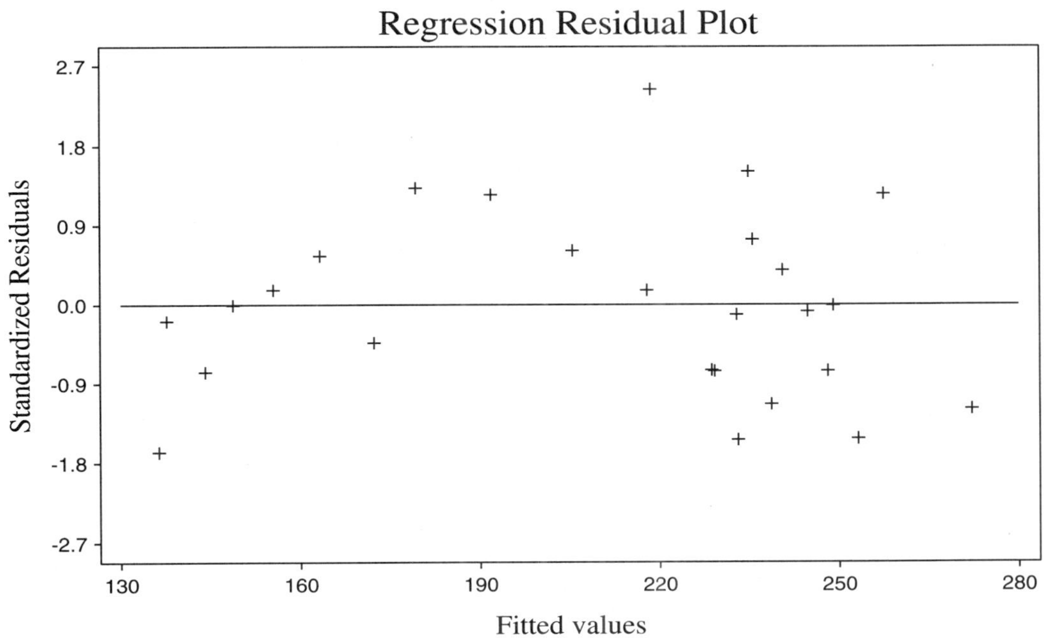

This gives a slight indication of structure, but given the very high R^2 here, it is unlikely that any corrective action would make much of a difference.

As a final check, let's look at the regression diagnostics; that is, standardized residuals, leverage and Cook's distance. They look reasonable:

CASE	YR	G	STDRES	LEVERAGE	COOKSD
1	60	129.7	M	M	M
2	61	131.3	-1.6829	0.1691	0.1441
3	62	137.1	-0.1830	0.1635	0.0016
4	63	141.6	-0.7685	0.1416	0.0244
5	64	148.8	-0.0005	0.1266	0.0000
6	65	155.9	0.1656	0.1077	0.0008
7	66	164.9	0.5478	0.0907	0.0075
8	67	171.0	-0.4435	0.0752	0.0040
9	68	183.4	1.3168	0.0680	0.0316
10	69	195.8	1.2556	0.0617	0.0259
11	70	207.4	0.6033	0.0695	0.0068
12	71	218.3	0.1578	0.0879	0.0006
13	72	226.8	-0.7643	0.1138	0.0188
14	73	237.9	0.7343	0.1236	0.0190
15	74	235.1	-1.1288	0.1454	0.0542
16	75	232.4	-0.1182	0.0695	0.0003
17	76	241.7	0.3945	0.0815	0.0035
18	77	249.2	-0.0036	0.0995	0.0000
19	78	261.3	1.2522	0.1194	0.0532
20	79	248.9	-1.5290	0.2275	0.1721
21	80	226.8	-0.7555	0.4569	0.1201
22	81	225.6	2.4416	0.2570	0.5154
23	82	228.8	-1.5394	0.2220	0.1691
24	83	239.6	1.5225	0.1769	0.1245
25	84	244.7	-0.0749	0.1466	0.0002
26	85	245.8	-0.7588	0.1378	0.0230
27	86	269.4	-1.1889	0.4614	0.3028

This new gasoline demand function has an appealing intuitive justification. Given the last two years' prices, gasoline demand is directly related to last year's demand, with the regression coefficient for GLAG indicating a small increase of about 6%. Given last year's demand and price, this year's demand is inversely related to this year's price, which is the inverse relationship between demand and price expected from economic theory. Further, given this year's price and last year's demand, this year's demand is directly related to last year's price. This also makes sense, since a higher value of last year's price, given this year's price is fixed, is consistent with a decreasing price trend, which would encourage additional consumption.

Summary

The presence of autocorrelation in the residuals complicates the construction of an empirical demand curve for gasoline. This problem can be addressed by using the previous year's gasoline consumption and gasoline price index as predictors, in addition to the current year's gasoline price index. The fit of such a model shows that the year of the first oil shock, 1974, is an outlier, with unusually low gasoline consumption. Adjusting for this, gasoline consumption is inversely related to the current year's price (the usual relationship between demand and price), directly related to the previous year's consumption (with a small annual increase indicated), and directly related to the previous year's price (reflecting sensitivity to a trend in prices), holding all else in the model fixed.

A comparison of the Dow Jones Industrial Average and the S & P 500 index

Topics covered: Modeling time series data.

Key words: Autocorrelation. Durbin–Watson statistic. Lagged variable. Residual plots. Time trend.

Data File: `djsp.dat`

The Dow Jones Industrial Average (`DJIA`) is an index of security (stock) prices issued by Dow Jones & Company that is used to study the movement of stock prices in the New York Stock Exchange. It is based on the stock prices of only 30 companies. The index was started in 1884 and at that time was based on 11 stocks, mostly of railway companies. The index was reorganized in 1928 to include 30 stocks and was given the base value of 100. The lowest value the index has taken was on July 2, 1932, when it was 41. The composition has remained more or less the same since 1928, except that companies that ceased to exist or were changed by merger or acquisition by another company were replaced by a similar company. `DJIA` is the most well–known index of stock market activity and is often used in the popular press to discuss the level of stock prices in the U.S. For example, historic highs of `DJIA` are viewed as evidence of strength in the market as a whole. There are three other Dow Jones indices representing price movements in U.S. home bonds, utilities, and transportation stocks. We will confine our attention to the `DJIA` since it is the most widely used indicator of stock prices.

How representative are the current values of `DJIA` of general stock price movement?

To give a partial answer to the question let's look at the relationship between the `DJIA` and another stock index that reflects a wider range of securities. Standard and Poor's 500 Stock Index (commonly called S & P 500, abbreviated here as `SP`) is constituted by using the prices of securities of 425 U.S. industrial companies and 75 railway and public–utility corporations. `SP` is certainly more broadly based than `DJIA`. Do these two indices track the price movements differently?

This case examines weekly values of `DJIA` and `SP` for the weeks ending February 1, 1991 through February 25, 1994 (no special significance is to be attached to these dates). `DJIA` and `SP` data are easily available and you may want to supplement these data with the latest available data. Calculate the correlation coefficient between `DJIA` and `SP`. The correlation is well above 0.95. `DJIA`, even though based on only 30 securities, is highly correlated with `SP`, which is based on 500 securities. It appears that we can track stock prices movements using `DJIA` even though it is based on only 30 stocks.

Can we describe the movement of `DJIA` over time? What is the best possible way to describe the movement of `DJIA`?

A plot of the data is always a good start for an analysis. A plot of `DJIA` against time shows that `DJIA` increases over time and a straight line is suggested. Regress `DJIA` against time to see if a linear trend line describes the behavior of `DJIA` over time. It appears that a linear trend line is not satisfactory. The coefficients are significant, and the value of R^2 is large, but a plot of the residuals shows the inadequacy of the linear time trend model. The residuals are autocorrelated; positive residuals being followed by positive residuals, and negative ones by negative ones. This is seen by plotting the standardized residuals against time (which is essentially the sequence in which the observations occur). The low value of the Durbin–Watson statistic (0.259) confirms this finding.

Regressing the variable against its lagged value often can remove time dependency of the residuals. We will therefore regress $\mathtt{DJIA}_t$ against $\mathtt{DJIA}_{t-1}$. Regressing $\mathtt{DJIA}_t$ against $\mathtt{DJIA}_{t-1}$, we find that the standardized residuals no longer have any time dependence. You should examine residual plots and the Durbin–Watson statistic to verify this.

Among the 161 observations there are three with large standardized residuals. These occur for the weeks ending 11/15/91, 12/27/91, and 2/5/93. The values of the standardized residuals are $-2.50, 3.45$, and 2.78 respectively. On examining these points we find that in these weeks the DJIA had very large changes of $-102, 167$, and 132, respectively. Our model describes the movement of DJIA adequately, as long as there is no drastic change in one particular week. The average weekly change in the DJIA for the period under consideration is 6.93. DJIA for one week can be described adequately using the DJIA of the previous week unless the stock market has a massive change (experiences a shock or jolt, as it were). The regression model has no mechanism to predict when these jolts will occur. There is no theory for predicting these jolts, although much folklore prevails.

The constant for this regression equation is not significant, but nevertheless it should be retained in the model. The constant should be interpreted as a general level (a base value) and not as the estimated value of DJIA when the previous value of DJIA is zero, as such a value for DJIA is not meaningful. In most regression models the constant term should be retained to represent a base value even though it may not be statistically significant. The fitted regression model is

$$\mathtt{DJIA}_t = 54.207 + 0.9856 \times \mathtt{DJIA}_{t-1}.$$

From our analysis we see that the movement of DJIA can be described satisfactorily using the value of DJIA in the previous period, unless the stock market receives a massive jolt (when all bets are off!). This confirms a theory in finance that asserts that the DJIA follows a random walk (possibly with drift). The implications of this theory are that $\mathtt{DJIA}_t$ and $\mathtt{DJIA}_{t-1}$ are linearly related to each other, and the residuals from this linear fit should have a distribution that is roughly normal (Gaussian) with zero mean and constant variance. An examination of the residuals from the fitted model in the present case supports these assumptions.

From the fitted model we can predict the value of DJIA for next week by using the value of DJIA this week. For example, the value of DJIA predicted for March 4, 1994 using the value of DJIA for February 25, 1994 (the last date that we have used in our analysis) is 3837.8, and the 95% prediction interval is $(3746.9, 3928.8)$. The actual value was 3833.9. The model seems to do quite satisfactorily in the short run. The random walk model implies that the past values of DJIA are not very useful in predicting future values, as only the most recent one is relevant. This statement holds with one qualification, that is, that the stock market has not received a violent jolt (shock).

Comments to the Instructor

A regression of DJIA on time is highly significant ($R^2 = 89.74\%$, $F = 1391.29$), but the structure in the residuals is apparent.

The regression of $\mathtt{DJIA}_t$ on $\mathtt{DJIA}_{t-1}$ is extremely strong ($R^2 = 97.3\%$, $F = 5694.05$). The autocorrelation of the residuals is gone, as evidenced by the Durbin–Watson statistic ($DW = 2.01$).

To maintain consistency with the unannotated edition,
the next case begins on the following page.

Further investigation of the DJIA and the S & P 500 index

Topics covered: Analysis of time series data.

Data File: `djsp.dat`

The previous case examined the behavior of the Dow Jones Industrial Average over time.

Carry out a similar analysis using the Standard & Poor's 500 index.

Does your analysis lead to similar conclusions?

A simplified version of the random walk model is that, except for a possible constant adjustment, the best prediction for a stock market index at time t is simply its value at time $t-1$. One way to test whether this simplification is appropriate for `DJIA` is to test whether the coefficient of $\texttt{DJIA}_{t-1}$ is significantly different from 1 in a regression of $\texttt{DJIA}_t$ on $\texttt{DJIA}_{t-1}$. Does the simpler model seem appropriate here? What about for `SP`?

Another way to check this is as follows. Construct the differenced `DJIA` series (`DDJIA`) by taking successive differences; i.e.,

$$\texttt{DDJIA}_t = \texttt{DJIA}_t - \texttt{DJIA}_{t-1}.$$

Examine if $\texttt{DDJIA}_t$ can be regarded as a random sample of normally distributed random variables with common mean and constant variance.

What about the differenced `SP` series? Why would such a result be consistent with the simplified random walk model?

Comments to the Instructor

A regression of `SP` on time is very strong ($R^2 = 94.94\%$, $F = 2981.04$), but the correlation structure in the residuals is evident ($DW = 0.58$).

A regression of $\texttt{SP}_t$ on $\texttt{SP}_{t-1}$ is extremely strong ($R^2 = 97.03\%$, $F = 5157.43$), with fitted model

$$\widehat{\texttt{SP}}_t = 11.795 + .9737 \times \texttt{SP}_{t-1}.$$

Autocorrelation has been removed ($DW = 2.14$). Thus, we obtain similar conclusions about `SP` as were found in the analysis of `DJIA`.

Tests for whether the slopes of the models for `DJIA` and `SP` are different from 1 are simply t–tests, as follows:

$$t = \frac{.98563 - 1}{.01306} = -1.10$$

for `DJIA` ($p = .273$), and

$$t = \frac{.9737 - 1}{.01356} = -1.94$$

for `SP` ($p = .054$). Thus, there is no evidence against the simplified random walk model for `DJIA`, and weak evidence against it for `SP`, based on these tests.

The normality of the two differenced variables can be examined in many ways, including histograms, stem–and–leaf plots, normal plots and control charts (to verify that the differenced series are in control). Tests for autocorrelation can also be performed on the differenced series. None of these methods leads to rejection of the normality assumption, supporting the use of the simplified random walk model.

It is straightforward to verify that normality of the differenced series is equivalent to the simplified random walk model, since the latter model (for `DJIA`, say) is consistent with

$$\texttt{DJIA}_t = \beta_0 + \beta_1 \times \texttt{DJIA}_{t-1} + \epsilon_t$$

with $\beta_1 = 1$ and ϵ_t being independent and identically distributed Gaussian random variables with zero mean.

This case can be used to discuss the general question of how to ascertain normality.

To maintain compatibility with the unannotated edition,
the appendix begins on the following page.

Descriptions of the data files

In this appendix, the data files used in the cases are described.

Included are the names of the files, a list of the variables in the file, the source of the data, and a printout of the first five observations in the file. The latter is included so that the reader can see the layout of the file, making it easier to input the data into a statistical package. We have not included a data source when the data were generally available in newspapers (e.g., stock prices). Missing values are represented by `M` in all data files.

All of the data files are included in the diskette that comes with this book. In addition, they can be obtained electronically over the Internet from the `statlib` server located at Carnegie–Mellon University. If the reader has electronic mail access to the Internet, one approach is to send the message

send chscase from datasets

to the Internet address `statlib@lib.stat.cmu.edu`. Alternatively, the data are available via **gopher** at the address

swis.stern.nyu.edu.

The information is filed under:

↪ Academic Departments & Research Centers
↪ Statistics and Operations Research
↪ Publications
↪ A Casebook for a First Course in Statistics and Data Analysis
↪ Welcome!

Finally, the files can be accessed from the World Wide Web (WWW) using a WWW browser (e.g., **mosaic**) starting from the URL address

gopher://swis.stern.nyu.edu:70

and following the same path. This will provide a short description of the information available. The latter two interactive methods (gopher and mosaic) also provide access to a growing archive of new cases (in Postscript format) and updated information concerning the cases in the book.

The data files are written in plain ASCII (character) text, so it should be possible to import them into virtually any statistical, database management, or spreadsheet package.

It is possible that, for some of the cases, results using these data sets will differ slightly from those in the text. The reason for this is that non–integer values are not necessarily provided to the full level of machine arithmetic in the data sets.

File name

`adopt.dat`

Description of data

Visas issued by the United States Immigration and Naturalization Service for the purpose of adoption by U.S. residents.

Variables in file

(1) Country or region of origin of child.
(2) The number of visas issued in 1988 (called `VISA88` in the case).
(3) The number of visas issued in 1991 (called `VISA91` in the case).
(4) The number of visas issued in 1992 (called `VISA92` in the case).

Data source

U.S. Immigration and Naturalization Service, published in *Ours* (May/June 1993), copyright Adoptive Families of America.

First five observations

```
Africa            28   41   63
Belize             6    4    8
Bolivia           21   51   74
Brazil           164  178  139
Cambodia           0   59   16
```

File name

`census1.dat`

Description of data

Incomes of full–time year–round female workers.

Variables in file

(1) Female income (in thousands of dollars) (called `INCOME` in the case).

Data source

Based on the *U.S. Census*, 1990.

First five observations

```
11.652
23.015
 5.604
 6.710
 7.293
```

File name

`census2.dat`

Description of data

Sample means of samples from the incomes in `census1.dat` (for different sample sizes).

Variables in file

Sample means for samples of size:

(1) 1 (called `SIZE1` in the case).
(2) 2 (called `SIZE2` in the case).
(3) 4 (called `SIZE4` in the case).
(4) 10 (called `SIZE10` in the case).
(5) 25 (called `SIZE25` in the case).
(6) 50 (called `SIZE50` in the case).
(7) 100 (called `SIZE100` in the case).
(8) 625 (called `SIZE625` in the case).

Data source

Based on the *U.S. Census*, 1990.

First five observations

```
16.7540 21.7430 16.0140 19.5413 17.5730 20.0110 19.3907 19.2227
17.4190 15.2250 12.5705 25.0069 20.4635 18.7705 19.4249 19.7238
42.0630 15.7570 30.8815 16.1605 20.8111 19.0999 19.5469 20.5771
15.9070 24.8285 16.0473 15.4880 23.9404 19.7499 18.7775 20.2231
35.7870 20.1530 21.5765 21.0381 20.6379 20.7242 18.6020 20.4897
```

File name

`census3.dat`

Description of data

Sample means of samples from the logarithm of the incomes in `census1.dat` (for different sample sizes).

Variables in file

Sample means for samples from the log–incomes of size:

(1) 1 (called `SIZE1L` in the case).
(2) 2 (called `SIZE2L` in the case).
(3) 4 (called `SIZE4L` in the case).
(4) 10 (called `SIZE10L` in the case).
(5) 25 (called `SIZE25L` in the case).
(6) 50 (called `SIZE50L` in the case).
(7) 100 (called `SIZE100L` in the case).
(8) 625 (called `SIZE625L` in the case).

Data source

Based on the *U.S. Census,* 1990.

First five observations

```
4.3830 4.4000 4.2788 4.2876 4.2397 4.2427 4.2538 4.3830
4.4170 4.2145 4.3759 4.1595 4.2758 4.2820 4.2182 4.4170
4.2894 4.3868 4.1404 4.2997 4.3123 4.2684 4.1946 4.2894
4.1397 4.1053 4.1357 4.1653 4.2470 4.2036 4.1990 4.1397
4.2282 4.3022 4.1870 4.3764 4.1921 4.2431 4.2441 4.2282
```

File name

`census4.dat`

Description of data

Sample medians of samples from the incomes in `census1.dat` (for different sample sizes).

Variables in file

Sample medians for samples of size

(1) 1 (called `SIZE1M` in the case).
(2) 2 (called `SIZE2M` in the case).
(3) 4 (called `SIZE4M` in the case).
(4) 10 (called `SIZE10M` in the case).
(5) 25 (called `SIZE25M` in the case).
(6) 50 (called `SIZE50M` in the case).
(7) 100 (called `SIZE100M` in the case).
(8) 625 (called `SIZE625M` in the case).

Data source

Based on the *U.S. Census*, 1990.

First five observations

```
34.6530 60.6355 13.8600 14.7120 17.9600 15.7530 18.3870 16.3150
 6.6770 47.6745 14.1830 15.0710 19.4240 17.2285 17.1905 16.8900
13.8720 11.9515 19.1690 24.3490 17.7810 18.2385 16.2030 17.3890
14.1250 27.8420 26.4380 12.7520 14.5420 18.3580 17.0740 17.3260
42.5080 15.4850 14.8085 13.1935 14.4110 15.3065 18.3760 17.0660
```

File name

`census5.dat`

Description of data

Sample medians of samples from the logarithm of the incomes in `census1.dat` (for different sample sizes).

Variables in file

Sample medians for samples from the log–incomes of size
(1) 1 (called `SIZE1ML` in the case).
(2) 2 (called `SIZE2ML` in the case).
(3) 4 (called `SIZE4ML` in the case).
(4) 10 (called `SIZE10ML` in the case).
(5) 25 (called `SIZE25ML` in the case).
(6) 50 (called `SIZE50ML` in the case).
(7) 100 (called `SIZE100ML` in the case).
(8) 625 (called `SIZE625ML` in the case).

Data source

Based on the *U.S. Census,* 1990.

First five observations

```
4.5397 4.7827 4.1418 4.1677 4.2543 4.1974 4.2645 4.2270
3.8246 4.6783 4.1518 4.1781 4.2883 4.2362 4.2353 4.2293
4.1421 4.0774 4.2826 4.3865 4.2500 4.2610 4.2096 4.2186
4.1500 4.4447 4.4222 4.1056 4.1626 4.2638 4.2323 4.2309
4.6285 4.1899 4.1705 4.1204 4.1587 4.1849 4.2643 4.2411
```

File name

`census6.dat`

Description of data

Standard deviations of samples from the incomes in `census1.dat` (for different sample sizes).

Variables in file

Sample standard deviations for samples of size
(1) 2 (called `SIZE2S` in the case).
(2) 4 (called `SIZE4S` in the case).
(3) 10 (called `SIZE10S` in the case).
(4) 25 (called `SIZE25S` in the case).
(5) 50 (called `SIZE50S` in the case).
(6) 100 (called `SIZE100S` in the case).
(7) 625 (called `SIZE625S` in the case).

Data source

Based on the *U.S. Census,* 1990.

First five observations

```
21.5250 21.5760 10.0227 10.4698 15.1795 10.3388 10.5899
22.8007  3.6888 13.3630  9.3414  9.3884 10.5529 10.6673
 0.7715 12.3097 11.0676  7.7973  9.7320  8.8305 14.0619
26.0130 14.4631 11.8456  9.9338 11.1705 11.5244 12.5909
 6.9692  6.9114  8.4057 14.6470  8.4625  8.8706 15.7175
```

File name

`chal.dat`

Description of data

Damage of O-rings at different temperatures for space shuttle flights.

Variables in file

(1) Ambient temperature at launch (called `TEMP` in the case).
(2) Number of O-rings damaged.

Data source

Report of the Presidential Commission on the Space Shuttle Challenger Accident, 1986. Data summarized in "Risk analysis of the space shuttle: Pre–Challenger prediction of failure," by S.R. Dalal, E.B. Fowlkes and B.A. Hoadley, *Journal of the American Statistical Association*, 84, 945–957 (1989). Reprinted with permission from the *Journal of the American Statistical Association*.

First five observations

```
53   2
57   1
58   1
63   1
66   0
```

File name

`demand.dat`

Description of data

Price indices and gasoline consumption in the United States over time.

Variables in file

(1) Year (called `YR` in the cases).
(2) Total gas consumption (called `G` in the cases).
(3) Price index for gasoline (called `PG` in the cases).
(4) Per capita real disposable income (called `I` in the cases).
(5) Price index for new cars (called `PNC` in the case).
(6) Price index for used cars (called `PUC` in the cases).
(7) Price index for public transportation (called `PPT` in the case).
(8) Aggregate price index for durable goods (called `PD` in the case).
(9) Aggregate price index for non–durables (called `PN` in the case).
(10) Aggregate price index for services (called `PS` in the case).
(11) Year squared (called `YRSQ` in the case).

Data source

Reprinted with the permission of Macmillan College Publishing Company from *Econometric Analysis 2nd ed.* by William H. Greene. Copyright ©1993 by Macmillan College Publishing Company, Inc.

First five observations

```
60 129.7 0.925  6036 1.045 0.836 0.810 0.444 0.331 0.302 3600
61 131.3 0.914  6113 1.045 0.869 0.846 0.448 0.335 0.307 3721
62 137.1 0.919  6271 1.041 0.948 0.874 0.457 0.338 0.314 3844
63 141.6 0.918  6378 1.035 0.960 0.885 0.463 0.343 0.320 3969
64 148.8 0.914  6727 1.032 1.001 0.901 0.470 0.347 0.325 4096
```

File name

`djsp.dat`

Description of data

Dow Jones Industrial Average and the S & P 500 index values weekly.

Variables in file

(1) Date
(2) Dow Jones Industrial Average at the close of the day (Called `DJI` in the case).
(3) Standard and Poor's 500 Stock Index at the close of the day (called `SP` in the case).

First five observations

```
2/ 1/91     2730.69   343.05
2/ 8/91     2830.69   359.35
2/15/91     2934.65   369.06
2/22/91     2889.36   365.65
3/ 1/91     2909.90   370.47
```

File name

`elusage.dat`

Description of data

Monthly domestic electricity consumption at different temperatures.

Variables in file

(1) Month of observation (1 = January, 2 = February, ..., 12 = December).
(2) Year of observation.
(3) Average daily usage (kilowatt–hours) (called `KWH` in the case).
(4) Average daily temperature (degrees) (called `TEMPERATURE` in the case).

Data source

Handcock family archives, August 1989 – February 1994.

First five observations

```
 8  1989 24.828   73
 9  1989 24.688   67
10  1989 19.310   57
11  1989 59.706   43
12  1989 99.667   26
```

File name

`empl.dat`

Description of data

Yearly employment rates in the United States of 25– to 34–year old males with 9–11 years of schooling (by year).

Variables in file

(1) Year.
(2) Percent of males employed.

Data source

The Condition of Education, 1991, U.S. Department of Education.

First five observations

```
1971    87.9
1972    88.5
1973    88.8
1974    90.2
1975    78.1
```

File name

`ers.dat`

Description of data

Emergency calls to the New York Auto Club.

Variables in file

(1) Date (January day/year).
(2) Emergency road service calls answered (called `CALLS` in the cases).
(3) Forecast high temperature (called `FHIGH` in the cases).
(4) Forecast low temperature (called `FLOW` in the cases).
(5) Daily high temperature (called `HIGH` in the case).
(6) Daily low temperature (called `LOW` in the case).
(7) Rain forecast (0 = no rain, 1 = rain) (called `RAIN` in the case).
(8) Snow forecast (0 = no snow, 1 = snow) (called `SNOW` in the case).
(9) Type of day (weekend/holiday = 0, normal business day = 1) (called `WEEKDAY` in the case).
(10) Year (1993=0, 1994=1).
(11) Sunday (false=0, true=1) (called `SUNDAY` in the case).
(12) Subzero temperature (false=0, true=1)

Data source

New York Motorist, March 1994 (page 3). ©Automobile Club of New York. Used with permission.

First five observations

```
16/93  2298  38  31  39  31  0  0  0  0  0  0
17/93  1709  41  27  41  30  0  0  0  0  1  0
18/93  2395  33  26  38  24  0  0  0  0  0  0
19/93  2486  29  19  36  21  0  0  1  0  0  0
20/93  1849  40  19  43  27  0  0  1  0  0  0
```

File name

`foot.dat`

Description of data

Betting on professional football results.

Variables in file

(1) Name of favored team.
(2) Name of underdog team.
(3) Betting result (Didn't cover = -1, push = 0, covered = 1) (called `BETTING` in the case).
(4) Day and time of game (0 = Sunday afternoon, 1 = Sunday night, 2 = Monday night, 3 = Thursday, 4 = Saturday) (called `DAY` in the case).
(5) Favored team at home or away (0 = away, 1 = home) (called `HOMEAWAY` in the case).
(6) Week of season (called `WEEK` in the case).
(7) Year (called `YEAR` in the case).

Data source

Compiled by Hal Stern. Submitted to the `statlib` facility by Robin Lock.

First five observations

```
BUF  MIA  -1  0  0   1  89
CHI  CIN   0  0  1   1  89
CLE  PIT   1  0  0   1  89
NO   DAL   1  0  1   1  89
MIN  HOU   1  0  1   1  89
```

File name

`free.dat`

Description of data

Performances of baseball players in their free agent year and the following year. The values affected by the baseball strike in 1981 were not analysed in the cases.

Variables in file

(1) Player name.
(2) Batting average (hits per at bat) in free agent year (called `BAFA` in the case).
(3) Batting average in the next year (called `BANEXT` in the case).
(4) Home runs in free agent year (called `HRFA` in the case).
(5) Home runs in the next year (called `HRNEXT` in the case).
(6) Runs batted in in free agent year (called `RBIFA` in the case).
(7) Runs batted in in the next year (called `RBINEXT` in the case).
(8) At bats in free agent year (called `ABFA` in the case).
(9) At bats in the next year (called `ABNEXT` in the case).
(10) Slugging average in free agent year.
(11) Slugging average in the next year.
(12) Runs scored in free agent year.

(13) Runs scored in the next year.
(14) Affected by the strike in 1981 (0 = no, 1 = year before becoming a free agent, 2 = year after becoming a free agent).

Data source

J.L Reichler, ed., *The Baseball Encyclopedia, 8th ed.*, Macmillan, New York (1991). Used with permission.

First five observations

R. Jackson (76-77)	0.277	0.286	27	32	91	110	498	525	0.502	0.550	84	93	0
L. Hisle	0.302	0.290	28	34	119	115	546	520	0.533	0.533	95	96	0
O. Gamble (77-78)	0.297	0.275	31	7	83	47	408	375	0.588	0.387	75	46	0
P. Rose (78-79)	0.302	0.331	7	4	52	59	655	628	0.421	0.430	103	90	0
B. Watson	0.303	0.307	16	13	71	68	457	469	0.569	0.456	63	62	0

File name

`funds.dat`

Description of data

Performance of different types of stock mutual funds.

Variables in file

(1) Fund name.
(2) Type of fund (Balanced = 1, Equity–Income = 2, Growth–and–Income = 3, Growth = 4, Aggressive–Growth = 5, Small–Company = 6, International–and–Global = 7).
(3) Five year investment performance (called `FIVEYR` in the case).
(4) 1992 return (called `RETURN92` in the case).

Data source

Copyright 1993 by Consumers Union of U.S., Inc., Yonkers, NY 10703-1057. Reprinted by permission from *Consumer Reports*, May 1993.

First five observations

CGM Mutual	1	15600	6
Fidelity Balanced	1	14650	8
MainStay Total Ret.	1	15001	4
Kemper Inv. Tot. Ret	1	15628	4
Pax World	1	14222	1

File names

`geyser1.dat`, `geyser2.dat`

Description of data

Eruption durations and intereruption times for the "Old Faithful" geyser in Yellowstone National Park.

Variables in file

(1) An index of the date the observation was taken (1 through 16). This appears in `geyser1.dat` only.
(2) The duration of an eruption of the geyser, in minutes. In `geyser2.dat`, some of these values are noted only as `Short`, `Medium` or `Long` (called `DURATION` in the case).
(3) The time until the next eruption (the intereruption time), in minutes (called `INTERVAL` in the case).

Data sources

`geyser1.dat`: S. Weisberg, *Applied Linear Regression, 2nd. ed.*, John Wiley, New York (1985). These data were taken in August 1978 and August 1979. ©John Wiley and Sons. Used with permission.

`geyser2.dat`: A. Azzalini and A.W. Bowman, "A look at some data on the Old Faithful geyser," *Applied Statistics*, **39**, 357–365 (1990). These data were taken in August 1985. ©The Royal Statistical Society. Used with permission.

First five observations

`geyser1.dat`

```
1   4.4   78
1   3.9   74
1   4.0   68
1   4.0   76
1   3.5   80
```

`geyser2.dat`

```
4.0    71
2.2    57
Long   80
Long   75
Long   55
```

File name

`health.dat`

Description of data

Comparison of health care spending across the United States.

Variables in file

(1) Name of the State
(2) Census Bureau region of the state.
(3) Census Bureau region number.
(4) Per capita health spending.
(5) Percent of per capita income spent on health.

Data source

The New York Times, October 15, 1993.

First five observations

Alabama	East South Central	5	1833	11.8
Alaska	Pacific	9	1801	8.2
Arizona	Mountain	8	1712	10.4
Arkansas	West South Central	6	1673	11.3
California	Pacific	9	1914	9.1

File name

`hockey1.dat`, `hockey2.dat`

Description of data

Success of National Hockey League Teams.

Variables in file

(1) Team name.
(2) Points (called `POINTS` in the case).
(3) Goals scored (called `GF` in the case).
(4) Goals given up (called `GA` in the case).
(5) Year. This appears in `hockey2.dat` only.

Data source

The National Hockey League Official Guide and Record Book, 1992–1993, Triumph Books, Chicago (1992).

First five observations

`hockey1.dat`

Pittsburgh Penguins	119	367	268
Boston Bruins	109	332	268
Chicago Blackhawks	106	279	230
Quebec Nordiques	104	351	300
Detroit Red Wings	103	369	280

`hockey2.dat`

Edmonton Oilers	106	372	284	1988

Philadelphia Flyers	100	310	245	1988
Calgary Flames	95	318	289	1988
Hartford Whalers	93	287	270	1988
Montreal Canadiens	92	277	241	1988

File name

`liinc.dat`

Description of data

Average incomes of Long Island communities in 1979 and 1989, in 1989 dollars.

Variables in file

(1) Community name (called `COMMUNITY` in the case).
(2) Average 1979 income in 1989 dollars (called `INCOME79` in the case).
(3) Average 1989 income (called `INCOME89` in the case).

Data source

Newsday, August 31, 1992.

First five observations

Albertson	47838	49676
Amagansett	27919	34705
Amityville	39294	46442
Asharoken	95159	84064
Atlantic Beach	54167	72812

File name

`lischool.dat`

Description of data

Distribution of white student enrollment in Nassau County school districts.

Variables in file

(1) School district (called `DISTRICT` in the case).
(2) Proposed legislative district (called `LEGISLAT` in the case).
(3) Total public school enrollment.
(4) White student enrollment.

Data source

Newsday, May 20, 1994

First five observations

Baldwin	5	4658	3549
Bellmore	19	950	912
Bellmore - Merrick	19	4706	4470
Bethpage	17	2460	2300
Carle Place	11	1439	1273

File name

`mort.dat`

Description of data

Interest rates for fixed and adjustable rate mortgages.

Variables in file

(1) Name of lender.
(2) Type of mortgage (Fixed = 0, Adjustable = 1).
(3) Mortgage rate.
(4) Points (called `POINTS` in the case).

Data source

Newsday, March 27, 1993.

First five observations

```
Anchor Savings          0   7.875   0.00
Bank of the Hamptons    0   7.125   1.50
Bay Ridge Federal       1   5.000   2.00
Brooklyn Federal        1   4.500   1.50
Chemical Bank           0   7.500   2.00
```

File name

`nba.dat`

Description of data

Performance of National Basketball Association guards.

Variables in file

(1) Player's name.
(2) Player's height (cm) (called `HEIGHT` in the case).
(3) Number of games appeared in (called `GAMES` in the case).
(4) Total minutes played (called `MINUTES` in the case).
(5) Player's age (called `AGE` in the case).
(6) Points scored per game.
(7) Assists per game.
(8) Rebounds per game.
(9) Percent of field goals made (called `FGP` in the case).
(10) Percent of free throws made (called `FTP` in the cases).

The variable measuring points per minute (called `PPM` in the cases) is derived from these variables by the expression:

$$\mathtt{PPM} = \mathtt{PPG} \times \mathtt{GAMES} / \mathtt{MINUTES}$$

The minutes per game (called `MPG` in the cases), assists per minute (called `APM` in the cases) and rebounds per minute are calculated similarly.

Data source

J. Cohn, *The Pro Basketball Bible*, Basketball Books Ltd., San Diego, CA. © 1994 by Jordan Cohn.

First five observations

R. Miller	201	82	2954	28	21.2	3.2	3.1	47.9	88.0
M. Jordan	198	78	3067	30	32.6	5.5	6.7	49.5	83.7
D. West	198	80	3104	26	19.3	2.9	3.1	51.7	84.1
J. Dumars	191	76	3094	30	23.5	4.0	1.9	46.6	86.4
M. Richmond	196	45	1728	28	21.9	4.9	3.4	47.4	84.5

File name

`nyseotc.dat`

Description of data

The return on stocks in the Over the Counter market and New York Stock Exchange, May 9 - May 13, 1994.

Variables in file

(1) Weekly return of NASDAQ stocks (called `NASDAQRET` in the case).
(2) Weekly return of NYSE stocks.

First five observations

-0.1053	-0.0246
0.0667	-0.0212
-0.0714	0.0789
-0.0294	-0.0250
0.0423	-0.0208

File name

`pcb.dat`

Description of data

Concentration of PCBs in different bays for 1984 and 1985.

Variables in file

(1) Bay name (called `BAY` in the case).
(2) 1984 PCB concentration (parts/billion) (called `PCB84` in the case).
(3) 1985 PCB concentration (parts/billion) (called `PCB85` in the case).

Data source

Environmental Quality 1987-1988.

First five observations

Casco Bay	95.28	77.55
Merrimack River	52.97	29.23
Salem Harbor	533.58	403.10
Boston Harbor	17104.86	736.00
Buzzards' Bay	308.46	192.15

File name

ppp.dat

Description of data

Comparison of changes in exchange rates and differences in inflation rates for various countries.

Variables in file

(1) Country name.
(2) Change in exchange rate 1975–1990 (called CHNGEX75 in the case).
(3) Change in exchange rate 1985–1990 (called CHNGEX85 in the case).
(4) Difference in inflation rates 1975–1990 (called CHNGIN75 in the case).
(5) Difference in inflation rates 1985–1990 (called CHNGIN85 in the case).

Data source

International Financial Statistics Yearbook.

First five observations

Australia	3.44697	-2.22615	3.62836	3.99243
Austria	-2.84312	-11.97250	-2.14455	-3.00037
Belgium	-0.63888	-11.49660	-2.17764	-4.45400
Canada	0.91474	-3.14511	0.95945	-0.37619
Chile	27.52827	12.77215	27.02101	12.64657

File name

prdq.dat

Description of data

Comparison of productivity and quality in Japanese and non–Japanese automobile manufacturing.

Variables in file

(1) Assembly defects per 100 cars (called QUALITY in the case).
(2) Hours per vehicle (called PRODUCTIV in the case).
(3) National origin of facility (non–Japan = 0, Japan = 1).
(4) Assembly defects per 100 cars (non–Japanese origin) (called QUALNONJ in the case).
(5) Assembly defects per 100 cars (Japanese origin) (called QUALJAPN in the case).
(6) Hours per vehicle (non–Japanese origin) (called PRODNONJ in the case).
(7) Hours per vehicle (Japanese origin) (called PRODJAPN in the case).

Data source

J.P. Womack, D.T. Jones, D. Roos, *The Machine That Changed the World*, Rawson Associates, New York (1990).

First five observations

29	31	0	29	39	31	27
38	26	0	38	38	26	23
39	27	1	68	42	21	15
38	23	1	67	48	26	20
42	15	1	69	50	30	17

File name

`return.dat`

Description of data

Annual return rates in the stock market.

Variables in file

(1) Year.
(2) Standard and Poor's Index year end value.
(3) Vanguard Index Trust 500 Portfolio year end value.

Data source

Vanguard Market Index Trust 500–Portfolio Annual Report, 1993 (page 7).

First five observations

```
1976   10000.000   10000.000
1977   9281.2618   9215.6490
1978   9885.7197   9756.9082
1979   11710.427   11517.425
1980   15509.067   15193.239
```

File name

`rock.dat`

Description of data

Sales and airplay rankings of popular music singles.

Variables in file

(1) Title of the song.
(2) Sales (rank in top 100).
(3) Sales last week (rank in top 100).
(4) Number of weeks on the sales top 100 list.
(5) Airplay (rank in top 100).
(6) Airplay last week (rank in top 100).
(7) Number of weeks on the airplay top 100 list.
(8) Change in sales ranking from last week.
(9) Change in airplay ranking from last week.

Data source

First five observations

```
100% Pure Love        26  31  16  18  17  14  -1   5
Action                43  42   4  43  45   8   2  -1
All I Wanna Do        40  61   2   7  15   5   8  21
Always                65  57  19  33  31  20  -2  -8
Always In My Heart    32  29  10  42  34  14  -8  -3
```

File name

`sex.dat`

Description of data

Numbers of reported sexual partners of a sample of males and females.

Variables in file

(1) Male total lifetime number of female partners (called `MALE` in the case).
(2) Female total lifetime number of male partners (called `FEMALE` in the case).

Data source

The General Social Survey 1989-1991.

First five observations

```
 1   1
 3   1
 6   5
 6   2
47   2
```

File name

`subway.dat`

Description of data

Perceptions of the New York City subway system.

Variables in file

For variables (1) – (18), 1 = Very unsatisfactory, 2 = Unsatisfactory, 3 = Neutral, 4 = Satisfactory, 5 = Very satisfactory.

(1) Usage of subway (0 = no use of subway, 1 = use subway).
(2) Cleanliness of stations (called `CLNSTAT` in the case).
(3) Cleanliness of trains (called `CLNTRAIN` in the case).
(4) Safety in stations (called `SAFSTAT` in the case).
(5) Safety on trains (called `SAFTRAIN` in the case).
(6) Rush hour crowding in stations.
(7) Rush hour crowding on trains.
(8) In–station information.
(9) On–train announcements.
(10) Convenience of train stops.
(11) Convenience of train schedule.
(12) Speed of travel.
(13) Frequency of trains.
(14) Ease of token purchase.
(15) Ease of token collection.
(16) Police presence in stations.
(17) Police presence on trains.
(18) Availability of maps.
(19) Number of uses per week.

Data source

Survey conducted at the Leonard N. Stern School of Business, Spring 1994.

First five observations

```
1  1  3  3  3  1  1  2  2  3  3  3  2  3  4  4  4  5  20
1  2  2  3  3  3  2  1  1  4  3  5  4  2  3  2  3  2   3
1  2  5  4  2  2  1  5  3  5  3  4  4  1  3  3  3  4  10
1  2  4  5  5  4  3  2  2  5  4  5  5  5  4  5  4  5   8
1  1  1  1  1  1  1  3  3  2  2  2  2  5  5  2  2  5  10
```

File name

`vine1.dat`

Description of data

Lug counts from vineyard harvests by row and year of harvest.

Variables in file

(1) Row number.
(2) Number of lugs for 1983.
(3) Number of lugs for 1984.
(4) Number of lugs for 1985.
(5) Number of lugs for 1986.
(6) Number of lugs for 1987.
(7) Number of lugs for 1988.
(8) Number of lugs for 1989.
(9) Number of lugs for 1990.
(10) Number of lugs for 1991.

Data source

Barnhill family archives 1976 – 1991.

First five observations

```
1    5.00   5.0   2.0   4.5   1.5   1.5   1.0   5.0   9.5
2   10.00   5.0   5.0   9.0   4.0   4.0   3.0   8.0  17.5
3   11.00  11.0   4.0   9.0   6.0   6.0   3.0  11.0  18.0
4   11.25  10.0   5.0   9.5   7.0   5.0   3.0   9.0  20.0
5   11.00   7.0   4.0  10.0   7.0   6.0   5.0   9.5  20.5
```

File name

`vine2.dat`

Description of data

Yields from vineyard harvests by row number and year of harvest.

Variables in file

(1) Harvest year (called `YEAR` in the case).
(2) Row of vines (called `ROW` in the case).
(3) Yield of grapes (lbs) (called `YIELD` in the case).

Data source

Barnhill family archives, 1976 – 1991.

First five observations

```
1983   1   158.672
1983   2   317.345
1983   3   349.079
1983   4   357.013
1983   5   349.079
```

File name

`vine3.dat`

Description of data

Yields from vineyard harvests for each year.

Variables in file

(1) Harvest Year (called `YEAR` in the case).
(2) Total number of collected lugs (called `LUGS` in the case).
(3) Total weight of grapes (lbs) (called `YIELD` in the case).
(4) Sum of the lug counts from rows 3 – 6 (called `INITIAL` in the case).

Data source

Barnhill family archives, 1976 – 1991.

First five observations

```
1976     11.00     313.0       M
1977    476.00   13299.0       M
1978     60.00    1848.0       M
1979    102.00    2894.0       M
1980     71.00    1792.0       M
```

File name

`vote.dat`

Description of data

Absentee and machine ballot votes in Philadelphia elections.

Variables in file

(1) Year of election.
(2) District number.
(3) Democrat absentee vote in district.
(4) Republican absentee vote in district.
(5) Democrat machine vote in district.
(6) Republican machine vote in district.

Data source

Provided by Orley Ashenfelter.

First five observations

```
1982   2    551    205   47767   21340
1982   4    594    312   44437   28533
1982   8    338    115   55662   13214
1984   1   1357    764   58327   38883
1984   3    716    144   78270    6473
```

File name

`whale.dat`

Description of data

The early lives of two beluga whale calves.

Variables in file

(1) Time period since birth for Hudson (called `PERIOD1` in the case).
(2) Number of nursing bouts in period for Hudson (called `BOUTS1` in the case).
(3) Total seconds spent nursing in period for Hudson (called `NURSING1` in the case).
(4) Number of lockons in period for Hudson (called `LOCKONS1` in the case).
(5) Time of day (night time = 0, daytime=1) for Hudson (called `DAYNIGHT1` in the case).
(6) Time period since birth for Casey.
(7) Number of nursing bouts in period for Casey.
(8) Total seconds spent nursing in period for Casey.
(9) Number of lockons in period for Casey.
(10) Time of day for Casey.

Data source

J.M. McLaughlin-Russell, *The Nursing Behaviors of Beluga Whales: A Study of Two Calves at the New York Aquarium*, unpublished Master's Thesis, New York University (1993).

First five observations

1	0	0	0	0	1	0	0	0	0
2	0	0	0	1	2	0	0	0	1
3	0	0	0	1	3	0	0	0	1
4	0	0	0	0	4	9	59	12	0
5	0	0	0	0	5	10	228	28	0

Index

Added variable plot, 191
  definition, 200
Adoption, 13–20, 189
  data description, 287
Air bags, 134–136
Amniocentesis, 41–42
Association in subgroups, 25
Autocorrelation, 224, 256, 276, 283
  definition, 225

Baseball free agency, 101–111, 112
  data description, 294
Beluga whale, 256–275
  data description, 306
Betting on professional football, 142–143
  data description, 294
Bias
  in estimate of mean, 50
  in estimate of median, 61
  in estimate of standard deviation, 68
Binomial random variable, 72
  definition, 81
Boxplot, 5, 6, 13, 88
  definition, 11
  side–by–side, 5, 8, 21, 22, 28, 72, 90, 224
  stacked, *see* Boxplot, side–by–side

Census incomes, 48–58, 59–66, 67–71
  data description, 287–290
Central Limit Theorem, 48–58, 59–66, 67–71, 72, 113, 129
  definition, 58
  for racial imbalance, 72–81
χ^2 test, 134, 137
  definition, 136
Collinearity, 200
Comparison
  of Binomial proportions, 132
  of location
    for two groups, 131
    of subgroups, 32
    to a standard, 99
  of proportions, 134, 137
  of volatility, 123
Computer experiment, 48, 59, 67, 129
Conditional probability, 37
  definition, 40
Condom use, 132–133
Confidence interval, 86, 156
  for a fitted value

definition, 163
for Binomial proportions, 132
for difference of means, 113, 123, 130, 131
for regression coefficients, 191, 250
for the mean, 82, 99, 100, 101, 129
definition, 85
testing assumptions, 82, 99
Constant shift model, 222
Contingency table, 43, 134, 137, 142, 144
definition, 136
definition of independence, 45
Cook's distance, 164
definition, 176
Correlation, 35, 191
coefficient, 283
definition, 31
Cross–classified data, *see* Contingency table

Density estimator, 24
Diagnostics, *see* Regression, diagnostics
Distributional shape, 48, 59, 67
Dow Jones Industrial Average, 283–284, 285
Down's syndrome, *see* Amniocentesis
Durbin–Watson statistic, 224, 256, 276, 283
definition, 225

Electricity usage, 177–184, 234
data description, 292
Emergency calls to the New York Auto Club, 145–152, 233
data description, 293
Employment rates, 190
data description, 293
Estimating a demand function, 191–200, 276–282
data description, 291
Examining subgroups, 224
Extra–Binomial variation, 72
definition, 81

F–statistic, 145, 153, 191
F–test
definition, 151
False negative, 37, 41
False positive, 37, 41
Fitted value, 145, 153
definition, 151
Forecasting, 227
Frequency estimate, 33
definition, 36

Gaussian, *see* Normal

Health care spending, 32

data description, 297
Heteroscedasticity, *see* Non–constant variance
Histogram, 5, 13, 59, 67
bin width, 24
definition, 11
of prediction errors, 225
Hypothesis
alternative
definition, 111
null
definition, 111
testing, 99, 237, 250, 253, *see also* Tests

Illusory correlation, 25, 235
definition, 31
Incomes of Long Island communities, 247–249, 252
data description, 298
Independence, 43, 47
definition, 45
test of, 134
Independent samples, 59, 67, 113, 129, 137
Indicator variables, 222, 234, 235, 236
Influential point
definition, 176
Inter–quartile range, 12, 89

Joint distribution, 43, 47
definition, 45

Lagged variable, 224, 233, 256, 276, 283
definition, 226
Leverage, 177
point, 164
value
definition, 176
Logarithm, *see* Transformation
Logistic regression, 35
Long–tailed data, 48, 59, 67, 101, 113, 129

Mann–Whitney rank sum test, *see* Wilcoxon rank sum test
Marginal distribution, 43, 47
definition, 45
Masking effect, 164, 185
definition, 176
Mean, 5, 13, 21, 48, 49, 82, 86, 177, 247
definition, 11
population, 49
Median, 5, 13, 48, 59, 89, 177, 201, 247
definition, 11
test, 123
definition, 128
Midhinge, 230

Mode, 5
  bimodal, 10, 11
  definition, 11
  unimodal, 11
Model selection, 191, 233
Mortgage rates, 131, 236
  data description, 299
Multicollinearity, 191
  definition, 200

National Basketball Association, 201–212, 230–231, 232
  data description, 299
National Hockey League, 227–229
  data description, 297
New York Stock Exchange, 100, 123–128, 283–284, 285
  data description, 292
Non–constant variance, 162, 237, 250, 253
Nonparametric tests, 101, 112, 113, 123
Normal
  approximation, 54
  distribution, 50
  random variable
    definition, 58
Normal plot, 72, 74
  definition, 81
  for returns on stocks, 82
  of standardized residuals, 149, 157

"Old Faithful" geyser, 5–12, 224, 130–226
  data description, 296
Outliers, 13, 15, 101, 153, 164, 177, 185, 201, 213, 247
  definition, 20
  in boxplot, 14, 22, 206
  in histogram, 103
  in normal plot, 81
  in regression, 158, 161
    definition, 163
  in scatter plot, 17, 23, 230, 247
  in standardized residual plot, 182, 183, 216, 218, 278
  in stem–and–leaf display, 205
Over the Counter (NASDAQ), 82–85, 123–128
  data description, 300

p–value, 101, 113, 145, 153
  definition, 111
Paired samples, 112
  t–test, 101
    definition, 111
Partial correlation, 191
Partial regression, 191
Partial regression plot

definition, 200
PCB contamination, 164–176
data description, 300
Popular music sales, 23
data description, 302
Population, 49, 114
definition, 58
Predicted value, 164, 201, 213
Predicting variable
definition, 152
Prediction, 145, 222, 253
Prediction interval, 82, 95, 97
for an individual case in regression, 145, 153, 156, 164, 201, 237, 250, 253
definition, 151
for the mean, 86, 95, 98, 99, 100
definition, 85
for the mean of the target variable in regression, 153
Probability, 34, 37, 41
definition, 36
Productivity versus quality, 25–31, 235
data description, 301
Purchasing power parity, 153–163, 222–223
data description, 301

Quality control, 25
Quartiles, 201
definition, 11

R^2, 145, 153
definition, 151
Racial imbalance, 72–81
data description, 298
Random drug and disease testing, 37–40
Random samples, 48
Random variable, 43
definition, 45
Rankits, *see* Normal plot
Regression, 164, 177, 186, 201, 247
coefficients, 191
diagnostics, 164, 168, 185, 191, 213, 276
multiple, 191, 213, 222, 224, 227, 230, 232, 233, 234, 235, 236, 247, 252, 256, 276, 283, 285
regression line, 25
definition, 31
simple, 145, 153, 164, 177, 185, 186, 189, 190, 237, 250, 253
testing assumptions, 148, 238, 241, 247, 284
through the origin, 253, 283
Removing autocorrelation, 224
Reporting bias, 16, 120
Reporting of sexual partners, 113–122, 129, 137–141

data description, 303
Residual, 145, 153, 164, 177, 201, 213
Residual plots, 145, 153, 164
definition, 151
Response bias, *see* Reporting bias
Return on stocks
NASDAQ, 82–85
NYSE, 100
Return rates of investments, 186–188
data description, 302
Right–skewed data
definition, 20
Robust, 15

Sample
definition, 58
Sample bias, 119
Sample proportions, 33, 43, 72
Sampling distribution
mean, 48, 50
median, 59–66
standard deviation, 67–71
Scatter plot, 5, 7, 13, 21
definition, 12
Semilog model, 186, 190
Sensitivity, 37, 41
Sign test, 101
definition, 111
Simple versus complicated models, 227, 237, 253
Skewness, 13, 50, 86, 90, 94, 97, 164, 201, 247
Space shuttle, 33–36
data description, 291
Spatial–Temporal data, 86
Specificity, 37, 41
Standard and Poor's 500 (S&P 500), 21–22, 99, 186–188, 283–284, 285
Standard deviation, 5, 13, 21, 48, 59, 67, 86
definition, 12
Standard error
definition, 58
of the coefficient, 194, 241
of the estimate, 148
definition, 151
of the mean, 51, 57
of the median, 62
of the regression line, 95, 97
of the standard deviation, 69
Standardized residual
definition, 151
Standardizing sample proportions, 72
Stem–and–leaf display, 5, 13, 94, 96, 201

definition, 12
Stock mutual funds, 21–22, 99, 185
data description, 295
Subway perceptions, 43–46, 47, 144
data description, 303

t–statistic, 153, 191, 222
t–test, 102, 123, 237, 250, 253
paired samples, 101
definition, 111
testing assumptions, 103, 117, 123
two–sample, 113, 123, 129
definition, 122
Tail probability, *see* p–value
Target variable
definition, 151
Tests
for differences in distributional shape, 113
for differences in location, 112, 113, 129
of independence, 137, 142, 144
Time series, 145, 149, 256, 276, 285
differencing, 256
Time series plot, 177, 191, 224
definition, 152
Time trend, 283
Transformation, 164, 177, 213, 233, 237
definition, 20
logarithm, 15, 54, 97, 166, 186, 247
Type I error, 145, 153

Unbiasedness, *see* Bias

Variance inflation factor, 191
definition, 200
Vineyard, 86–98, 237–246, 250–251, 253–255
data description, 304–305
Volatility of stock exchanges, 123–128
Voting fraud, 213–221
data description, 306

Weighted linear regression, 237, 247, 250, 252
definition, 246
Wilcoxon (Mann–Whitney) rank sum test, 113, 123
definition, 122
Wilcoxon signed rank test, 101
definition, 111

z–score, 72
z–test, 132
definition, 133
two–sample, 132